Gödel's Incompleteness Theorems

Dirk W. Hoffmann

Gödel's Incompleteness Theorems

A Guided Tour Through Kurt Gödel's Historic Proof

Dirk W. Hoffmann
Fakultät für Informatik und
Wirtschaftsinformatik
Hochschule Karlsruhe
Karlsruhe, Baden-Württemberg, Germany

ISBN 978-3-662-69549-4   ISBN 978-3-662-69550-0 (eBook)
https://doi.org/10.1007/978-3-662-69550-0

This book is a translation of the original German edition "Die Gödel'schen Unvollständigkeitssätze," 2nd edition, by Dirk W. Hoffmann, published by Springer-Verlag GmbH, DE in 2017. The translation was done with the help of an artificial intelligence machine translation tool. A subsequent human revision was done primarily in terms of content, so that the book will read stylistically differently from a conventional translation. Springer Nature works continuously to further the development of tools for the production of books and on the related technologies to support the authors.

Translation from the German language edition: "Die Gödel'schen Unvollständigkeitssätze" by Dirk W. Hoffmann, © Springer-Verlag GmbH Deutschland 2017. Published by Springer Berlin Heidelberg. All Rights Reserved.

© The Editor(s) (if applicable) and The Author(s), under exclusive license to Springer-Verlag GmbH, DE, part of Springer Nature 2024

This work is subject to copyright. All rights are solely and exclusively licensed by the Publisher, whether the whole or part of the material is concerned, specifically the rights of translation, reprinting, reuse of illustrations, recitation, broadcasting, reproduction on microfilms or in any other physical way, and transmission or information storage and retrieval, electronic adaptation, computer software, or by similar or dissimilar methodology now known or hereafter developed.
The use of general descriptive names, registered names, trademarks, service marks, etc. in this publication does not imply, even in the absence of a specific statement, that such names are exempt from the relevant protective laws and regulations and therefore free for general use.
The publisher, the authors, and the editors are safe to assume that the advice and information in this book are believed to be true and accurate at the date of publication. Neither the publisher nor the authors or the editors give a warranty, expressed or implied, with respect to the material contained herein or for any errors or omissions that may have been made. The publisher remains neutral with regard to jurisdictional claims in published maps and institutional affiliations.

This Springer imprint is published by the registered company Springer-Verlag GmbH, DE, part of Springer Nature.
The registered company address is: Heidelberger Platz 3, 14197 Berlin, Germany

If disposing of this product, please recycle the paper.

"Logic will never be the same again."

John von Neumann

Preface

For thousands of years, it was the unspoken assumption of mathematics that every mathematical statement can be either proved or disproved. In 1931, Kurt Gödel laid this long-held dream to rest. The young mathematician had discovered that the notion of truth and the notion of provability cannot coincide. In any sufficiently expressive formal system, statements exist that can neither be proved nor disproved within the system.

Gödel's work has fundamentally changed our view of mathematics. It is a jewel of our cultural heritage, ranking on par with Einstein's work on the theory of relativity [18] or Heisenberg's work on the uncertainty principle [39]. All three define fundamental limits we cannot overcome.

Since their discovery, many authors have dealt with Gödel's incompleteness theorems and shed light on their mathematical and philosophical implications. For the first time, I read about the incompleteness theorems in Douglas Hofstadter's famous book *Gödel, Escher, Bach* [61] in my senior year of high school. I was quickly captivated by the theorems and wanted to delve deeper into the subject: I wanted to know what Gödel had really proved. Yet, even my first glance at the original paper made me resign. The entire presentation was so formal that I could not understand it, even rudimentarily. I wished for a book that explained the original proof in comprehendible words, but such a book did not exist.

Since then, many years have passed in which I could never detach myself mentally from the incompleteness theorems. As a result, two books eventually came into being. One is the book I missed for so long and is now in your hands; it is my very personal attempt to fill the gap I just described. The other is *Limits of Mathematics* (*Grenzen der Mathematik* [59]), which addresses a similar subject with a different objective. While the present book deals in detail with Gödel's historical proof, the other is closer in style to a classical textbook. It covers a broader array of themes and explores diverse ideas and thoughts that are somewhat distant from the epicenter of the incompleteness theorems. It also introduces the fundamental principles of classical mathematical logic, crucial for understanding the present text.

I hope you find this book enjoyable, and I welcome any comments or suggestions from all readers.

Karlsruhe, 30 March 2024   Dirk W. Hoffmann

About this Book

The subsequent chapters feature a complete reproduction of Gödel's original article, divided into annotated sections. The original passages are printed on a shaded background to visually separate them from the surrounding text. Apart from that, the layout of the original work has been largely retained. Only the footnotes, which are numerous in Gödel's work, are treated differently. To create a smooth reading experience, they appear at the end of the referencing text fragments.

The original manuscript has been marginally modified at seven locations to rectify a few known errors. Specifically, the changes are those listed in Gödel's collected works under 'Textual Notes' [36]:

Page:Row	Original	Correction
*175:25	$\overline{n \varepsilon K}$	$\overline{q \varepsilon K}$
177:12	18a	19a
177:33	18a	19a
180:15	auch R	auch $\overline{R}$
184:7	$u * R\,(n\,G\,l\,x)\,v$	$u * R\,(n\,G\,l\,x) * v$
187:5	rekursiv	*rekursiv*
*189:30	Existenz	Existenz von aus x

The entries in the first column correspond to the pages and lines in Gödel's original manuscript. Gödel himself made the two alterations marked with an asterisk. They originate from a manually corrected manuscript that was part of his estate.

Gödel had written his work in German. As a German-speaking author, this came in handy, given that the initial two editions of this book explicitly targeted the German market. With no language barrier present, I could concentrate entirely on the substantive aspects of Gödel's work. In 2023, my publisher asked me to translate the book into English, an opportunity I readily embraced as it offered access to a significantly broader audience. Nevertheless, the translation process was not solely a linguistic challenge for me; it also influenced the book's conception. First and foremost, I had to decide whether to retain Gödel's work in its original German form or to replace it with an English translation. On the one hand, I was hesitant to exclude the German original since the mathematical presentation of Gödel's proof is only one aspect of this book, albeit a crucial one.

From the outset, my objective has been to present not just the mathematical facets of the incompleteness theorems but also, and foremost, to convey Gödel's very own account. I regard his original work as a precious part of our scientific heritage, and this book is my modest contribution to helping readers access its breathtaking content.

On the other hand, I understood that the majority of English-speaking readers might grapple with the German language, even with additional linguistic explanations in the surrounding text. As a result, I've chosen to adopt a hybrid approach. This English edition of the book showcases Gödel's article in both its original German form and an English translation.

In the past, three translations emerged during Gödel's lifetime: the initial one crafted by Bernard Meltzer in 1962 [68], followed by Elliott Mendelson's rendition in 1965 [69], and the third by Jean van Heijenoort in 1967 [35]. I've chosen to present Mendelson's translation alongside the German original in this book. Minor adjustments have been made compared to the published version in [69], predominantly revolving around three aspects. First, I've slightly reformatted the text to better align with the layout of the German original, facilitating ease of reference between the texts. Second, I've restored the page references within Gödel's text to their original numbers to assure consistency with the German original. The third alteration pertains to typography: where Gödel utilizes italics to signify a specific semantic meaning, Mendelson's translation utilizes capitalized letters. In this book's reprint, I've reverted to using italics to mirror the style of the German original. Beyond these subtle adjustments in layout and typography, the wording of Mendelson's translation remains unchanged and is reproduced verbatim in this edition.

Acknowledgements

I take the opportunity to thank the Institute for Advanced Study in Princeton for granting permission to reprint Gödel's historic paper in the original German. Furthermore, I thank Dover Books for granting permission to reprint Elliott Mendelson's English translation. Finally, I thank Hal Prince for providing several helpful comments on my published logic books and writing his own on Gödel's historic paper [82]. Hal's book explores the same subject but sheds light on it from a different angle. I am grateful to him for having written this insightful piece.

Bird's-Eye View

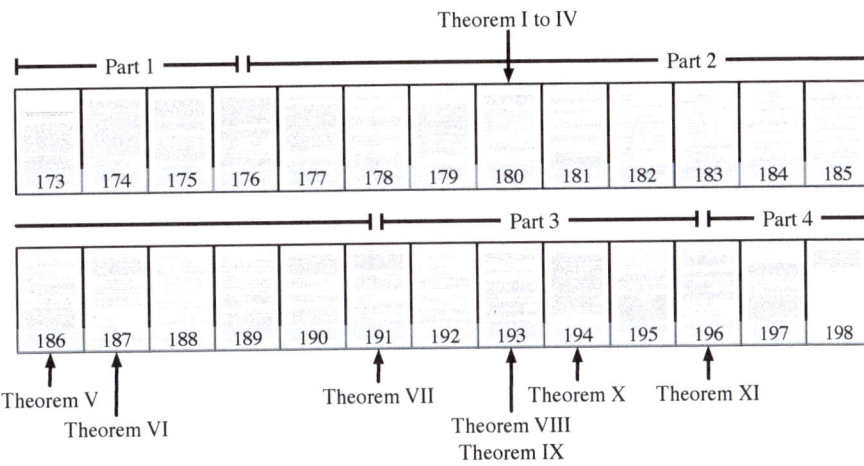

Gödel's original paper comprises 4 parts and establishes a total of 11 theorems. Part 1 features a proof sketch elucidating the fundamental concept behind his argument 🔖. 🔖 [173 – 176] Part 2 initiates the detailed exposition of the proof, commencing with the introduction of a formal system named P, which is central to the subsequent argument 🔖. Af- 🔖 [176 – 179] terward, Gödel introduces the notion of primitive-recursive functions and establishes elementary properties about this class of functions by proving Theorems I to IV 🔖. Subse- 🔖 [179 – 181] quently, Gödel meticulously demonstrates that many meta-mathematical concepts concerning formal systems are expressable using primitive-recursive definitions 🔖. He then 🔖 [182 – 186] continues to prove Theorem V, which reveals a significant relationship between formulas within a formal system and primitive-recursive relations 🔖. This is followed by The- 🔖 [186 – 187] orem VI, the main result of his work. Gödel calls it the general result on the existence of undecidable propositions 🔖. The third part draws various conclusions from the gen- 🔖 [187 – 191] eral result. Theorem VII establishes a connection between arithmetic and primitive-recursive relations, while Theorem VIII asserts the existence of undecidable statements within arithmetic 🔖. This theorem is commonly referred 🔖 [191 – 193] to as the first incompleteness theorem in modern terms. Through Theorems IX and X, Gödel concludes that the

[193 – 196] decision problem of first-order predicate logic remains unsolvable within P. The fourth section outlines the proof of Theorem XI, now recognized as the second incompleteness theorem.
[196 – 198]

Page Overview

The following overview indicates where the various pages of Gödel's historic paper are discussed in this book.

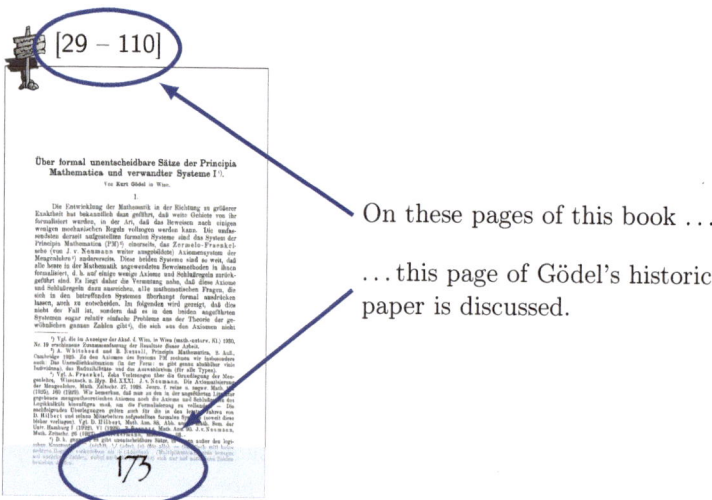

On these pages of this book ...

...this page of Gödel's historic paper is discussed.

Bird's-Eye View

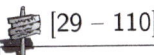

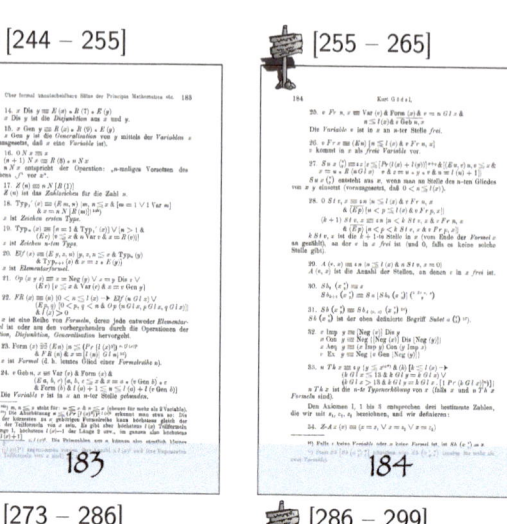

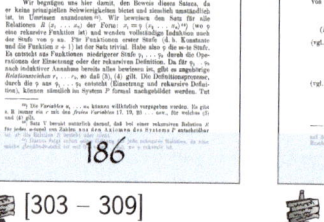
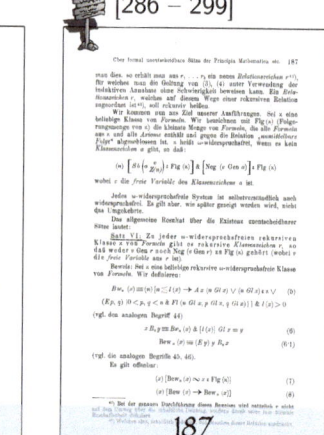

Bird's-Eye View

XV

[314 – 320]

[320 – 325]

[325 – 338]

[338 – 344]

[344 – 350]
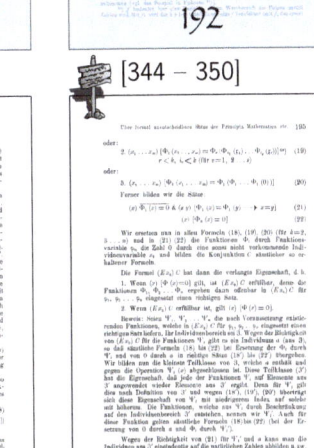

[350 – 355]
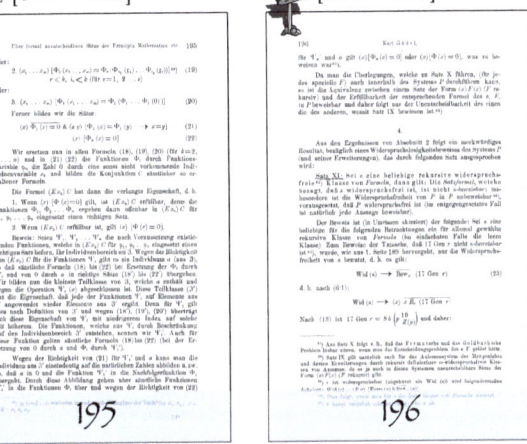

[355 – 360]

[360 – 362]
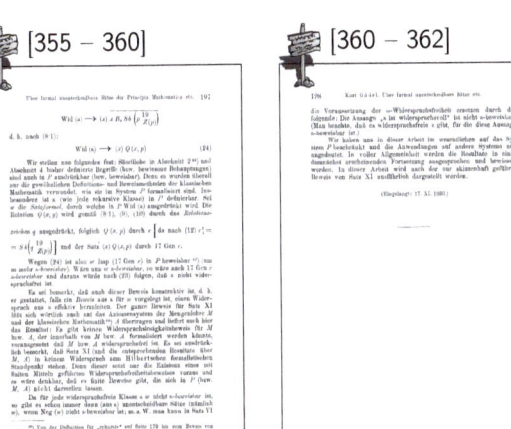

Contents

1 Introduction — 1
- 1.1 The Axiomatic Method — 1
- 1.2 Formal Systems — 6
- 1.3 Metamathematics — 13
  - 1.3.1 Consistency — 15
    - 1.3.1.1 Proof of Consistency at the Semantics Level — 16
    - 1.3.1.2 Proof of Consistency at the Syntax Level — 17
  - 1.3.2 Completeness — 18
  - 1.3.3 Hilbert's Program — 20
- 1.4 The Incompleteness Theorems — 26
  - 1.4.1 The First Incompleteness Theorem — 26
  - 1.4.2 The Second Incompleteness Theorem — 28
- 1.5 On Formally Undecidable Propositions — 28

2 Foundations of Mathematics — 33
- 2.1 The Logicist Program — 33
  - 2.1.1 *Begriffsschrift* — 34
  - 2.1.2 Axioms of the *Begriffsschrift* — 37
  - 2.1.3 Formalization of Arithmetic — 39
- 2.2 The Natural Numbers — 43
  - 2.2.1 *Arithmetices Principia* — 43
  - 2.2.2 Axioms of the *Arithmetices Principia* — 46
  - 2.2.3 Dedekind's Isomorphism Theorem — 50
- 2.3 *Principia Mathematica* — 56
  - 2.3.1 Cantor's Theorem — 60
  - 2.3.2 Russell's Antinomy — 65
    - 2.3.2.1 Logical Formulation of the Antinomy — 65
    - 2.3.2.2 Set-Theoretic Formulation of the Antinomy — 68
    - 2.3.2.3 In Ruins — 69
  - 2.3.3 Type Theory — 73
  - 2.3.4 The Logic of the *Principia Mathematica* — 77
- 2.4 Axiomatic Set Theory — 86
  - 2.4.1 Continuum Hypothesis — 87
  - 2.4.2 Well-Ordering Theorem — 90
  - 2.4.3 Criticism of Zermelo's Proof — 99
    - 2.4.3.1 The Axiom of Choice — 99
    - 2.4.3.2 Non-Predicative Definitions — 101

		2.4.4	Zermelo's Axioms	102

3 Proof Sketch — 109
- 3.1 Arithmetic Formulas — 110
- 3.2 The Arithmetization of Syntax — 117
- 3.3 I Am Unprovable! — 122
- 3.4 Gödel, Richard and the Liar — 129
  - 3.4.1 The Liar Paradox — 130
  - 3.4.2 Richard's Antinomy — 132
  - 3.4.3 When Is a Formal System Affected? — 134

4 System P — 139
- 4.1 Syntax — 140
  - 4.1.1 Terms and Formulas — 142
  - 4.1.2 Substitutions — 148
- 4.2 Semantics — 150
  - 4.2.1 Definition of Equality — 154
  - 4.2.2 Definition of Natural Numbers — 155
- 4.3 Axioms and Inference Rules — 156
- 4.4 Formal Proofs — 162
  - 4.4.1 Propositional Logic Theorems — 166
  - 4.4.2 Hypothesis-Based Proving — 178
  - 4.4.3 Predicate Logic Theorems — 182
  - 4.4.4 Theorems About Equality — 185
  - 4.4.5 Numerical Theorems — 190
- 4.5 The Arithmetization of Syntax — 200

5 Primitive-Recursive Functions — 209
- 5.1 Definition and Properties — 209
- 5.2 Primitive-Recursive Functions and Relations — 228
- 5.3 Decision Procedures — 274
- 5.4 Theorem V — 279

6 The Limits of Mathematics — 293
- 6.1 Gödel's Main Result — 293
  - 6.1.1 Incompleteness of System P — 296
  - 6.1.2 Consequences of the Main Result — 305
- 6.2 The First Incompleteness Theorem — 315
  - 6.2.1 Incompleteness of Arithmetic — 315
  - 6.2.2 Implications for the Restricted Function Calculus — 330
    - 6.2.2.1 Syntax of First-Order Predicate Logic — 330
    - 6.2.2.2 Semantics of First-Order Predicate Logic — 332
- 6.3 The Second Incompleteness Theorem — 352

7 Epilogue — 363

Bibliography	365
Image Credits	373
Name Index	375
Index	379

1 Introduction

> "Kurt Gödel's achievement in modern logic is singular and monumental – indeed it is more than a monument, it is a landmark which will remain visible far in space and time."
>
> Attributed to John von Neumann

For Rudolf Carnap, Herbert Feigl, Kurt Gödel, and Friedrich Waismann, a steamboat trip along the Baltic coast marked the end of a long journey from Vienna to Königsberg. Joined in Swinemünde by two other scientists, Kurt Grelling, and Hans Hahn, all six disembarked on September 4, 1930 [14]. The objective of their trip was to participate in the 2nd Conference on Epistemology of the Exact Sciences, hosted by the Berlin Society for Empirical Philosophy in the East Prussian metropolis from September 5 to 7. On this late summer day, there was nothing to suggest that September 7, 1930, would later be remembered as the day that changed mathematics forever.

This year, the conference centered around the foundations of mathematics. The philosophical-sounding topic of the conference must be put in its historical context. In the 1930s, mathematics was not yet the isolated discipline we know today, and mathematical and philosophical questions were tightly intertwined. Accordingly, there was a great demand for a conceptual foundation on which mathematics could be solidly built.

Three philosophical perspectives dominated the mathematics of the early twentieth century, presented in three 60-minute tutorials on the first day. First, the German philosopher Rudolf Carnap introduced *Logicism* [8]. Next, the Dutch mathematician Arend Heyting spoke about *Intuitionism* [41]. Last, the Austro-Hungarian-born mathematician John von Neumann lectured on *Formalism*. From today's perspective, Formalism was the most important of the three philosophical views, as it propagated the approach that would later shape all of modern mathematics: the axiomatic method. The historical roots of axiomatic thinking trace back to ancient Greece, where our journey continues.

1.1 The Axiomatic Method

The axiomatic method employs the idea of logically deducing statements from a set of a priori established assumptions. It is disputed among historians to whom the intellectual authorship of this several thousand-year-old idea should

Figure 1.1: An original fragment of Euclid's *Elements*

be attributed. Some believe that axiomatic thinking dates back to the Greek scholar Eudoxos of Knidos [40]; others trace it back to Plato and Aristotle [60].

The axiomatic method was popularized by a man about whom little is known: Euclid of Alexandria. Euclid was born around 360 BC and likely attended the Platonic Academy in Athens for a time. His contributions to Greek philosophy were comparatively small [3], unlike classical mathematics, where he made significant contributions. He became best known for his work *Elements*, which summarized the preceding three hundred years of Greek mathematics. His work consisted of 13 so-called books, which we would rather call chapters today. Euclid's *Elements* stands as the most successful writing in mathematical world literature. Over the years, it has been translated into countless languages, and new editions still appear at regular intervals.

First and foremost, Euclid's *Elements* is notable for its methodology. The author had taken an axiomatic approach to develop geometry, deriving all theorems from a set of elementary facts established a priori. Today, we are all in Euclid's tradition when we state the basic facts of a mathematical theory in advance and adhere to strict rules when deriving new theorems. In [40], Harro Heuser calls the axiomatic method the *lifeblood of mathematics* and the *greatest contribution the remarkable Greek people have made to mathematics*. He does not exaggerate in any way.

The most famous passages of Euclid's *Elements* are part of the first book. Among the numerous definitions, postulates, and axioms, the five postulates listed in Figure 1.2 are the most important [19]. They are what we call *theory axioms* today.

Viewed from a modern perspective, two peculiarities of Euclid's work stand out:

- Despite its axiomatic character, Euclid's *Elements* is far less formal than modern mathematical writings. Although Euclid derived the geometry theorems deductively, his presentation was deficient in many respects. In various passages, his logical conclusions rely on unspoken facts that are intuitively correct but underivable from the axioms. Furthermore, the applied logical apparatus is not formally defined.

1.1 The Axiomatic Method

Let the following be postulated:

Postulate 1: *"To draw a straight line from any point to any point."*

Postulate 2: *"To produce a finite straight line continuously in a straight line."*

Postulate 3: *"To describe a circle with any centre and distance."*

Postulate 4: *"That all right angles are equal to one another."*

Postulate 5: *"That, if a straight line falling on two straight lines make the interior angles on the same side less than two right angles, the two straight lines, if produced indefinitely, meet on that side on which are the angles less than two right angles."*

Figure 1.2: The five postulates of Euclidean geometry (from [19])

- Euclid did not utilize a dedicated formula language. He formulated all definitions, postulates, and axioms colloquially, as well as the derived theorems. Consequently, he represented all mathematical objects in the same language he used to talk about these objects.

Euclid's axiomatic method remained almost unchanged for over 2000 years. It was not until the end of the nineteenth century that it awakened from its slumber. At that time, all of a sudden, it underwent a metamorphosis that was about to change its face completely in just a few years.

The development was significantly driven by the German mathematician David Hilbert (Figure 1.3). Born in Königsberg in 1862, Hilbert was unusually versatile, shifting his scientific focus several times. In addition to significant contributions to logic and the foundations of mathematics, he made momentous discoveries in algebraic geometry, analysis, number theory, and theoretical physics. His scientific legacy is among the most valuable ever left by a single mathematician.

Hilbert spent most of his professional career at the mathematics faculty in Göttingen, where Gauss, Dirichlet, and Riemann had achieved great achievements years before. Hilbert's appointment in 1895 marked a new beginning for the faculty. It allowed Göttingen to regain its old glory, which had begun to fade towards the end of the nineteenth century.

Hilbert was an advocate of the axiomatic method and the figurehead of the formalists. He unfailingly articulated his perspective in numerous publications and lectures using easily comprehensible language. Hilbert's exceptional clarity of expression makes his writings a pleasure to read, even decades after his death.

Figure 1.3

DAVID HILBERT
1862 – 1943

At this point, we want to let him speak for himself and quote from a lecture he gave in 1917 to the Swiss Mathematical Society. He explained his formalistic position with the following words:

> "When we assemble the facts of a definite, more-or-less comprehensive field of knowledge, we soon notice that these facts are capable of being ordered. This ordering always comes about with the help of a certain *framework of concepts* in the following way: a concept of this framework corresponds to each individual object of the field of knowledge, and a logical relation between concepts corresponds to every fact within the field of knowledge. This framework of concepts is nothing other than the *theory* of the field of knowledge. [...] If we consider a particular theory more closely, we always see that a few distinguished propositions of the field of knowledge underlie the construction of the framework of concepts, and these propositions then suffice by themselves for the construction, in accordance with logical principles, of the entire framework. [...] These fundamental propositions can be regarded from an initial standpoint as *the axioms of the individual fields of knowledge.*"

"Wenn wir die Tatsachen eines bestimmten Wissensgebietes zusammenstellen, so bemerken wir bald, daß diese Tatsachen einer Ordnung fähig sind. Diese Ordnung erfolgt jedesmal mit Hilfe eines gewissen *Fachwerkes von Begriffen* in der Weise, daß dem einzelnen Gegenstande des Wissensgebietes ein Begriff dieses Fachwerkes und jeder Tatsache innerhalb des Wissensgebietes eine logische Beziehung zwischen den Begriffen entspricht. Das Fachwerk der Begriffe ist nichts anderes als die *Theorie des Wissensgebietes.* [...] Wenn wir eine bestimmte Theorie näher betrachten, so erkennen wir allemal, daß der Konstruktion des Fachwerkes von Begriffen einige wenige ausgezeichnete Sätze des Wissensgebi-

1.1 The Axiomatic Method

etes zugrunde liegen und diese dann allein ausreichen, um aus ihnen nach logischen Prinzipien das ganze Fachwerk aufzubauen. [...] Diese grundlegenden Sätze können von einem ersten Standpunkte aus als die Axiome der einzelnen Wissensgebiete angesehen werden."

David Hilbert [49, 42]

Towards the end of the nineteenth century, Hilbert impressively demonstrated the fertility of the ground he had chosen for the refoundation of mathematics. In his book *Foundations of Geometry*, published in 1899, he introduced several axioms that allowed all the propositions of Euclidean geometry to be derived with an accuracy way beyond that of Euclid's *Elements*. However, to achieve this goal, Hilbert had to opt for an axiomatic system far more complex than its historical counterpart. While the theorems of Euclid's *Elements* are derived from five main postulates, Hilbert's system comprises a total of 21 axioms divided into five groups. There are

- 8 axioms of connection (Group I),
- 4 axioms of order (Group II),
- 1 axiom of parallels (Group III),
- 6 axioms of congruence (Group IV), and
- 2 axioms of continuity (Group V).

It would be short-sighted to consider Hilbert's axioms merely as a refinement of Euclid's postulates, for in one crucial aspect, they were entirely new. For thousands of years, people understood axioms to express basic facts about the real world, which, according to Aristotle, are *"neither capable nor in need of proof,"* and for just as long, they were used to define mathematical objects. For instance, the seventh book of Euclid's *Elements* contains the following definition of the natural numbers [19]:

- *"A unit is that by virtue of which each of the things that exist is called one."*
- *"A number is a multitude composed of units."*

For Hilbert, all attempts to define mathematical objects by virtue of their nature were doomed to fail. Each definition merely reduces a concept to other concepts that in turn require a definition. In Euclid's axioms, these are concepts such as *unit*, *thing*, and *multitude*. Since the chain of definitions cannot be continued indefinitely, one has to stop at a certain level and accept its terms and concepts as given. But which level is the proper one? Is it Euclid's level of things, units, and multitudes, or perhaps the level of the natural numbers themselves?

Hilbert tackled the problem simply by refraining from defining mathematical objects by virtue of their nature. Consequently, his axiom systems never describe these objects as such; instead, they solely concentrate on their relationships and the resulting logical conclusions. More precisely, in Hilbert's axiomatization of geometry, it does not matter what the terms *point*, *line*, and *plane* mean by their nature. All that matters is that they relate to each other as the axioms dictate. Thus, even if the terms *point*, *line* and *plane* were replaced by the terms *beer mug*, *bench* and *table*, the result may still be called an axiomatization of Euclidean geometry. It is said that Hilbert once explained his formalistic standpoint with this example, but no reliable source is known.

1.2 Formal Systems

Hilbert's formalistic approach has established a perspective in mathematics that distinguishes between an object level and a meaning level. For the object level, artificial formula languages were created, conforming to well-defined construction rules. Later, logical reasoning was added to the object level through textual transformation rules. From now on, it was possible to treat every axiom and every derived theorem as a mere chain of symbols that could be transformed according to a particular set of rules. In the resulting *formal systems*, mathematics had become a mechanical game akin to a game like chess.

As a result of this formalization, vaguely defined concepts, such as conducting a proof, could suddenly be grasped with mathematical precision, giving rise to a new mathematical branch called *proof theory*. In 1923, Hilbert described this new branch of mathematics with the following words:

> "Everything that previously made up mathematics is to be rigorously formalized, so that mathematics proper or mathematics in the strict sense becomes a stock of formulae. [...] Certain formulae that serve as building-blocks for the formal edifice of mathematics are called axioms. A proof is a figure that must intuitively appear to us as such; it consists of inferences using the inference-schema
>
> $$\frac{\mathfrak{S},\ \mathfrak{S} \to \mathfrak{T}}{\mathfrak{T}}$$
>
> where in every case each of the premises – that is, the formulas $\mathfrak{S}$ and $\mathfrak{S} \to \mathfrak{T}$ – either is an axiom, or results directly from an axiom by substitution, or agrees with the end-formula $\mathfrak{T}$ of an inference that appears earlier in the proof, or results from such an end-formula by substitution. A formula shall be called provable if it either is an axiom, or results from an axiom by substitution, or is the end-formula of a proof."

> "Alles, was im bisherigen Sinne die Mathematik ausmacht, wird streng formalisiert, so daß die eigentliche Mathematik oder die Mathematik

1.2 Formal Systems

in engerem Sinne zu einem Bestande an Formeln wird. [...] Gewisse Formeln, die als Bausteine des formalen Gebäudes der Mathematik dienen, werden Axiome genannt. Ein Beweis ist eine Figur, die uns als solche anschaulich vorliegen muss; er besteht aus Schlüssen vermöge des Schlußschemas

$$\frac{\mathfrak{S} \quad \mathfrak{S} \to \mathfrak{T}}{\mathfrak{T}}$$

wo jedesmal jede der Prämissen, d. h. der betreffenden Formeln $\mathfrak{S}$ und $\mathfrak{S} \to \mathfrak{T}$, entweder ein Axiom ist bzw. direkt durch Einsetzung aus einem Axiom entsteht oder mit der Endformel $\mathfrak{T}$ eines Schlusses übereinstimmt, der vorher im Beweis vorkommt bzw. durch Einsetzung aus einer solchen Endformel entsteht. Eine Formel soll beweisbar heißen, wenn sie entweder ein Axiom ist bzw. durch Einsetzen aus einem Axiom entsteht oder die Endformel eines Beweises ist."

David Hilbert [51, 43]

Hilbert's proof theory is of such central importance for understanding Gödel's work that we want to examine two specific examples closely. We introduce both of them in two steps. In the first step, we define the syntax of the formal system, that is, we agree on a set of symbols, together with a set of rules that determine how these symbols can be chained together to symbol strings called formulas. In the second step, we introduce the axioms and the inference rules. They define how new theorems can be obtained from the axioms and what has already been proven.

System B

The first example, called system B, is based on an artificial language that has little in common with ordinary mathematics. The syntax is given by the alphabet $\{\square, \blacksquare, (,)\}$ and the following set of rules:

> **Definition 1.1** — Syntax of system B
>
> The language of system B is defined as follows:
>
> 1. $\square$ and $\blacksquare$ are formulas.
> 2. If φ and ψ are formulas, then so is $(\varphi\psi)$.

For instance, the symbol strings

$$\square, \blacksquare, (\blacksquare\square), (\square\blacksquare), ((\square\square)\blacksquare), (\square(\square\blacksquare)), ((\square\square)(\blacksquare\blacksquare)), ((\blacksquare(\blacksquare\blacksquare))((\square\square)\square))$$

Table 1.1: Axioms and inference rules of system B

Axioms of system B			
■			(B1)
Inference rules of system B			
$\dfrac{■}{(■□)}$	(S1)	$\dfrac{(\varphi\psi)\chi}{\varphi(\psi\chi)}$	(S4)
$\dfrac{□}{(□□)}$	(S2)	$\dfrac{\varphi(\psi\chi)}{(\varphi\psi)\chi}$	(S5)
$\dfrac{□}{(■■)}$	(S3)	$\dfrac{((\varphi\psi)(\varphi\psi))}{□}$	(S6)

are formulas, but the following are not:

$$□■, (■)□, ■■■, (□■□), (□□)(■)$$

Table 1.1 summarizes the axioms and inference rules of system B. With ■, there exists a single axiom. Thus, every proof must begin with this formula. Six rules are available for deriving new theorems. Rules (S1) to (S3) permit the replacement of the symbols ■ and □ with other symbol combinations, whereas rules (S4) and (S5) allow brackets to be shifted back and forth. Note that the three placeholders φ, ψ, and χ must be appropriately substituted with other formulas before the rule is applied. Rule (S6) is the only one capable of shortening a formula. It states that repeating subexpressions of a certain kind may be replaced with the symbol □. All inference rules are meant to be used as it is common in so-called *rewriting systems*, that is, replacements may be applied not only to whole formulas but also to any subexpression.

It is time to look at some examples:

■ **Example 1:** Derivation of $((■(■■))((□□)□))$

1. ⊢ ■ (B1)
2. ⊢ (■□) (S1,1)
3. ⊢ ((■□)□) (S1,2)
4. ⊢ ((■(■■))□) (S3,3)
5. ⊢ ((■(■■))(□□)) (S2,4)
6. ⊢ ((■(■■))((□□)□)) (S2,5)

1.2 Formal Systems

■ **Example 2:** Derivation of (□■)

1. ⊢ ■   (B1)
2. ⊢ (■□)   (S1,1)
3. ⊢ (■(■■))   (S3,2)
4. ⊢ ((■■)■) $[\varphi, \psi, \chi = \blacksquare]$ (S5,3)
5. ⊢ (((■□)■)■)   (S1,4)
6. ⊢ (((■□)(■□))■)   (S1,5)
7. ⊢ (□■) $[\varphi = \blacksquare, \psi = \square]$ (S6,6)

Even though system B is far off from any practical use, it nicely illustrates the core ideas of Hilbert's proof theory. The following definition recaps what we have covered so far:

Definition 1.2 — Formal System, Proof

A *formal system* consists of

- a set of axioms and
- a set of inference rules.

A *formal proof* is a chain of formulas $\varphi_1, \varphi_2, \ldots, \varphi_n$, which is formed according to the following rules:

- φ_i is an axiom, or
- φ_i is derived from preceding formulas by applying an inference rule.

The last formula of the proof chain is called the proven *theorem*. The symbolic expression ⊢ φ indicates that φ is a theorem.

The definition also clarifies the meaning of the symbol '⊢', which you might have already spotted in the derivation sequences. It expresses that a formula is provable, that is, it can be derived from the axioms by repeated application of inference rules. Always remember that every axiom is also a theorem, proven trivially through a chain with a single element, where the axiom serves as both the starting and ending formula.

Before proceeding to the next example, let us consider whether the formula (■■) is derivable from the axioms. Please take a moment and try to construct a suitable derivation sequence on your own. We will provide the answer to this question in just a few pages.

System E

The structure of the next example system quite closely resembles the one Hilbert had in mind for the refoundation of mathematics. System E utilizes the alphabet $\{0, \mathsf{s}, (,), =, >, \neg, \rightarrow\}$ and adheres to the following syntax rules:

Definition 1.3 — Syntax of system E

The language of system E is defined as follows:

- 0 is a term.
- If σ is a term, then so is $\mathsf{s}(\sigma)$.
- If σ and τ are terms, then the following expressions are formulas:
$$(\sigma = \tau), (\sigma > \tau), \neg(\sigma = \tau), \neg(\sigma > \tau)$$
- If φ and ψ are formulas, then so is $\varphi \rightarrow \psi$.

In contrast to the first example, system E distinguishes between *terms* and *formulas*. Every string of the form

$$0, \mathsf{s}(0), \mathsf{s}(\mathsf{s}(0)), \mathsf{s}(\mathsf{s}(\mathsf{s}(0))), \mathsf{s}(\mathsf{s}(\mathsf{s}(\mathsf{s}(0)))), \ldots$$

is a term, but not a formula. Formulas are created by combining two terms with a symbol from the set $\{=, >, \neg\}$. Among others, we can construct the following formulas according to the syntax definition provided above:

$$(0 = 0), (0 > 0), \neg(0 = 0), \neg(0 > 0), (\mathsf{s}(\mathsf{s}(0)) = \mathsf{s}(0)), (0 = 0) \rightarrow (0 > 0)$$

Table 1.2 summarizes the axioms and inference rules. System E provides six *axiom schemata*, from which the axioms are obtained by replacing each of the two placeholders, σ and τ, with a term. Since these terms can be chosen arbitrarily, each schema generates an infinite number of axioms.

The inference apparatus of system E is comparatively sparse. New theorems can only be deduced via the *law of detachment*, more commonly known as *modus ponens* (MP). It is the primary inference rule of classical logic and states in words that ψ is true if φ is true and ψ can be inferred from φ.

Once again, let's look at some examples:

- **Example 1:** Derivation of $(\mathsf{s}(\mathsf{s}(\mathsf{s}(0))) > \mathsf{s}(0))$

1. $\vdash (\mathsf{s}(0) = \mathsf{s}(0))$ $\qquad [\sigma = \mathsf{s}(0)]$ (A1)
2. $\vdash (\mathsf{s}(0) = \mathsf{s}(0)) \rightarrow (\mathsf{s}(\mathsf{s}(0)) > \mathsf{s}(0))$ $\qquad [\sigma = \mathsf{s}(0)]$ (A2)

1.2 Formal Systems

Table 1.2: Axioms and inference rules of system E

Axioms of system E			
$(\sigma = \sigma)$	(A1)	$(\sigma > \tau) \to \neg(\sigma = \tau)$	(A4)
$(\sigma = \sigma) \to (\mathsf{s}(\sigma) > \sigma)$	(A2)	$(\sigma > \tau) \to \neg(\tau = \sigma)$	(A5)
$(\sigma > \tau) \to (\mathsf{s}(\sigma) > \tau)$	(A3)	$(\sigma > \tau) \to \neg(\tau > \sigma)$	(A6)

Inference rules of system E	
$\dfrac{\varphi,\ \varphi \to \psi}{\psi}$	(MP)

3. $\vdash\ (\mathsf{s}(\mathsf{s}(0)) > \mathsf{s}(0))$ \hfill (MP, 1,2)
4. $\vdash\ (\mathsf{s}(\mathsf{s}(0)) > \mathsf{s}(0)) \to (\mathsf{s}(\mathsf{s}(\mathsf{s}(0))) > \mathsf{s}(0))$ \hfill $[\sigma = \mathsf{s}(\mathsf{s}(0)), \tau = \mathsf{s}(0)]$ (A3)
5. $\vdash\ (\mathsf{s}(\mathsf{s}(\mathsf{s}(0))) > \mathsf{s}(0))$ \hfill (MP, 3,4)

■ **Example 2:** Derivation of $\neg(\mathsf{s}(\mathsf{s}(0)) = \mathsf{s}(\mathsf{s}(\mathsf{s}(0))))$

1. $\vdash\ (\mathsf{s}(\mathsf{s}(0)) = \mathsf{s}(\mathsf{s}(0)))$ \hfill $[\sigma = \mathsf{s}(\mathsf{s}(0))]$ (A1)
2. $\vdash\ (\mathsf{s}(\mathsf{s}(0)) = \mathsf{s}(\mathsf{s}(0))) \to (\mathsf{s}(\mathsf{s}(\mathsf{s}(0))) > \mathsf{s}(\mathsf{s}(0)))$ \hfill $[\sigma = \mathsf{s}(\mathsf{s}(0))]$ (A2)
3. $\vdash\ (\mathsf{s}(\mathsf{s}(\mathsf{s}(0))) > \mathsf{s}(\mathsf{s}(0)))$ \hfill (MP, 1,2)
4. $\vdash\ (\mathsf{s}(\mathsf{s}(\mathsf{s}(0))) > \mathsf{s}(\mathsf{s}(0))) \to \neg(\mathsf{s}(\mathsf{s}(0)) = \mathsf{s}(\mathsf{s}(\mathsf{s}(0))))$ \hfill (A5)
   \hfill $[\sigma = \mathsf{s}(\mathsf{s}(\mathsf{s}(0))), \tau = \mathsf{s}(\mathsf{s}(0))]$
5. $\vdash\ \neg(\mathsf{s}(\mathsf{s}(0)) = \mathsf{s}(\mathsf{s}(\mathsf{s}(0))))$ \hfill (MP, 3,4)

Up to this point, we have studied system E solely at the syntactic level, where theorems are nothing more than strings of symbols manipulated mechanically according to a predetermined set of rules.

We will now introduce a semantic level by assigning a substantive meaning to the individual formula components. For the sake of simplicity, let's agree on the abbreviation

$$\overline{n} := \underbrace{\mathsf{s}(\mathsf{s}(\ldots \mathsf{s}\,(0)\ldots))}_{n \text{ times}} \tag{1.1}$$

which allows us to write down the proven theorems in a compact manner:

$$(\overline{3} > \overline{1})\ \text{ denotes }\ (\mathsf{s}(\mathsf{s}(\mathsf{s}(0))) > \mathsf{s}(0))$$

$\neg(\overline{2} = \overline{3})$ denotes $\neg(\mathsf{s}(\mathsf{s}(0)) = \mathsf{s}(\mathsf{s}(\mathsf{s}(0))))$

Now, each formula can be readily assigned a substantive meaning as follows:

$$\begin{aligned}
\overline{n} & \quad \text{states} \quad n \in \mathbb{N} \\
(\overline{n} = \overline{m}) & \quad \text{states} \quad n = m \\
(\overline{n} > \overline{m}) & \quad \text{states} \quad n > m \\
\neg(\overline{n} = \overline{m}) & \quad \text{states} \quad n \neq m \\
\neg(\overline{n} > \overline{m}) & \quad \text{states} \quad n \leq m \\
\varphi \to \psi & \quad \text{states} \quad \text{``}\varphi \text{ implies } \psi\text{''}
\end{aligned}$$

On the meaning level, it is possible to speak about true and false statements. In the following, the symbol '$\models$' labels a formula as substantively true, in analogy to the already introduced symbol '$\vdash$', which expresses that a formula is formally provable.

Definition 1.4 — Provability relation, Model relation

The *provability relation* '$\vdash$' has the following meaning:

$$\vdash \varphi \;:\Leftrightarrow\; \text{Formula } \varphi \text{ is formally provable}$$
$$\nvdash \varphi \;:\Leftrightarrow\; \text{Formula } \varphi \text{ is not formally provable}$$

The *model relation* '$\models$' has the following meaning:

$$\models \varphi \;:\Leftrightarrow\; \text{Formula } \varphi \text{ is substantively true}$$
$$\not\models \varphi \;:\Leftrightarrow\; \text{Formula } \varphi \text{ is substantively false}$$

Always keep in mind that the relations '$\vdash$' and '$\models$' depend on the underlying formal system and the selected interpretation, respectively. A formula that is provable in a particular formal system may be unprovable in another, just like a true formula can become false when the substantive meanings of its symbols change. In short, provability and truth are two unrelated notions, each independently determined by the choice of the formal system and the substantive interpretation of the formula symbols.

For each formula φ, the following four cases must be distinguished:

- φ is substantively true and formally provable ($\models \varphi$ and $\vdash \varphi$)
- φ is substantively true but formally unprovable ($\models \varphi$ and $\nvdash \varphi$)
- φ is substantively false but formally provable ($\not\models \varphi$ and $\vdash \varphi$)
- φ is substantively false and formally unprovable ($\not\models \varphi$ and $\nvdash \varphi$)

1.3 Metamathematics

	True formulas ($\models \varphi$)	False formulas ($\not\models \varphi$)
Provable formulas ($\vdash \varphi$)	$(0 = 0)$ $(s(0) = s(0))$ $(s(0) > 0)$ $(s(s(0)) > 0)$ $(s(s(s(0))) > s(0))$ $\neg(s(s(0)) = s(s(s(0))))$ $\ldots$	
Unprovable formulas ($\not\vdash \varphi$)	$\neg(0 > 0)$ $\neg(s(0) > s(0))$ $\neg(s(s(0)) > s(s(0)))$ $\neg(s(s(s(0))) > s(s(s(0))))$ $\neg(s(s(s(s(0)))) > s(s(s(s(0)))))$ $\ldots$	$\neg(0 = 0)$ $\neg(s(0) = s(0))$ $\neg(s(0) > 0)$ $\neg(s(s(0)) > 0)$ $\neg(s(s(s(0))) > s(0))$ $(s(s(0)) = s(s(s(0))))$ $\ldots$

Figure 1.4: Quadrant representation for the formal system E

Figure 1.4 graphically visualizes the four cases by distributing the formulas across four quadrants. All provable formulas appear in the upper two quadrants, and all unprovable formulas in the lower two. Thus, a formula's vertical position is solely determined by the axioms and inference rules of the underlying formal system. Similarly, the chosen interpretation determines the horizontal position. True formulas appear on the left and false formulas on the right.

1.3 Metamathematics

Formal systems allow mathematics to be performed with utmost precision. Proofs become mechanically verifiable, thus eliminating all doubts about their validity or invalidity. Moreover, formal systems exhibit a peculiarity that opens up an entirely new perspective on the mathematical method. Due to their strictly formal character, these systems can be made the subject of mathematical investigations themselves, that is, mathematical reasoning can be employed to prove statements about a formal system. Along these lines, a metamathematics emerges, existing side by side with ordinary mathematics.

Do you remember the exercise we gave you above? We asked you to clarify whether the formula (■■) is a theorem of system B. We are dealing here with a classical question of metamathematics since it makes a statement *about* the system that cannot be answered *within*. The language of B does not even offer the necessary means to formulate this question.

From a metatheoretical perspective, a surprisingly simple answer can be given based on the following observation:

In every theorem of system B,
the symbol ■ occurs an odd number of times.

A quick analysis of the axioms and inference rules verifies the property. The single axiom ■ has an odd number of ■'s, and the inference rules inherit this property from the premise to the conclusion. Because the formula (■■) contains an even number of ■'s, it can never be the end formula of a proof chain. Quod erat demonstrandum: We have just conducted our first metamathematical proof, albeit a very simple one.

Hilbert described the purpose of metamathematics in the following way:

> "In addition to this formalized mathematics proper, we have a mathematics that is to some extend new: a metamathematics that is necessary for securing mathematics, and in which – in contrast to the purely formal modes of inference in mathematics proper – one applies contentual inference, but only to prove the consistency of the axioms. In this metamathematics we operate with the proofs of mathematics proper, and these proofs are themselves the object of contentual investigation."

> "Zu der eigentlichen so formalisierten Mathematik kommt eine gewissermaßen neue Mathematik, eine Metamathematik, die zur Sicherung jener notwendig ist, in der – im Gegensatz zu den rein formalen Schlußweisen der eigentlichen Mathematik – das inhaltliche Schließen zur Anwendung kommt, aber lediglich zum Nachweis der Widerspruchsfreiheit der Axiome. In dieser Metamathematik wird mit den Beweisen der eigentlichen Mathematik operiert, und diese letzteren bilden selbst den Gegenstand der inhaltlichen Untersuchung."

David Hilbert [51, 43]

In this passage, Hilbert explicitly addressed consistency, one of the four questions of particular interest in metamathematics:

1.3 Metamathematics

Definition 1.5 — Meta-properties of formal systems

- **Consistency** $\quad(\text{\reflectbox{$\mathbb{F}$}}\ \nvdash \varphi \text{ or } \nvdash \neg\varphi)$

  A formal system is *consistent* or *non-contradictory* if no formula is provable along with the negation of that formula.

- **Negation completeness** $\quad(\text{\reflectbox{$\mathbb{F}$}}\ \vdash \varphi \text{ or } \vdash \neg\varphi)$

  A formal system is *negation complete* if, for each formula, the formula itself or its negation is provable.

- **Correctness** $\quad(\text{\reflectbox{$\mathbb{F}$}}\ \vdash \varphi \text{ implies } \models \varphi)$

  A formal system is *correct* or *sound* if every provable formula is substantively true.

- **Completeness** $\quad(\text{\reflectbox{$\mathbb{F}$}}\ \models \varphi \text{ implies } \vdash \varphi)$

  A formal system is *complete* if every substantively true formula is provable.

Keep in mind that the properties of consistency and negation completeness refer exclusively to the syntactic level of a formal system. As they do not rely on the notion of truth, they are meaningful even for uninterpreted formulas. In particular, these notions can be applied to all formal systems capable of negating formulas at the symbolic level, and nearly all logics in use today utilize the symbol '¬' for this purpose. The example system E already offered this symbol for precisely this reason. The first example, system B, had no means of negating a formula, rendering the question of whether the formal system is consistent or negation complete meaningless.

The properties of correctness and completeness establish a connection between the syntactic and the semantic level. They are only meaningful when the symbols of a formal system can be interpreted such that each formula represents a substantively true or substantively false statement.

The quadrant representation in Figure 1.4 offers a graphical interpretation for both properties. If a formal system is correct, no substantively false formula is provable, resulting in an empty upper right quadrant. If a system is complete, every substantively true formula is provable, thus leaving the lower left quadrant empty.

1.3.1 Consistency

In the past, two approaches for conducting consistency proofs have been established. We will apply both to demonstrate system E's consistency in the following two sections.

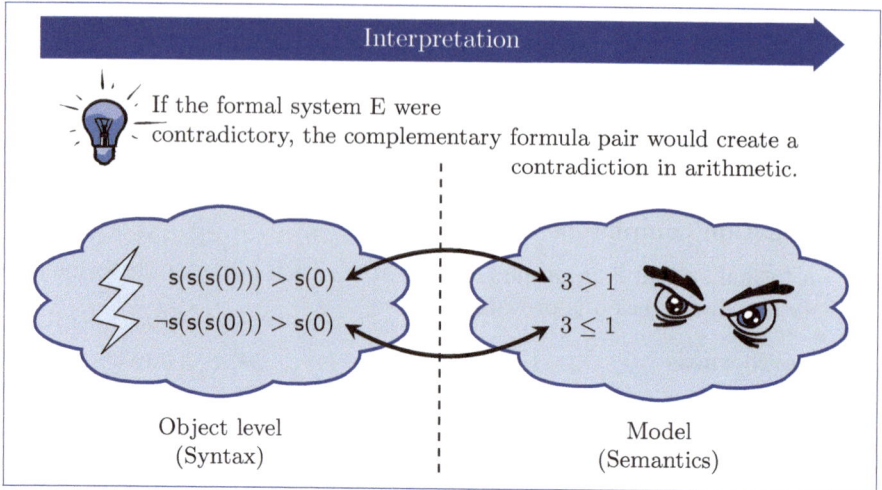

Figure 1.5: A relative consistency proof for the formal system E. By specifying a model, the model's consistency transfers to the object level.

1.3.1.1 Proof of Consistency at the Semantics Level

One way to prove the consistency of a formal system is to provide a *model*, that is, an interpretation that renders all theorems of the formal system substantively true. Above, we have already constructed a model for system E by identifying the terms 0, s(0), s(s(0)), etc., with the natural numbers and assigning the symbols '=' and '>' their usual meaning. That way, each axiom became a true statement of number theory, and the modus ponens inference rule inherited the truth of the premises to the conclusion. Consequently, all theorems of system E became true statements of number theory.

Now, if there were a formula φ with the property that both φ and $\neg\varphi$ were derivable from the axioms, a contradiction in the realm of natural numbers would arise (Figure 1.5). Conversely, if we trust arithmetic, the consistency of the formal system E is the inevitable consequence.

Hilbert had similarly proven the consistency of his axiom system for Euclidean geometry. He constructed a specific range of numbers such that any provable relationship between two geometric objects corresponded to a provable relationship between two elements of this number range and vice versa. Consequently, every derivable contradiction would have become visible as a contradiction in arithmetic.

What is crucial about this method is that it does not prove the consistency of a formal system in an absolute sense. The proof relies on the assumption that arithmetic is free of contradictions and transfers this property to the formal system. Proofs of this kind are thus called *relative* consistency proofs.

1.3 Metamathematics

But is a relative consistency proof for system E meaningful at all? A brief look at the axioms and inference rules reveals that E is a small fraction of arithmetic, capable of proving no more than elementary statements about the arrangement of natural numbers. Hence, we have backed up our consistency proof with an argument that builds upon the consistency of a more complicated and, therefore, less reliable system.

In reality, the situation is even worse. If we tried to formalize the model-theoretic arguments, we would have to rely on knowledge and reasoning from set theory. In Chapter 2, we will discover that this puts us on less stable ground than it might seem.

Consequently, if a relative consistency proof is insufficient to back up the consistency of system E, there is only one way out: to prove consistency in an absolute sense. Fortunately, providing such a proof for system E is not too challenging, as we will now observe.

1.3.1.2 Proof of Consistency at the Syntax Level

Our goal is to prove the consistency of system E without referring to interpretations, models, or any other semantic concept. For this purpose, we perform a classical proof of contradiction by assuming the existence of a formula φ such that both φ and $\neg \varphi$ are derivable from the axioms. We will now show that this assumption is incompatible with the structure of the formal system. As the language rules do not allow to form a formula of the form $\neg(\varphi \to \psi)$, it is sufficient to distinguish two cases:

- **Case 1:** Assume $\vdash (\sigma = \tau)$ and $\vdash \neg(\sigma = \tau)$

    Any formula of the form $(\sigma = \tau)$ must have been created by instantiating axiom schema (A1). In that case, however, σ and τ are identical terms, so it remains to resolve whether a formula of the form

    $$\neg(\sigma = \sigma) \qquad (1.2)$$

    can be derived from the axioms. The modus ponens rule can output this formula only if the formula

    $$(\sigma > \sigma) \qquad (1.3)$$

    has been proven already. Then, and only then, formula (1.2) could be derived via an instance of (A4) or (A5). However, a formula of type (1.3) must have been produced via axiom schema (A3). Consequently, a formula of the form

    $$(\sigma > \mathsf{s}(\sigma)) \qquad (1.4)$$

    has to be part of the proof chain. To prove (1.4), we need a theorem of the form $(\sigma > \mathsf{s}(\mathsf{s}(\sigma)))$, and we can repeat this argument as many times as we like. For each number n, the formula

$$(\sigma > \underbrace{s(s(\ldots s(\sigma)\ldots)))}_{n\text{-mal}} \tag{1.5}$$

would have to be proven already, contradicting the finiteness of a proof chain.

■ **Case 2:** Assume $\vdash (\sigma > \tau)$ and $\vdash \neg(\sigma > \tau)$

The formula $\neg(\sigma > \tau)$ can only be derived via axiom schema (A6) and the formula $(\tau > \sigma)$. Thus, $(\tau > \sigma)$ must appear somewhere in the proof chain. If σ equals τ, this formula is identical to (1.3) and, as stated above, is unprovable. If σ and τ are different terms, one of the two formulas $(\sigma > \tau)$ and $(\tau > \sigma)$ must have the form (1.5) and, as stated above, is also unprovable.

This proves the consistency of E. □

In fact, our example system fulfills an even stronger property. Since all provable formulas are substantively true, system E is not only consistent but also correct.

1.3.2 Completeness

Now that consistency is assured, let us investigate whether system E is complete. For this purpose, let's take another look at Figure 1.4. The entries in the lower left quadrant already suggest that E is incomplete, as true formulas exist that are not derivable from the axioms. Indeed, the unprovability of those formulas can be easily recognized. A formula of the type

$$\neg(\sigma > \sigma) \tag{1.6}$$

is only derivable if the formula $(\sigma > \sigma)$ appears earlier in the proof chain. However, it was stated above that this formula is unprovable in E.

This can be solved by adding the schema

$$(\sigma = \tau) \rightarrow \neg(\tau > \sigma) \tag{A7}$$

to the axioms. Since the formula $(\sigma = \sigma)$ is a theorem for every term σ, all formulas of the type (1.6) are now provable.

However, the extended system is still incomplete, as with

$$(0 = 0) \rightarrow (0 = 0)$$

there is yet a formula that is substantively true but underivable from the axioms.

1.3 Metamathematics

Figure 1.6

GOTTFRIED WILHELM LEIBNIZ
1646 – 1716

Formal systems that are simultaneously correct and complete are the dream of most mathematicians. Every true mathematical statement expressible in the formal system's artificial languages would be provable while ensuring no substantively false statement would ever be derivable from the axioms. In such systems, the notions of provability and truth perfectly align. There, and only there, the provable formulas *are* the true statements.

The idea of linking concepts and thoughts with objects of a formal language originates back to the seventeenth century. It traces back to Gottfried Wilhelm Leibniz, undoubtedly one of the most extraordinary scholars of the late seventeenth and early eighteenth centuries (Figure 1.6). It would be narrow-minded to reduce his person to a scientist, a philosopher, or a jurist. Leibniz, born 1646 in Leipzig, Germany, made significant contributions in all these areas, and conventional standards can hardly measure his life's work. Therefore, we want to join all those who, for lack of alternatives, refer to Leibniz as a universal scholar. This title is not very telling, but none could be more honorable.

Leibniz based his views on several *great principles*, with the *Principle of Contradiction* and the *Principle of Sufficient Reason* being the most prominent. The latter expresses the belief that every effect is preceded by a cause and that nothing in the world happens without reason. In Leibniz's words, it is the principle,

> "[...] in virtue of which we hold that there can be no fact real or existing, no statement true, unless there be a sufficient reason, why it should be so and not otherwise, although these reasons usually cannot be known by us."

> "[...] en vertu du qvel nous considerons qv'aucun fait ne sauroit se trouver vray ou existent, aucune Enonciation veritable, sans qv'il y ait une raison

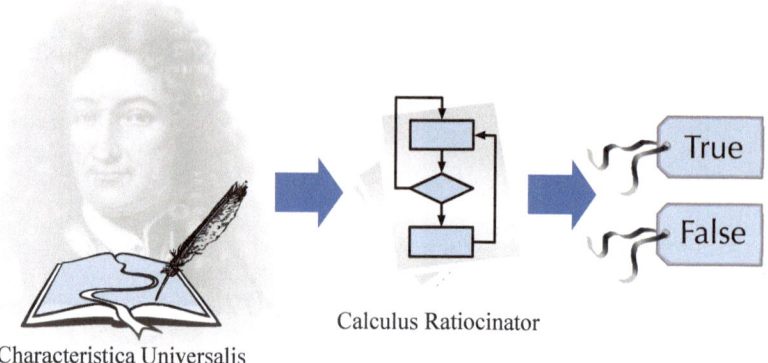

Figure 1.7: Throughout his life, Leibniz believed in the possibility of creating a *calculus ratiocinator* capable of mechanically determining the truth value of a formalized statement.

> *suffisante, pour qvoy il en soit ainsi, et non pas autrement; quoyque ces raisons le plus souvent ne puissent point nous etre connues."*
>
> Gottfried Wilhelm Leibniz [65]

Leibniz envisioned a universal language expressive enough to describe all of human knowledge. This *characteristica universalis* was to be a symbolic language, very similar to the one we use in modern mathematics. Within this language, he wanted to encode the objects and concepts of the real world to make their relationships visible at the syntactic level.

Leibniz firmly believed that the truth value of any formalized statement was calculable. This was to be done with the *calculus ratiocinator*, a fixed set of rules meant to operate much like the algorithmically working computers of our age (Figure 1.7). With such a set of rules, we could tackle all outstanding problems without fear. We would be ready to answer each of them, starting with the famous Leibnizian saying: *Calculemus – Let us calculate!*

With his visionary idea, Leibniz was far ahead of his time, but until his death, neither he nor any other scholar succeeded in bringing it to life. But then, more than 200 years later, signs of an imminent incarnation began to emerge. Hilbert's axiomatic method seemed to bring a *characteristica universalis* within reach – at least for the field of mathematics.

1.3.3 Hilbert's Program

With the development of the axiomatic method, Hilbert also pursued a practical interest. He sought to resolve a dispute that arose at the beginning of the

1.3 Metamathematics

Figure 1.8

LEOPOLD KRONECKER
1823 – 1891

twentieth century, revolving around which concepts and methods should be permitted in mathematics and which should not.

Leopold Kronecker is credited with saying, *"The good Lord made the integers, all else is the work of man."*[a] The German mathematician was one of the most prominent and also one of the most vehement opponents of the new mathematics invented by his student Georg Cantor. Towards the end of the nineteenth century, Cantor had begun experimenting with certain sets, formerly called *manifolds*, which stretched the concept of infinity in a way that was considered improper at the time. By repeatedly combining infinite sets of accumulation points into new sets, Cantor obtained significant results in the field of Fourier series. [6, 84]. However, his constructions seemed so adventurous that several renowned mathematicians rejected them.

Most critics believed that his wildly constructed sets must never be regarded as a self-contained whole, but this is precisely what Cantor had in mind. He had formed his manifolds according to the same principles used today for the construction of *ordinal numbers* and treated each set as an independent, self-contained individual. What is today considered a standard method of mathematics seemed strange to many mathematicians at the time. The severity of his alleged wrongdoing was assessed quite differently back then. For some, Cantor's mathematics was seen as a frivolous but harmless liaison with the *actual infinite*, while others considered it a dangerous play with fire.

Chapter 2 will reveal why the expressed concerns were partly justified. This much in advance: Cantor went a step too far. Just like the German mathematician Gottlob Frege, whose work will be discussed in detail later, Cantor had, initially unnoticed, opened a gateway to logical antinomies. What had

---

[a] *"Die ganzen Zahlen hat der liebe Gott gemacht, alles andere ist Menschenwerk"* [98]

Figure 1.9

GEORG CANTOR
1845 – 1918

begun as a play with fire was about to become a conflagration, and the new mathematics was on the verge of engulfing a blazing inferno.

At the beginning of the twentieth century, the Dutch mathematician Luitzen Brouwer initiated another attack. He not only criticized the reckless handling of infinity but also questioned long-established basic principles of classical mathematics. At the core of Brouwer's philosophy was the idea of constructive mathematics. The intuitionists accepted a statement as true only if constructive proof would back it up. Likewise, they only considered those mathematical objects to exist that could be explicitly constructed. In this respect, intuitionism is a counterpoint to *Platonism*, which grants mathematical objects an independent existence in the realm of thought. There, the truth or falsity of a statement is a static property that exists independently of the real world. Platonists see mathematics merely as a tool for unveiling the truth value of a proposition through deductive reasoning.

The intuitionists behind Brouwer vehemently refused to assign a truth value to a statement if it could not be determined constructively. In doing so, they openly opposed the *principle of excluded middle* (*tertium non datur*), which asserts that either a statement or its negation must be true. With his intuitionistic program, Brouwer attacked mathematics at its core since all derivations that proved a statement by excluding its opposite were now rejected.

Throughout his life, Hilbert was bothered by the intuitionists, and he did everything to protect mathematics against their attacks. In his famous treatise *On the Infinite* from 1926, he defended the methods that entered mathematics through Cantor's way of thinking, with the well-known quote:

> "No one shall drive us out of the paradise which Cantor has created for us."

1.3 Metamathematics

> "Aus dem Paradies, das Cantor uns geschaffen, soll uns niemand vertreiben können."
>
> David Hilbert [46, 44]

For Hilbert, the *principle of excluded middle* was integral to the mathematical method, and the idea of removing it was simply inconceivable to him. In 1928, he expressed his view with the famous words:

> "Taking the principle of excluded middle from the mathematician would be the same, say, as proscribing the telescope to the astronomer or to the boxer the use of his fists."
>
> "Dieses Tertium non datur dem Mathematiker zu nehmen, wäre etwa, wie wenn man dem Astronomen das Fernrohr oder dem Boxer den Gebrauch der Fäuste untersagen wollte."
>
> David Hilbert [48, 45]

Hilbert regarded the axiomatic method as an instrument for settling the ongoing dispute. For him, the correctness of the criticized concepts was beyond question, and he was confident that he could back them up with formal arguments. *Hilbert's program* began in the 1920s and is motivated in [43] with the following words:

> "My investigations in the new grounding of mathematics have as their goal nothing less than this: to eliminate, once and for all, the general doubt about the reliability of mathematical inference. We can see how necessary such an investigation is, if we think of how changeable and imprecise the intuitions of even the most distinguished mathematicians have been in this area, or if we remember that the inferences that were previously regarded as the most certain in mathematics have been challenged by some of the most renowned mathematicians of modern times"
>
> "Meine Untersuchungen zur Neubegründung der Mathematik bezwecken nichts Geringeres, als die allgemeinen Zweifel an der Sicherheit des mathematischen Schließens definitiv aus der Welt zu schaffen. Wie nötig eine solche Untersuchung ist, gewahren wir, wenn wir bedenken, wie wechselnd und unpräzise die diesbezüglichen Anschauungen oft selbst der hervorragendsten Mathematiker waren, oder wenn wir uns erinnern, daß von einigen der namhaftesten Mathematiker der neuesten Zeit die bisher für die sichersten gehaltenen Schlüsse in der Mathematik verworfen werden."
>
> David Hilbert [51, 43]

The realization of his project would have required constructing a formal system capable of representing all concepts and methods of classical mathematics. The natural, rational, and real numbers would be present, along with the concept

Hilbert envisioned constructing a formal system
expressive enough to formalize classical mathematics.

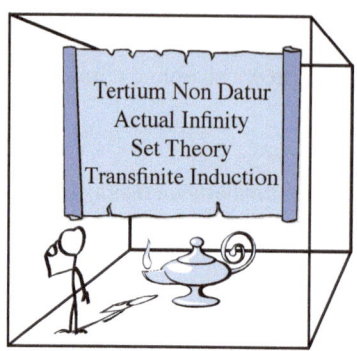 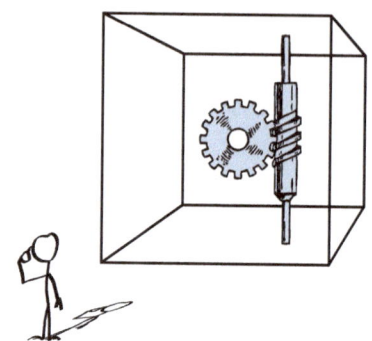

From the inside, the system would represent classical mathematics, encompassing all concepts that were controversially discussed at the time and rejected by some mathematicians as inadmissible.

From the outside, the system would appear as a set of rules to derive theorems mechanically. If its consistency were provable with finite means, all methods existing within the system would be safeguarded against contradictions.

Figure 1.10: Hilbert's program

of sets, which was controversial at the time. Proof methods would also be available, including accepted methods such as direct proof, as well as controversial principles like transfinite induction or the law of excluded middle. From the inside, the system would resemble the mathematical framework we know, embedded within a complete formal system in the sense of Definition 1.5.

From the outside, the system would appear as a complex set of rules, operating on the same basic principles as our example systems B and E. What would appear as a proof in classical mathematics, when viewed from the inside, would appear as a sequence of mechanical operations on symbol strings when viewed from the outside.

Hilbert planned to exploit this dual view to prove the consistency of classical mathematics. Such a proof should be conducted *from the outside*, that is, a mathematically precise analysis of all axioms and inference rules should ensure that no contradictions are derivable within the system. We have shown above how such a proof can be performed for simple formal systems like our example system E. If a similar proof succeeded for the system Hilbert had in mind, all methods *within* the system would be safeguarded against contradictions. In short, this was the objective of what is today known as *Hilbert's program* (Figure 1.10).

1.3 Metamathematics

Figure 1.11

JOHN VON NEUMANN
1903 – 1957

Of course, we would gain nothing if such a proof utilized the same potentially unsafe methods it was meant to back up. Hilbert intended to carry out the proof exclusively with *finite means*. Roughly speaking, this term refers to all proof methods the disputants considered legitimate at the time. Arguments that relied, for example, on the law of excluded middle had to be avoided, just as any argument that treated infinite collections of objects as a totality.

Initially, Hilbert's program went according to plan since it turned out that consistency could indeed be proved for distinct areas of mathematics. Based on the initial successes, it seemed only a matter of time before the consistency of all classical mathematics could be assured by finite means. Hilbert's program seriously threatened Brouwer and his followers, for they knew: If Hilbert were to present flawless proof in the intuitionistic sense, it would be the death blow to their philosophical movement.

At this juncture, we pause our journey through the history of mathematics and return to where it all began. It was noon in Königsberg when John von Neumann (Fig. 1.11) concluded his talk about formalism with a summary of the current state of Hilbert's program:

> "Although the consistency of classical mathematics has not yet been proved, such a proof has been found for a somewhat narrower mathematical system. [...] Thus Hilbert's system has passed the first test of strength: the validity of a non-finitary, not purely constructive mathematical system has been established through finitary constructive means. Whether someone will succeed in extending this validation to the more difficult and more important system of classical mathematics, only the future will tell."

> "Der gegenwärtige Stand der Dinge ist dadurch gekennzeichnet, daß

> *die Widerspruchsfreiheit der klassischen Mathematik immer noch unbewiesen ist, dagegen dieser Beweis für ein etwas engeres mathematisches System bereits geglückt ist. [...] Dadurch hat Hilberts System die erste Kraftprobe bestanden: die Rechtfertigung eines nicht finiten und nicht rein konstruktiven mathematischen Systems ist mit finit-konstruktiven Mitteln geglückt. Ob es gelingen wird, diese Rechtfertigung am schwierigeren und wesentlicheren System der klassischen Mathematik zu wiederholen, wird die Zukunft lehren."*
>
> John von Neumann [74, 72]

Von Neumann had no idea how close the answer to this question already was.

1.4 The Incompleteness Theorems

The second day began with lectures by Hans Reichenbach and Werner Heisenberg about the impact of quantum mechanics on the concepts of physical truth and causality. On the agenda for the afternoon was a 60-minute lecture on the history of pre-Greek mathematics and three 20-minute short lectures on the foundations of mathematics. The short lectures were given by Arnold Scholz, Walter Dubislav, and Kurt Gödel.

In his talk *On the Completeness of the Logic Calculus*, Gödel discussed what is now known as the *completeness theorem*. He had proven the theorem in his dissertation, thereby solving a critical foundational question in mathematical logic. The completeness theorem makes a statement about *first-order predicate logic*, or PL1 for short, thoroughly discussed in Section 6.2.2. Gödel proved that PL1 is complete if this term is restricted to the derivability of *universally valid formulas*. To anticipate: A universally valid formula becomes a true statement under all possible interpretations of its predicate and function symbols.

For the formalists, Gödel's completeness theorem was a significant milestone. It seemed to bring the realization of Hilbert's program within reach, and no one suspected the hope for its early completion to be shattered by the very next morning.

1.4.1 The First Incompleteness Theorem

On the agenda for the third and final day was a discussion on the foundations of mathematics, opened up by a longer lecture by Hans Hahn. Alongside Rudolf Carnap, John von Neumann, Arend Heyting, and Kurt Gödel were also present. Our protagonist only spoke up towards the end of the session in his typical reserved and precise manner:

1.4 The Incompleteness Theorems

Figure 1.12

KURT GÖDEL
1906 – 1978

"One can – assuming the consistency of classical mathematics – even give examples of sentences [...] that are indeed correct in content, but unprovable in the formal system of classical mathematics."

"Man kann – unter Voraussetzung der Widerspruchsfreiheit der klassischen Mathematik – sogar Beispiele für Sätze [...] angeben, die zwar inhaltlich richtig, aber im formalen System der klassischen Mathematik unbeweisbar sind."

Kurt Gödel [9]

What the young mathematician expressed that morning was the first public formulation of what is now known as the *first incompleteness theorem*. With the terminology from Definition 1.5, Gödel's statement can be formulated as follows:

Theorem 1.6 — Gödel, 1930

Every consistent formal system expressive enough to formalize ordinary mathematics is incomplete.

Gödel discovered that in every sufficiently expressive formal system, it is possible to formulate true statements that are not provable within the system. In short, the concepts of truth and provability cannot coincide. This result destroyed the hope of all those who, like Hilbert, believed in the existence of a consistent and, at the same time, complete formal system for mathematics. A *characteristica universalis*, as Leibniz envisioned it, cannot exist. Born centuries ago as a visionary dream, it will remain as such forever.

1.4.2 The Second Incompleteness Theorem

Immediately after the discussion, John von Neumann sought a conversation with Gödel. Unlike most other listeners, who received Gödel's comment rather impassively, he appeared to have immediately grasped the significance hidden in the somewhat casual remark.

Several weeks later, von Neumann wrote a letter to Gödel. After the Königsberg conference, he extensively studied the incompleteness theorem and made another shattering discovery. His letter of November 20, 1930, begins with the following words:

> "Dear Mr. Gödel!
>
> *I have recently been dealing with logic again, using the methods that you have so successfully used to reveal undecidable properties. In doing so, I have achieved a result that seems remarkable to me. I was able to show that the consistency of mathematics is unprovable."*
>
> "Lieber Herr Gödel!
>
> *Ich habe mich in der letzten Zeit wieder mit Logik beschäftigt, unter Verwendung der Methoden, die Sie zum Aufweisen unentscheidbarer Eigenschaften so erfolgreich benützt haben. Dabei habe ich ein Resultat erzielt, das mir bemerkenswert erscheint. Ich konnte nämlich zeigen, dass die Widerspruchsfreiheit der Mathematik unbeweisbar ist."*
>
> John von Neumann [37]

Von Neumann had discovered *Gödel's second incompleteness theorem*. Simply put, this theorem states that a formal system strong enough to formalize the first incompleteness theorem cannot prove its own consistency.

The first incompleteness theorem was a hard blow for Hilbert's program, but what the second incompleteness theorem stated was like a walk to the scaffold. Here is the reason: If it is impossible in the system of classical mathematics to prove the consistency of classical mathematics, a fortiori such a proof cannot succeed if we restrict ourselves to a limited number of proof methods. But this was the core of Hilbert's ambitious program: proving the consistency of classical mathematics with finite means.

1.5 On Formally Undecidable Propositions

Von Neumann's letter came too late, as shortly after the conference in Königsberg, Gödel had independently discovered the second incompleteness theorem himself. As early as October 23, 1930, he sent a summary to the *Vienna*

1.5 On Formally Undecidable Propositions

Academy of Sciences and submitted the completed article on November 17, 1930 [31, 14].

In 1931, the *Monthly Journal for Mathematics and Physics* published Gödel's article under the somewhat awkward-sounding title

> **Über formal unentscheidbare Sätze der Principia Mathematica und verwandter Systeme I**[1]).
>
> Von Kurt Gödel in Wien.
>
> ---
>
> [1]) Vgl. die im Anzeiger der Akad. d. Wiss. in Wien (math.-naturw. Kl.) 1930, Nr. 19 erschienene Zusammenfassung der Resultate dieser Arbeit.

> Kurt Gödel
>
> **On Formally Undecidable Propositions of Principia Mathematica and Related Systeme I**[1])
>
> ---
>
> [1]) Cf. the summary of the results of this paper which appeared in the Anzeiger der Akad. d. Wiss. in Wien (math.-naturw. Kl.) 1930, Nr. 19.

Gödel intended the article to be the first of two parts. In the announced second part, he planned to provide a detailed proof of the second incompleteness theorem, which he had only outlined. However, this never came to fruition. Most mathematicians found Gödel's arguments in the presented form so convincing that he saw no need to publish a sequel. The gaps were later filled by David Hilbert and Paul Bernays [56].

Let's postpone the details for now and let Gödel speak instead:

> 1.
>
> Die Entwicklung der Mathematik in der Richtung zu größerer Exaktheit hat bekanntlich dazu geführt, daß weite Gebiete von ihr formalisiert wurden, in der Art, daß das Beweisen nach einigen wenigen mechanischen Regeln vollzogen werden kann. Die umfassendsten derzeit aufgestellten formalen Systeme sind das System der Principia Mathematica (PM)[2] einerseits, das Zermelo-Fraenkelsche (von J. v. Neumann weiter ausgebildete) Axiomensystem der Mengenlehre[3] andererseits. Diese beiden Systeme sind so weit, daß alle heute in der Mathematik angewendeten Beweismethoden in ihnen formalisiert, d. h. auf einige wenige Axiome und Schlußregeln zurückgeführt sind. Es liegt daher die Vermutung nahe, daß diese Axiome

1.

It is well known that the development of mathematics in the direction of greater precision has led to the formalization of extensive mathematical domains, in the sense that proofs can be carried out according to a few mechanical rules. The most extensive formal systems constructed up to the present time are the system of Principia Mathematica (PM)[2], on the one hand, and, on the other hand, the Zermelo-Fraenkel axiom system for set theory[3] (which has been developed further by J. v. Neumann). Both of these systems are so broad that all methods of proof used in mathematics today can be formalized in them, i.e. can be reduced to a few axioms and rules of inference.

---

[2] A. Whitehead and B. Russell, Principia Mathematica, 2nd edition. Cambridge, 1925. Among the axioms of the system PM we also include, in particular, the axiom of infinity (in the form: there exist precisely denumerably many individuals), the axiom of reducability and the axiom of choice (for all types).

[3] Cf. A. Fraenkel, "Zehn Vorlesungen über die Grundlegung der Mengenlehre." Wissensch. u. Hyp., Vol. XXXI. J. v. Neumann, "Die Axiomatisierung der Mengenlehre." Math. Zeitschr. 27 (1928). Journ. f. reine u. angew. Math. 154 (1925), 160 (1929). We note that, in order to complete the formalization, one must add the axioms and rules of inference of the logical calculus to the set-theoretic axioms and rules of inference of the logical calculus to the set-theoretic axioms given in the literature just cited. The arguments that follow also hold for the formal systems constructed recently by D. Hilbert and his co-workers (so far as these have been published up to the present). Cf. D. Hilbert, Math. Ann. 88, Abh. aus d. math. Sem. der Univ. Hamburg I (1922), VI (1928); P. Bernays, Math. Ann. 90; J. v. Neumann, Math. Zeitschr. 26 (1927); W. Ackermann, Math. Ann. 93.

The article begins with a brief survey of early twentieth-century mathematics. In particular, Gödel mentions the *Principia Mathematica* and the *Zermelo-Fraenkel set theory* as the two major formal foundations of mathematics. Both systems were born out of necessity, dating back to when mathematics experienced one of its greatest crises. It was the time when mathematics became tangled in the webs of set-theoretic antinomies, with little hope of an easy way out.

1.5 On Formally Undecidable Propositions

Gödel refers to the mentioned systems multiple times, and his opening remarks make one very clear: To understand Gödel's work, one must understand history. For this reason, let us leave the article behind and travel into the past once again. This time, the destination of our journey is the mathematics of the late nineteenth and early twentieth centuries.

2 Foundations of Mathematics

> *"The fundamental thesis [...], that mathematics and logic are identical, is one which I have never since seen any reason to modify."*
>
> Bertrand Russell [92]

In this chapter, we will delve deeper into the history of mathematical logic and introduce various concepts crucial for comprehending Gödel's work. Our journey starts in Section 2.1 with a short visit to the late nineteenth century. Here, we will acquaint ourselves with Gottlob Frege, who not only made significant contributions to the development of modern logic but also stands as one of the tragic figures in the annals of science. Section 2.2 will discuss the contributions of Giuseppe Peano, in particular, his axiomatic foundation of natural numbers. The endeavors of Frege and Peano were pivotal in the life of our next protagonist, Bertrand Russell. Section 2.3 will derive *Russell's antinomy* and explain why it damaged mathematics at its core. Subsequently, we will open up the monumental work that Gödel mentions in the title of his paper: the *Principia Mathematica*. Finally, Section 2.4 will discuss modern set theory and provide an overview of the various axiomatic systems invented to put mathematics on solid ground.

2.1 The Logicist Program

The history of mathematical logic is closely intertwined with the life of Gottlob Frege, who was born on November 8, 1848, in the Mecklenburg town of Wismar. After completing his early education at his birthplace, Frege enrolled at the University of Jena in 1869, where he crossed paths with Ernst Abbe. The former director of Carl Zeiss AG not only served as an influential teacher but also became a lifelong supporter. It may have been one of Abbe's suggestions that led Frege to transfer from Jena to the prestigious mathematics faculty in Göttingen in 1871. Here, Frege specialized in Geometry and earned his doctoral degree in 1873. In 1874, he submitted his habilitation thesis in Jena and secured an Extraordinarius position in 1879 after several years of private lecturing.

Figure 2.1

GOTTLOB FREGE
1848 – 1925

2.1.1 *Begriffsschrift*

That very same year, Frege published his first major work titled *Begriffsschrift, eine der arithmetischen nachgebildete Formelsprache des reinen Denkens* [28]. The booklet of only 88 pages was fundamental to developing mathematical logic. It is no exaggeration when some authors refer to 1879 as *"the most important date in the history of logic since Aristotle"* [a].

Frege's *Begriffsschrift* initiated a philosophical movement that is now referred to as *Logicism*. It adopts the perspective that mathematics is a branch of logic rather than logic being a branch of mathematics. Frege firmly believed that all mathematics could be developed within logic and envisioned defining mathematical entities, such as natural numbers, by breaking them down into fundamental logical principles.

Frege understood that he needed to create an artificial language many times more precise than all natural languages developed throughout human history. The creation of this language is the content of the *Begriffsschrift*. In the preface, he writes:

> "To prevent anything intuitive from penetrating here unnoticed, I had to bend every effort to keep the chain of inferences free of gaps. In attempting to comply with this requirement in the strictest possible way I found the inadequacy of language to be an obstacle; no matter how unwieldy the expressions I was ready to accept, I was less and less able, as the relations became more and more complex, to attain the precision that my purpose required. This deficiency led me to the idea of the present ideography."

---

[a] *"das wichtigste Datum in der Geschichte der Logik seit Aristoteles"* [3]

2.1 The Logicist Program

"[es] musste alles auf die Lückenlosigkeit der Schlusskette ankommen. Indem ich diese Forderung auf das strengste zu erfüllen trachtete, fand ich ein Hindernis in der Unzulänglichkeit der Sprache, die bei aller entstehenden Schwerfälligkeit des Ausdruckes doch, je verwickelter die Beziehungen wurden, desto weniger die Genauigkeit erreichen ließ, welche mein Zweck verlangte. Aus diesem Bedürfnisse ging der Gedanke der vorliegenden Begriffsschrift hervor."

Gottlob Frege [28, 21]

Frege demanded mathematical proofs to exhibit a precision akin to Hilbert's formalistic view. Nevertheless, logicism and formalism differ in crucial aspects. Unlike Hilbert, Frege employed the axiomatic method in the Euclidean sense, that is, he viewed the axioms as being representatives of *"the true"*, implying that *"the false"* could never be deduced when only correct rules of inference were applied. From this philosophical perspective, it is understandable why Frege saw no purpose in seeking a consistency proof for his formal system. In a letter to Hilbert from December 27, 1899, he justified his position as follows:

"I call axioms sentences that are true, but which are not proven, because their knowledge flows from a source of knowledge different from the logical one, which can be called spatial intuition. From the truth of the axioms it follows that they do not contradict each other. This therefore requires no further proof."

"Axiome nenne ich Sätze, die wahr sind, die aber nicht bewiesen werden, weil ihre Erkenntnis aus einer von der logischen verschiedenen Erkenntnisquelle fliesst, die man Raumanschauung nennen kann. Aus der Wahrheit der Axiome folgt, dass sie einander nicht widersprechen. Das bedarf also keines weiteren Beweises."

Gottlob Frege [30]

The logical calculus of the *Begriffsschrift* corresponds, in essential parts, to what modern literature recognizes as a second-order predicate logic calculus with equality. Thus, Frege's contributions extended well beyond the work of George Boole and Augustus De Morgan, who had laid the foundations of modern propositional logic a few years earlier. Unlike the logic of Boole or De Morgan, Frege's logic was expressive enough to capture the entire core of classical mathematics.

Contemporary readers may find it challenging to recognize the similarities between Frege's logic and modern predicate logic, primarily because of the unique notation, which significantly diverges from current standards. Frege pioneered an entirely novel system, organizing the elements of formulas in two dimensions. Figure 2.2 summarizes how the elementary logic operations are represented in this notation.

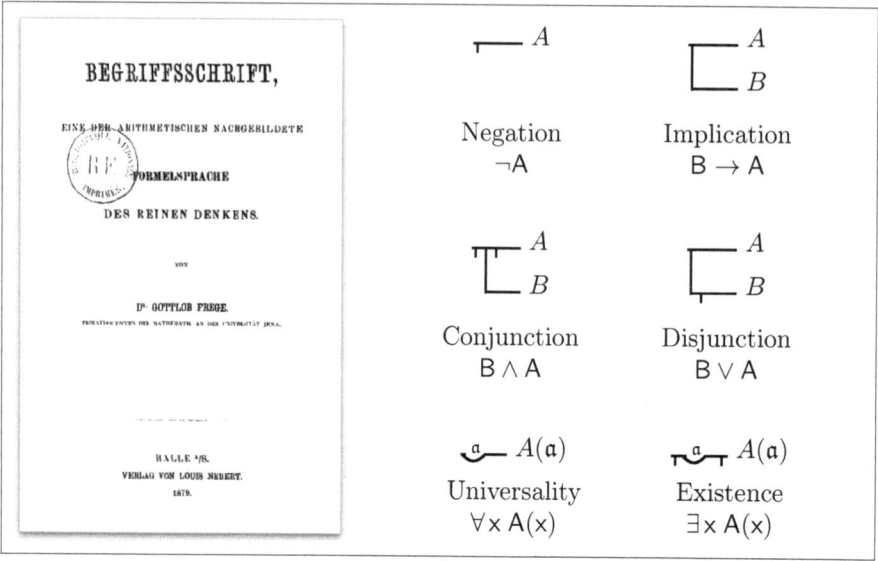

Figure 2.2: The symbolic notation of the *Begriffsschrift*

While Frege's notation did not stand the test of time, his way of formalizing logical facts did. The *Begriffsschrift* shaped the perspective of considering predicates as *sentence functions* with subjects as their arguments. Accordingly, a statement like "Socrates is a human" is represented in the form Human(Sokrates). The statement is created by applying the sentence function Human(x) to the argument Socrates.

To express "All humans are mortal" in the notation of the *Begriffsschrift*, one would write:

$$\vcenter{\hbox{$\displaystyle\begin{array}{l} \text{Mortal}(\mathfrak{a}) \\ \text{Human}(\mathfrak{a}) \end{array}$}} \qquad (2.1)$$

Frege was well aware of the significance of his novel concept and believed it would take a firm place in logic. In the preface of his Begriffsschrift, we find:

> "In particular, I believe that the replacement of the concepts *subject* and *predicate* by *argument* and *function*, respectively, will stand the test of time."

> "Insbesondere glaube ich, dass die Ersetzung der Begriffe *Subject* und *Praedicat* durch *Argument* und *Funktion* sich auf die Dauer bewähren wird."

Gottlob Frege [28, 21]

Frege was right. In modern notation, formula (2.1) looks quite familiar to us:

$$\forall x\, (\text{Human}(x) \to \text{Mortal}(x))$$

2.1 The Logicist Program

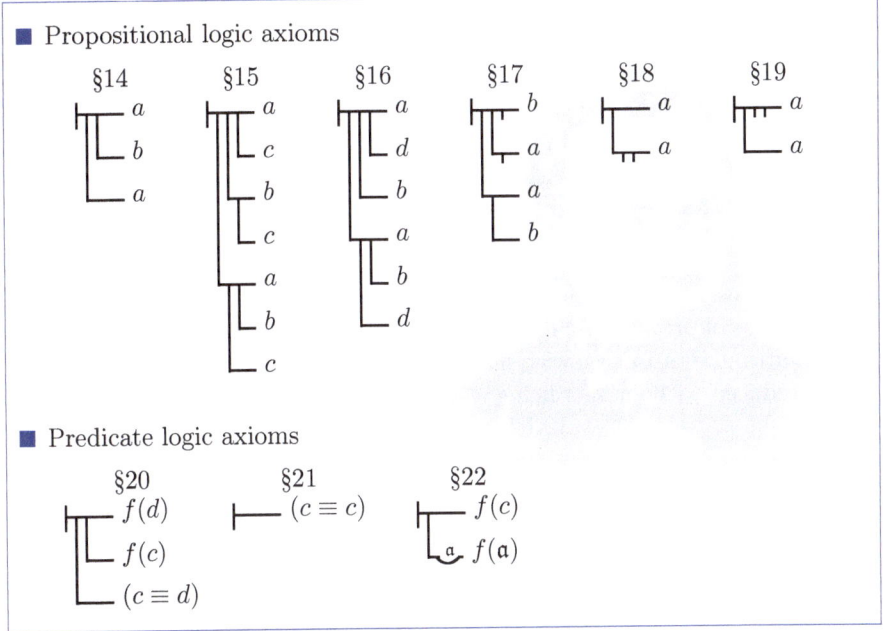

Figure 2.3: The axioms of the *Begriffsschrift*

Today, we are so accustomed to this structural style that we rarely consider its origin. Few are aware that we resort to the formal framework introduced with Frege's *Begriffsschrift* when writing down logical facts.

2.1.2 Axioms of the *Begriffsschrift*

In §14 to §22 of the *Begriffsschrift*, Frege introduces nine axioms, summarized in Figure 2.3. They are the fundamental building blocks of his logical framework and break down into two groups. The first group consists of six propositional logic axioms. The remaining three axioms relate to predicate logic; they describe the basic properties of sentence functions and equality. To reveal their meaning, we first translate them into modern notation:

$$\S14 : A \to (B \to A) \tag{F.1}$$
$$\S15 : (C \to (B \to A)) \to ((C \to B) \to (C \to A)) \tag{F.2}$$
$$\S16 : (D \to (B \to A)) \to (B \to (D \to A)) \tag{F.3}$$
$$\S17 : (B \to A) \to (\neg A \to \neg B) \tag{F.4}$$
$$\S18 : \neg\neg A \to A \tag{F.5}$$
$$\S19 : A \to \neg\neg A \tag{F.6}$$
$$\S20 : c = d \to (F(c) \to F(d)) \tag{F.7}$$
$$\S21 : c = c \tag{F.8}$$

§22 : $\forall x\, F(x) \rightarrow F(c)$ (F.9)

Frege provided two inference rules for the deduction of theorems: the *rule of detachment* and the *rule of substitution*. We are already familiar with the former: it is identical to the modus ponens we've extensively used in Section 1.2 for deriving theorems within the formal system E. The rule of substitution is novel, stating that replacing a variable of an axiom or a theorem with another expression yields a new theorem.

Frege's propositional axioms are *complete*, as they are sufficient to derive all true formulas of propositional logic. They are not *minimal*, though. The Polish mathematician Jan Łukasiewicz has demonstrated that the number of axioms can be reduced without sacrificing a single theorem. In [67] he writes:

> "Frege is the founder of the modern propositional calculus. His system, which does not even seem to be known in Germany, is based on the following 6 axioms: 'CpCqp', 'CCpCqrCCpqCpr', 'CCpCqrCqCpr', 'CCqpCNqNp', 'CNNpp', 'CpNNp'. The third axiom is superfluous, for it is derivable from the first two. The last three axioms can be replaced by the sentence 'CCNpNqCqp'."

> "Frege ist der Begründer des modernen Aussagenkalküls. Sein System, das nicht einmal in Deutschland bekannt zu sein scheint, ist auf folgenden 6 Axiomen aufgebaut: 'CpCqp', 'CCpCqrCCpqCpr', 'CCpCqrCqCpr', 'CCqpCNqNp', 'CNNpp', 'CpNNp'. Das dritte Axiom ist überflüssig, denn es ist aus den beiden ersten ableitbar. Die drei letzten Axiome können durch den Satz 'CCNpNqCqp' ersetzt werden."

Jan Łukasiewicz [67]

Apart from the quotation's content, the notation used for writing down the formulas is particularly remarkable. To achieve a compact representation, Łukasiewicz developed a distinct notation, called *Polish notation* today, in honor of the inventor's origin. In this notation, Cpq stands for $P \rightarrow Q$, and Np stands for $\neg P$, where the symbols p and q are placeholders for arbitrary formulas. With this knowledge, we can gradually translate Łukasiewicz's three axioms into modern notation:

$$Cp\,\underbrace{Cqp}_{(Q \rightarrow P)} = P \rightarrow (Q \rightarrow P)$$

$$C\underbrace{Cp\,\underbrace{Cqr}_{(Q \rightarrow R)}}_{P \rightarrow (Q \rightarrow R)}\,\underbrace{C\,\underbrace{Cpq}_{(P \rightarrow Q)}\,\underbrace{Cpr}_{(P \rightarrow R)}}_{(P \rightarrow Q) \rightarrow (P \rightarrow R)} = (P \rightarrow (Q \rightarrow R)) \rightarrow ((P \rightarrow Q) \rightarrow (P \rightarrow R))$$

$$C\underbrace{C\,\underbrace{Np}_{\neg P}\,\underbrace{Nq}_{\neg Q}}_{(\neg P \rightarrow \neg Q)}\,\underbrace{Cqp}_{(Q \rightarrow P)} = (\neg P \rightarrow \neg Q) \rightarrow (Q \rightarrow P)$$

2.1 The Logicist Program

The proposed axioms have passed the test of time. Many contemporary textbooks use these axioms as the axiomatic foundation for propositional logic.

2.1.3 Formalization of Arithmetic

Frege strongly believed that his *Begriffsschrift* would provide a suitable formal framework for executing his logicist program. Initially, everything proceeded according to plan, and with his second major work, *The Foundations of Arithmetic* [25] (*Die Grundlagen der Arithmetik* [22]), published in 1884, he reached an important milestone. In this book, he attempted to develop arithmetic within a logical framework, which he believed was only possible by solidly defining the concept of numbers. Frege had chosen a set-theoretic approach, thus basing his work on the conceptual framework that Cantor had invented. Among others, this enabled him to compare quantities based on their cardinalities.

Frege's original wording is rather challenging to understand because he did not use the term *set*. Instead, he employed the German word *Begriffsumfang*, which loosely translates to *conceptual scope* or *conceptual extent*. Therefore, we do not reproduce Frege's definition in its original but its the set-theoretic formulation given in [3]:

> "(i) The cardinality of a set X is the totality of all sets that have the same cardinality as X. (ii) n is a number if a set X exists such that n is the cardinality of X. (iii) 0 is the cardinality of the empty set. (iv) 1 is the cardinality of the set that consists only of 0. (v) The number n is the successor of the number m if there is a set X and an element a of X such that n is the cardinality of X and m is the cardinality of the set X without the element a (i.e., of $X\backslash\{a\}$). (vi) n is a finite (natural) number if n is an element of all sets Y for which the following holds: if 0 is an element of Y, and if k is an element of Y, then so is the successor of k."

> "(i) Die Mächtigkeit einer Menge X ist die Gesamtheit aller Mengen, die gleichmächtig zu X sind. (ii) n ist eine Zahl, wenn eine Menge X existiert, so dass n die Mächtigkeit von X ist. (iii) 0 ist die Mächtigkeit der leeren Menge. (iv) 1 ist die Mächtigkeit der Menge, die nur aus 0 besteht. (v) Die Zahl n ist der Nachfolger der Zahl m, wenn es eine Menge X und ein Element a von X gibt, so dass n die Mächtigkeit von X ist und m die Mächtigkeit der Menge X ohne das Element a (also von $X\backslash\{a\}$). (vi) n ist eine endliche (natürliche) Zahl, wenn n ein Element aller Mengen Y ist, für die gilt: 0 ist Element von Y, und ist k Element von Y, dann auch der Nachfolger von k."

> Thomas Bedürftig, Roman Murawski [3]

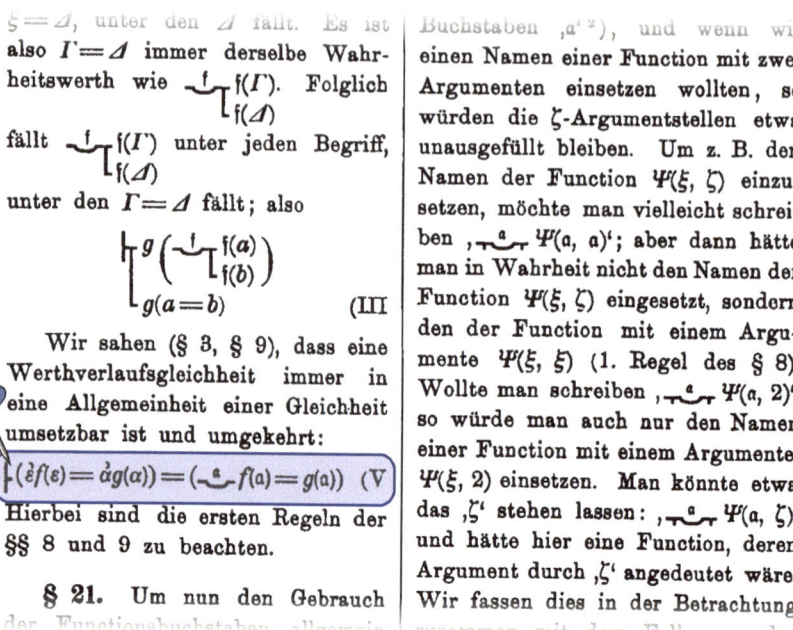

Figure 2.4: Excerpt from the 1st volume of *Grundgesetze der Arithmetik*. The displayed page contains the definition of law V, which made Frege a tragic figure in the history of science. This law opens a gateway for paradoxes.

In hindsight, reducing the natural numbers to sets may seem negligent. Today, we know that Frege relied on a construct less intuitive and considerably more uncertain than the object he was trying to justify. Back then, however, he knew nothing about the dangers hidden in naive set theory, still waiting to be discovered.

The Foundations of Arithmetic was written entirely in colloquial language. After publishing the book, Frege began translating the concepts into the logic of the *Begriffsschrift*. The result was *The Fundamental Laws of Arithmetic*, divided into two volumes. The first volume appeared in 1893, and the second in 1903 (cf. [26, 27]).

To put arithmetic on solid grounds, Frege supplemented the axioms of the *Begriffsschrift* with several basic arithmetic laws. Particularly well-known is the fifth basic law from the first volume (Fig. 2.4). It is called the *Basic law of course-of-values* and reads like this:

$$\vdash (\grave{\epsilon}f(\epsilon) = \grave{\alpha}g(\alpha)) = (\underset{\smile}{\mathfrak{a}}\; f(\mathfrak{a}) = g(\mathfrak{a})) \qquad (V)$$

2.1 The Logicist Program

The expression $\grave{\epsilon}f(\epsilon)$ denotes the *course-of-values* of function f. The term refers to the set representation of f, in which the arguments are combined with their function values to form ordered pairs:

$$\grave{\epsilon}f(\epsilon) := \{(x, y) \mid y = f(x)\}$$

Now, we can easily see what the fifth basic law is saying: Two functions f and g are identical if and only if they map each argument to the same value.

The law becomes particularly interesting when f and g represent what Frege called *concepts*. In this case, the functions map their arguments to "*the true*" or "*the false*" and are thus nothing but predicates. Frege referred to the course-of-values of a concept as the *conceptual scope* ("Begriffsumfang"). In the logic of the *Begriffsschrift*, each conceptual scope is uniquely associated with a set that includes precisely those objects mapped to "*the true*". In modern notation, this is the set $\{x \mid F(x)\}$, which allows us to rewrite basic law V as follows:

$$\{x \mid F(x)\} = \{x \mid G(x)\} \Leftrightarrow \forall z \, (F(z) \leftrightarrow G(z))$$

Frege was unaware of the explosive power hidden in this scheme. Without realizing it, he had compromised the foundation of his logic at a critical point and opened a gateway for paradoxes. This flaw remained unnoticed for many years, and an ever-growing edifice of thought was erected on unstable ground. But how could the contradictions in Frege's logic remain undiscovered for so long?

It happened mainly for two reasons. Firstly, set theory was a developing branch of mathematics, and even renowned mathematicians were still inexperienced in dealing with the novel structures. Much more important in this context, however, is the fact that the scientific community largely ignored Frege's contributions. They were neither noticed in the annual mathematical reports nor cited by Dedekind, who was also working on a formal justification of the natural numbers at that time. Kronecker did not mention them either, and even Cantor considered Frege's contributions largely insignificant.

From early on, Frege knew that finding an audience would be challenging for his ideas. In the preface of the *Begriffsschrift*, he explicitly pointed out:

> "I hope that logicians, if they do not allow themselves to be frightened off by an initial impression of strangeness, will not withhold their assent from the innovations that, by a necessity inherent in the subject matter itself, I was driven to make."
>
> "Ich hoffe, dass die Logiker, wenn sie sich durch den ersten Eindruck des Fremdartigen nicht zurückschrecken lassen, den Neuerungen, zu denen ich durch eine der Sache selbst innewohnende Notwendigkeit getrieben wurde, ihre Zustimmung nicht verweigern werden."
>
> Gottlob Frege [28, 21]

Figure 2.5

GIUSEPPE PEANO
1858 – 1932

Back then, he was still optimistic that his ideas would sooner or later find the acceptance they deserved. In the preface of the first volume of his *Foundations of Arithmetic*, however, he already sounded more pessimistic:

> "With this, I arrive at a second reason for the delay: the despondency that at times overcame me as a result of the cool reception, or rather, the lack of reception, by mathematicians of the writings mentioned above, and the unfavourable scientific currents against which my book will have to struggle."
>
> "Hiermit komme ich auf den zweiten Grund der Verspätung: die Muthlosigkeit, die mich zeitweilig überkam angesichts der kühlen Aufnahme, oder besser gesagt, des Mangels an Aufnahme meiner Schriften bei den Mathematikern und der Ungunst der wissenschaftlichen Strömungen, gegen die mein Buch zu kämpfen haben wird."
>
> Gottlob Frege [29, 23]

The great esteem in which we hold his work today came too late for Frege. Worn down by the long struggle for acceptance and acknowledgment, the uncovering of the paradoxes in 1902 was a blow from which he never recovered. He considered his life's work a failure and published only a few insignificant articles afterward. On July 26, 1925, Gottlob Frege died at 76 as a lonely and bitter man.

2.2 The Natural Numbers

The next stop on our journey is Spinetta, a small Italian village in southwestern Piedmont, where Giuseppe Peano was born on August 27, 1858. Aside from other mathematical contributions, Peano has influenced the modern way of phrasing mathematical facts like few others. Peano was born into a peasant family and spent his early years on a farm near Cuneo. After proving himself an exceptionally talented boy in school, his parents sent him to Turin at the age of 12, where he resided with his uncle for the next few years. During that time, Peano attended high school and studied mathematics at the university. His extraordinary performance paved the way for a smooth academic career. In 1880, Peano earned his doctorate from the University of Turin, and after working as a research assistant for several years, he was appointed a professor in 1889.

Peano devoted a large part of his scientific life to the *formulario project*. His goal was to describe mathematical knowledge in a symbolic language, precise enough to allow mathematical statements to be formally derived from a set of axioms defined a priori. The outcome of this work is the *Formulario Mathematica*, published in five volumes between 1895 and 1908.

Many of Peano's philosophical views trace back to Leibniz. The following quote from [3] is worth mentioning. It shows that Peano apparently believed to have realized Leibniz's dream of a *characteristica universalis* with his symbolic language:

> "After two centuries, this 'dream' of the inventor of calculus has become reality. We have indeed fulfilled the task set by Leibniz."
>
> "Nach zwei Jahrhunderten ist dieser 'Traum' des Erfinders der Infinitesimalrechnung Wirklichkeit geworden. Wir haben nämlich die von Leibniz gestellte Aufgabe erfüllt."
>
> Giuseppe Peano [3]

2.2.1 *Arithmetices Principia*

We will not delve into the details of the *Formulario Mathematica* in this book but instead focus on an earlier work: the *Arithmetices Principia* of 1889. It is one of Peano's most influential publications, which later appeared as an English translation titled *The principles of arithmetic*. The following quotes refer to the version reprinted in [79].

The *Arithmetices Principia* was Peano's first attempt to develop an axiomatic foundation for classical mathematics. He had come to believe that mathematics had reached a point where colloquial language was no longer sufficient to

	Peano	Russell	Hilbert, Ackermann		Present	
	1889 [77]	1910 [99]	1928 [52]	1958 [54]	Present	
Negation	$-\varphi$	$\sim \varphi$	$\overline{\varphi}$	$\rightharpoondown \varphi$	$\overline{\varphi}$	$\neg \varphi$
Disjunction	$\varphi \cup \psi$	$\varphi \vee \psi$		$\varphi \vee \psi$	$\varphi \vee \psi$	
Conjunction	$\varphi \cap \psi$ $\varphi . \psi$	$\varphi . \psi$	$\varphi \& \psi$	$\varphi \wedge \psi$	$\varphi \wedge \psi$	
Implication	$\varphi \supset \psi$	$\varphi \supset \psi$	$\varphi \rightarrow \psi$		$\varphi \rightarrow \psi$	
Equivalence	$\varphi = \psi$	$\varphi \equiv \psi$	$\varphi \sim \psi$	$\varphi \leftrightarrow \psi$	$\varphi \leftrightarrow \psi$	

Figure 2.6: Evolution of logic symbols over time

describe the increasingly complex concepts with the necessary precision, and just like Frege, Peano saw the solution in the creation of an artificial language:

> "Questions that pertain to the foundations of mathematics, although treated by many in recent times, still lack a satisfactory solution. The difficulty has its main source in the ambiguity of language. That is why it is of the utmost importance to examine attentively the vary words we use. My goal has been to undertake this examination, and in this paper I am presenting the results of my study, as well as some applications to arithmetic. I have denoted by signs all ideas that occur in the principles of arithmetic, so that every proposition is stated only by means of these signs."
>
> Giuseppe Peano [79]

The passage is reminiscent of the preface of the *Begriffsschrift*. However, Peano was unaware of this or any of Frege's other publications back then, as he cited Frege for the first time in 1891 [78].

Working independently for many years, the two mathematicians devised widely different solutions. While Frege created a highly developed logical apparatus in a difficult-to-understand notation, Peano achieved the opposite. On the one hand, Peano's logical framework was far less mature than Frege's. On the other hand, Peano had created a symbolic language capable of expressing logical facts in a remarkably elegant and precise manner. Many of Peano's symbols have withstood the test of time and are still used today in either their original or slightly modified form.

Figure 2.6 shows how some common logic symbols have matured over time. The symbols '∪' and '∩', which are used in the *Arithmetices Principia* to denote the disjunction (OR) and the conjunction (AND), respectively, had been introduced by Peano in 1888 for the union and the intersection of sets [76]. In set theory, they are still in use today. In logic, however, they were eventually replaced by '∨' and '∧'. For the implication operator, Peano wrote a rotated C ('⊃'). This

2.2 The Natural Numbers

symbol was later replaced by '⊃' and eventually by '→'. He also introduced the element symbol '∈', but with a different font, making it look like 'ϵ'. Finally, Peano invented the existential quantifier '∃', which is still used today in modern predicate logic.

In addition to the symbols mentioned above, Peano invented a distinct notion for grouping expressions. In particular, he proposed to replace the parenthesis symbols '(' and ')' with dots to structure formulas concisely. The core idea is so simple that a few words suffice to explain his dot notation:

> "To understand a formula divided by dots we first take together the signs that are not separated by any dot, next those separated by one dot, next those separated by two dots, and so on."
>
> Giuseppe Peano [79]

Peano illustrated his notation with the following example:

$$\text{ab . cd : ef . gh :. k} \qquad (2.2)$$

With the previously quoted explanation in our mind, it is easy to translate this expression into an ordinary parenthesized formula:

1. "We first take together the signs that are not separated by any dot,"
   ☞ (ab) . (cd) : (ef) . (gh) :. k

2. "next those separated by one dot,"
   ☞ ((ab)(cd)) : ((ef)(gh)) :. k

3. "next those separated by two dots,"
   ☞ (((ab)(cd))((ef)(gh))) :. k

4. "and so on."
   ☞ ((((ab)(cd))((ef)(gh)))k)

The following examples originate from the *Principia Mathematica*, which extensively uses Peano's symbolism. The symbol '⊃' is a typographical refinement of Peano's logical implication operator '⊃'.

■ **Example 1:** p ⊃ q . ⊃ : q ⊃ r . ⊃ . p ⊃ r

$$\begin{aligned}
\text{p} \supset \text{q} . \supset : \text{q} \supset \text{r} . \supset . \text{p} \supset \text{r} &= (\text{p} \supset \text{q}) . \supset : (\text{q} \supset \text{r}) . \supset . (\text{p} \supset \text{r}) \\
&= (\text{p} \supset \text{q}) \supset : ((\text{q} \supset \text{r}) \supset (\text{p} \supset \text{r})) \\
&= (\text{p} \supset \text{q}) \supset ((\text{q} \supset \text{r}) \supset (\text{p} \supset \text{r}))
\end{aligned}$$

☞ In modern notation: $(p \to q) \to ((q \to r) \to (p \to r))$

"if p implies q, then if q implies r, p implies r" [99]

- **Example 2:** p ∨ q . ⊃ :. p . ∨ . q ⊃ r : ⊃ . p ∨ r

$$\begin{aligned}
\text{p} \vee \text{q} . \supset :. \text{p} . \vee . \text{q} \supset \text{r} : \supset . \text{p} \vee \text{r} &= (\text{p} \vee \text{q}) . \supset :. \text{p} . \vee . (\text{q} \supset \text{r}) : \supset . (\text{p} \vee \text{r}) \\
&= (\text{p} \vee \text{q}) \supset :. (\text{p} \vee (\text{q} \supset \text{r})) : \supset (\text{p} \vee \text{r}) \\
&= (\text{p} \vee \text{q}) \supset :. ((\text{p} \vee (\text{q} \supset \text{r})) \supset (\text{p} \vee \text{r})) \\
&= (\text{p} \vee \text{q}) \supset ((\text{p} \vee (\text{q} \supset \text{r})) \supset (\text{p} \vee \text{r}))
\end{aligned}$$

☞ In modern notation: $(p \vee q) \rightarrow ((p \vee (q \rightarrow r)) \rightarrow (p \vee r))$

"if either p or q is true, then if either p or 'q implies r' is true, it follows that either p or r is true." [99]

The examples illustrate why Peano's dot notation did not stand the test of time. Unlike in the parenthesized expressions, the grouping is not immediately apparent in some dotted formulas. To make matters worse, Peano assigned the dot a double meaning. It served not only as a replacement for parentheses but also as a symbol for the AND operation.

Theoretically, this is not an issue, as it is unambiguous whether a dot acts as a grouping symbol or a logical operator. In practice, however, the double usage significantly affects readability, demonstrated vividly by the next example:

- **Example 3:** p ∨ q : p . ∨ . q ⊃ r : ⊃ . p ∨ r

$$\begin{aligned}
\text{p} \vee \text{q} : \text{p} . \vee . \text{q} \supset \text{r} : \supset . \text{p} \vee \text{r} &= (\text{p} \vee \text{q}) : \text{p} . \vee . (\text{q} \supset \text{r}) : \supset . (\text{p} \vee \text{r}) \\
&= (\text{p} \vee \text{q}) : (\text{p} \vee (\text{q} \supset \text{r})) : \supset (\text{p} \vee \text{r}) \\
&= ((\text{p} \vee \text{q}) : (\text{p} \vee (\text{q} \supset \text{r}))) \supset (\text{p} \vee \text{r}) \\
&= ((\text{p} \vee \text{q}) \wedge (\text{p} \vee (\text{q} \supset \text{r}))) \supset (\text{p} \vee \text{r})
\end{aligned}$$

☞ In modern notation: $((p \vee q) \wedge (p \vee (q \rightarrow r))) \rightarrow (p \vee r)$

"if either p or q is true, and either p or 'q implies r' is true, then either p or r is true." [99]

For most mathematicians, the dot notation is a relic of the past that hardly any young scientist can relate to. This is one of the reasons why it is challenging to read historical texts such as the *Principia Mathematica* today.

2.2.2 Axioms of the *Arithmetices Principia*

So far, we have primarily commented on Peano's notation, which, in reality, constitutes the less significant part of his 1889 contribution. The *Arithmetices*

2.2 The Natural Numbers

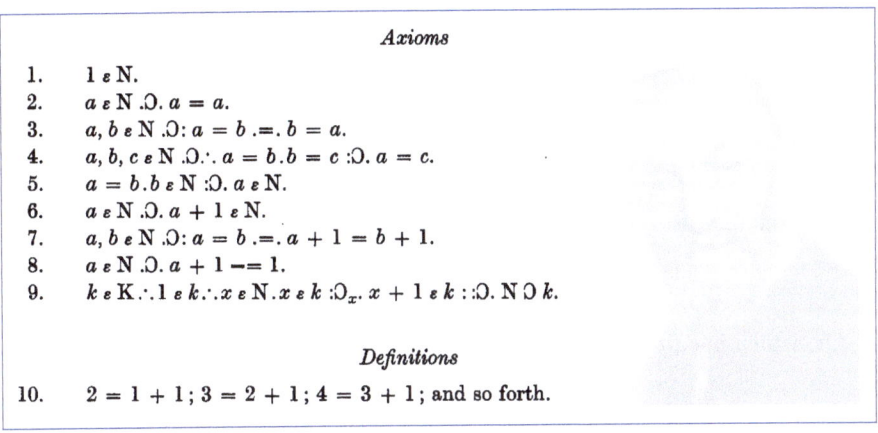

Figure 2.7: Characterization of the natural numbers in Peano's symbolic language [79]

Principia ranks among the most important pieces of mathematical world literature mainly because of their contents. Therein, Peano formulated the famous five axioms that unambiguously characterize the structure of the natural numbers. When contemporary mathematicians talk about the *Peano axioms*, they mean the axioms 1, 6, 8, 7, and 9 in Figure 2.7 (usually in this order). Let's take a closer look and sort out what these formulas are all about:

- **1.** 1 ε N.

  Peano uses the symbol 'N' to represent the natural numbers. Therefore, the axiom is the same as the formula

  $$1 \in \mathbb{N} \qquad \text{(PA.1)}$$

  and states:

  "*1 is a natural number.*"

  His choice to let the natural numbers begin with 1 instead of 0 holds little significance. The crucial aspect here is the fixed starting point of the number sequence, which is 1 in Peano's work. Substituting the symbol 1 with 0 in all axioms will make the sequence start at 0.

- **6.** a ε N .Ɔ. a + 1 ε N.

  In modern notation, the axiom appears in the following shape:

  $$a \in \mathbb{N} \Rightarrow a + 1 \in \mathbb{N} \qquad \text{(PA.2)}$$

  The axiom asserts that the successor of a natural number is also a natural number. Because '+' is a function, the successor is uniquely determined, allowing us to express the substantial meaning of the axiom as such:

  "*Every number has a unique successor.*"

- **8.** $a \, \varepsilon \, N \, .\supset. \, a + 1 \, -= \, 1.$

The minus symbol preceding the equal sign expresses logical negation. Thus, in modern notation, the axiom corresponds to the following formula:

$$a \in \mathbb{N} \Rightarrow a + 1 \neq 1 \qquad \text{(PA.3)}$$

The axiom states:

> "All successors are different from 1."

Or, which is equivalent and a more common formulation today:

> "1 is not the successor of any natural number."

- **7.** $a, b \, \varepsilon \, N \, .\supset: a = b \, .=. \, a + 1 = b + 1.$

In modern notation, this formula reads as follows:

$$a, b \in \mathbb{N} \Rightarrow (a = b \Leftrightarrow a + 1 = b + 1)$$

The axiom states that two numbers are equal if and only if their successors are. The direction from left to right follows from the definition of equality, making the following form equivalent:

$$a, b \in \mathbb{N} \Rightarrow (a + 1 = b + 1 \Rightarrow a = b)$$

Reversing the direction of the inner implication results in

$$a, b \in \mathbb{N} \Rightarrow (a \neq b \Rightarrow a + 1 \neq b + 1) \qquad \text{(PA.4)}$$

which is the form utilized by most contemporary textbooks:

> "Different numbers have different successors."

- **9.** $k \, \varepsilon \, K \, .\therefore . \, 1 \, \varepsilon \, k \, .\therefore . \, x \, \varepsilon \, N \, . \, x \, \varepsilon \, k \, :\supset_x. \, x + 1 \, \varepsilon \, k \, ::\supset. \, N \, \supset \, k.$

In Peano's terminology, the expression $k \, \epsilon \, K$ states that k is a set, allowing us to rephrase the axiom as follows:

$$1 \in M \wedge \forall x \, ((x \in \mathbb{N} \wedge x \in M) \Rightarrow x + 1 \in M) \Rightarrow \mathbb{N} \subseteq M \qquad \text{(PA.5)}$$

In colloquial terms, the axiom states:

> "If a set M contains the number 1 and for every natural number x from M also its successor $x + 1$, then all natural numbers are contained in M."

For systems such as Gödel's system P, where only natural numbers exist as individuals, the wording can be slightly simplified:

> "If a subset $M \subseteq \mathbb{N}$ contains the number 1 and for every element n also its successor $n + 1$, then $M = \mathbb{N}$."

2.2 The Natural Numbers

Next, let us consider an arbitrary property of the natural numbers, represented by the predicate symbol P, and write $P(x)$ if the number x has the property P. Then we may choose for M the set $\{x \in \mathbb{N} \mid P(x)\}$ and reformulate the axiom as follows:

> "If $P(1)$ holds and the property P inheres from every natural number x to its successor, then the property P applies to all natural numbers."

At this point, the meaning of Peano's fifth axiom becomes evident. It encapsulates the fundamental principle of mathematical induction.

Peano's approach to the natural numbers is appealing due to its simplicity. Unlike Frege, who developed a cumbersome conceptual framework for the same purpose, Peano achieved the same goal with only a few elementary definitions. The reason is simple: Unlike Frege, Peano refrained from justifying the natural numbers according to their nature. He primarily created his symbolic language to eliminate the ambiguities of everyday language.

> "In this way, one fixes a unique correspondence between thoughts and symbols, a correspondence not found in everyday language."
>
> "Auf diese Weise fixiert man eine eindeutige Korrespondenz zwischen Gedanken und Symbolen, eine Korrespondenz, die man in der Umgangssprache nicht findet."
>
> Giuseppe Peano [3]

Peano adopted the same pragmatic view that can be encountered in nearly all disciplines of modern mathematics, where ontological arguments are rarely made. Today, it is mainly philosophers who keep the discourse alive.

Peano's pragmatic approach led to a logical apparatus that separated the object level from the meta-level far less precisely than the one devised by Frege. As an example, let us consider the proof for the statement *"2 is a natural number"*, represented by the formula $2 \, \epsilon \, N$:

■ *Arithmetices Principia* ([79], page 94):

Proof:

P 1 .⊃:	$1 \, \epsilon \, N$	(1)
1 [a] (P 6) .⊃:	$1 \, \epsilon \, N \, . \supset . \, 1 + 1 \, \epsilon \, N$	(2)
(1) (2) .⊃:	$1 + 1 \, \epsilon \, N$	(3)
P 10 .⊃:	$2 = 1 + 1$	(4)
(4).(3).(2, 1 + 1) [a, b] (P 5) :⊃:	$2 \, \epsilon \, N$	(Theorem).

Peano handles axioms and theorems with a precision reminiscent of the proofs of Frege and Hilbert. In every line, he meticulously describes the substitutions

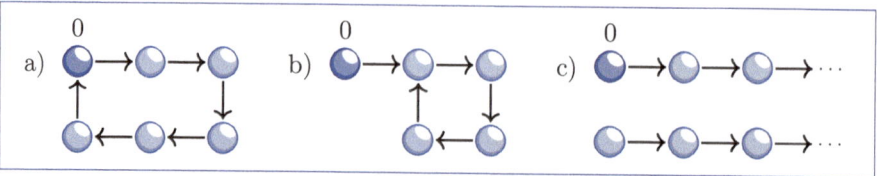

Figure 2.8: All three structures fulfill the first two Peano axioms.

necessary to obtain a particular formula from one of the axiom schemata. However, the derivation of new theorems happens by itself. Two different inference rules can be identified in this short proof, though not explicitly mentioned. One is the modus ponens, which is already well known to us and utilized to derive formula (3) from formulas (1) and (2). The other is needed to derive the formula in the last line and shaped as follows:

$$\frac{\varphi \quad \psi \quad \varphi \wedge \psi \to \chi}{\chi}$$

This rule is logically flawless but has yet to be defined anywhere. It becomes apparent that Peano conducted proofs just like they were conducted outside of mathematical logic, breaking them down into a series of basic logical steps that any serious mathematician would consider legitimate. However, unlike Frege and Hilbert, Peano refrained from defining a formal inference calculus.

2.2.3 Dedekind's Isomorphism Theorem

This section addresses two questions that naturally arise in the context of the Peano axioms. First, we aim to determine whether these axioms are adequate for uniquely defining the natural numbers. Second, we will explore whether all five of them are needed.

Each of us has a mental conception of the chain-like structure of the natural numbers, and at first glance, it may seem that the first two Peano axioms uniquely describe this structure. Figure 2.8 exposes this impression as deceptive by exhibiting three structures that satisfy the first two Peano axioms without being isomorphic to the natural numbers. To uniquely characterize the natural numbers, additional constraints must be satisfied. Figure 2.9 shows that all three remaining axioms are required to eliminate the non-isomorphic structures from Figure 2.8. Among the examples shown, only the chain-like structure of the natural numbers does indeed fulfill all five Peano axioms.

Adding axioms 3 to 5 eliminates all of our example structures containing a cycle, which raises the question of whether the axioms are strong enough to

2.2 The Natural Numbers

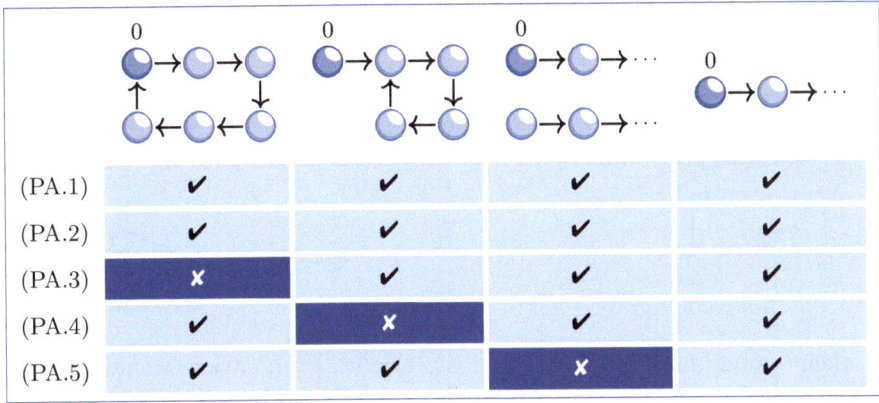

Figure 2.9: Only the structure of the natural numbers fulfills all five Peano axioms.

cancel out all cyclic structures. The analysis presented in Figure 2.10 answers this question positively.

The absence of cycles is a significant result and expressible in the form of an arithmetic property:

Theorem 2.1 — Absence of cycles in the natural numbers

The five Peano axioms imply:
$$x + n \neq x \qquad \text{(for all } n \geq 1\text{)}$$

Whether the Peano axioms uniquely characterize the natural numbers has yet to be settled. What has been said so far does not rule out the existence of cycle-free structures not listed in Figure 2.9, yet fulfilling all Peano axioms.

The German mathematician Richard Dedekind was the first to precisely answer this question (Figure 2.11). His interest in this matter arose in the second half of the nineteenth century when the discussion about the nature of numbers was in full swing. Most attempts of the time were so vague and imprecise that Dedekind saw the need to tackle the problem with mathematical precision. Over the next few years, his work on this matter was repeatedly interrupted, but 1888 he finally succeeded. That year, he published a paper titled "*Was sind und was sollen die Zahlen?*" [15] ("What are numbers and what should they be?" [16]), which is, in retrospect, one of his most significant scientific contributions.

In §9.126, he proved a theorem, which is reprinted in its original wording in Figure 2.12 and referred to by Dedekind as the *Theorem of Definition by Induction*. In this theorem, formula

$$\psi(N) \; 3 \; \Omega$$

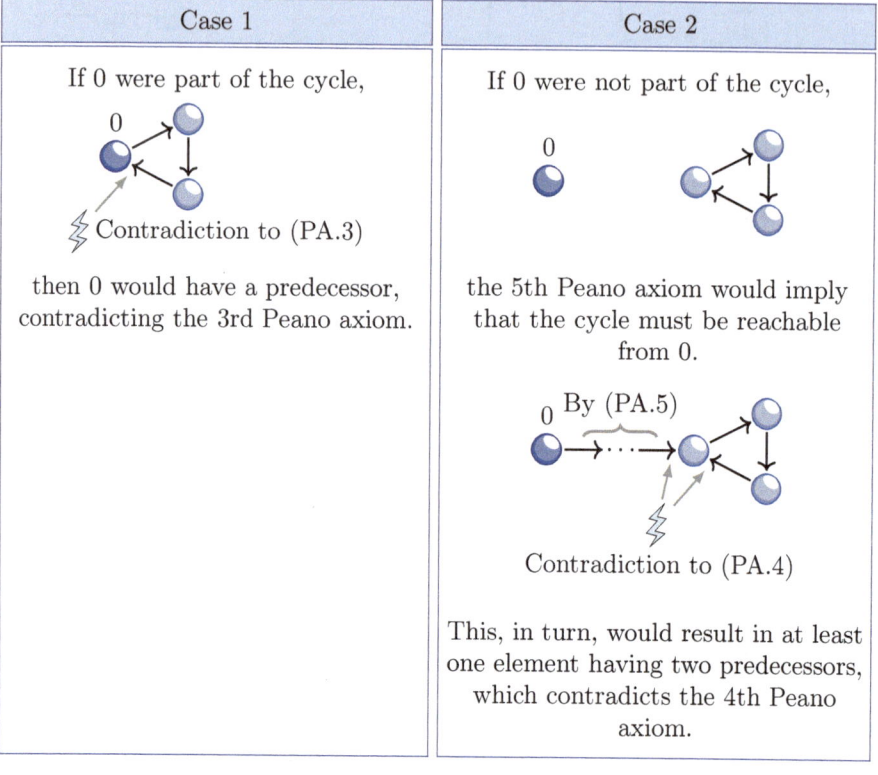

Figure 2.10: The five Peano axioms imply the absence of cycles in the structure of the natural numbers.

expressed that ψ maps the natural numbers into the set Ω, thus being a function of the following form:

$$\psi : \mathbb{N} \to \Omega$$

The formulas

$$\psi(1) = \omega$$
$$\psi(n') = \theta\,\psi(n)$$

denote a recursion scheme, which Dedekind later utilizes to define addition, multiplication, and exponentiation on the natural numbers. ω is the base element, and the function θ is the recursion rule, defining the function value $\psi(n+1)$ by the previous value $\psi(n)$. Dedekind writes n' for the successor of the natural number n.

Contemporary literature refers to the Theorem of Definition by Induction as *Dedekind's recursion theorem*, usually formulated as follows:

2.2 The Natural Numbers

Figure 2.11

RICHARD DEDEKIND
1831 – 1916

126. Satz der Definition durch Induktion. Ist eine beliebige (ähnliche oder unähnliche) Abbildung θ eines Systems Ω in sich selbst und außerdem ein bestimmtes Element ω in Ω gegeben, so gibt es eine und nur eine Abbildung ψ der Zahlenreihe N, welche den Bedingungen
  I. ψ(N) 3 Ω,
  II. ψ(1) = ω,
  III. ψ(n') = θ ψ(n) genügt, wo n jede Zahl bedeutet.
Beweis. Da, wenn es wirklich eine solche Abbildung ψ gibt,

Figure 2.12: Excerpt from *What are numbers and what should they be?*

Theorem 2.2 — Recursion Theorem (Dedekind)

If Ω is a set, $\omega \in \Omega$, and $\theta : \Omega \to \Omega$, then there is a unique $\psi : \mathbb{N} \to \Omega$ with:

$$\psi(0) = \omega \quad \text{and} \quad \psi(n+1) = \theta(\psi(n)) \text{ for all } n \in \mathbb{N}$$

Figure 2.13 illustrates the recursion theorem graphically. The way function ψ maps the natural numbers onto the elements of the set Ω makes it irrelevant whether the successor of a natural number n is determined before ψ is applied or if ψ is applied first, followed by an application of θ. The recursion theorem ensures that ψ always exists and is uniquely defined.

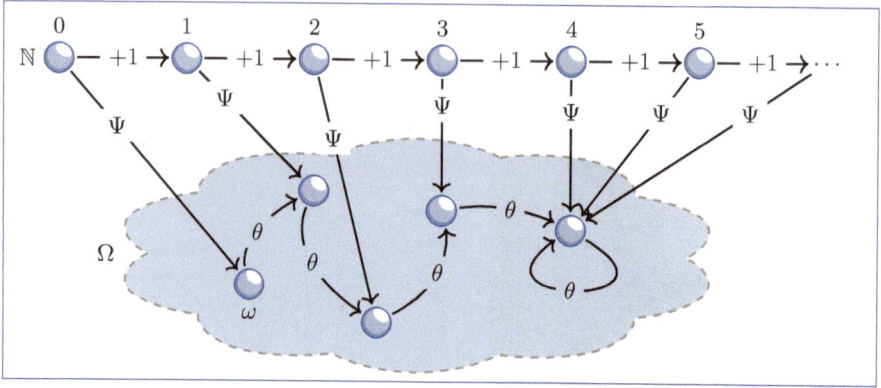

Figure 2.13: Illustration of Dedekind's recursion theorem

Next, we will discuss what Dedekind's recursion theorem can tell us about the Peano axioms. In the following consideration, let

$$(\mathbb{N}, 0, +1)$$

be the structure of the natural numbers, where $+1$ denotes the successor operation. Furthermore, let

$$(\Omega, 0_\Omega, +_\Omega 1)$$

be another structure with its own zero and successor operation, also fulfilling all five Peano axioms.

Now, let $\theta : \Omega \to \Omega$ be the following function:

$$\theta(x) = x +_\Omega 1$$

Dedekind's recursion theorem guarantees the existence of a mapping ψ with the following two properties:

$$\psi(0) = 0_\Omega \qquad (2.3)$$
$$\psi(x+1) = \psi(x) +_\Omega 1 \qquad (2.4)$$

Property (2.4) identifies the mapping ψ as a *homomorphism*, and it takes little effort to extend our understanding of ψ even further:

■ ψ is injective   ☞ $\forall x \, \forall y \, \psi(x) = \psi(y) \Rightarrow x = y$

If the contrary were true, there would be a natural number x and some $n \geq 1$ with $\psi(x) = \psi(x+n)$

2.2 The Natural Numbers

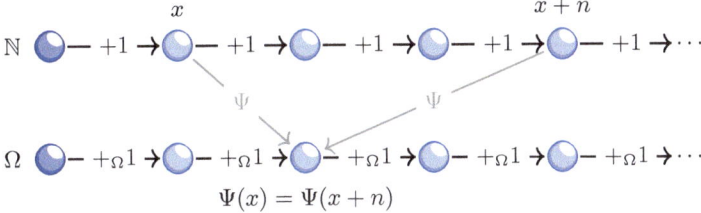

$$\Psi(x) = \Psi(x+n)$$

Because of $\psi(x+1) = \psi(x) +_\Omega 1$, it follows that $\psi(x+n) = \psi(x) +_\Omega n$. Thus,

$$\psi(x) = \psi(x) +_\Omega n,$$

contradicting Theorem 2.1. Consequently, ψ must be injective.

■ ψ ist surjective ☞ $\forall y\, \exists x\, \psi(x) = y$

Let U be the set of all $y \in \Omega$ with a preimage:

$$U := \{y \in \Omega \mid \psi(x_y) = y \text{ for some } x_y \in \mathbb{N}\}$$

Property (2.3) implies that the element 0_Ω belongs to U. Furthermore, for any $y \in U$, the relationship

$$y +_\Omega 1 = \psi(x_y) +_\Omega 1 = \psi(x_y + 1)$$

shows that $y +_\Omega 1$ also has a preimage. Consequently, for every $y \in U$, its successor $y +_\Omega 1$ also belongs to U. From the 5th Peano axiom, it follows that U comprises all elements of Ω, which implies by the definition of U that every element of Ω has a preimage.

A homomorphism being both injective and surjective is called an *isomorphism* (Figure 2.14). The existence of such a mapping implies that $\mathbb{N}$ and Ω are structurally identical, that is, they may only differ in the naming of their elements. Thus, we have just proved *Dedekind's isomorphism theorem*:

Theorem 2.3   Isomorphism Theorem (Dedekind)

Every structure that satisfies all five Peano axioms is isomorphic to the structure of the natural numbers.

Dedekind's isomorphism theorem proves that the five Peano axioms uniquely characterize the natural numbers. We will return to the Peano axioms in Chapter 4, where Gödel integrates them into his formal system P to restrict the domain of possible interpretations to the set of natural numbers.

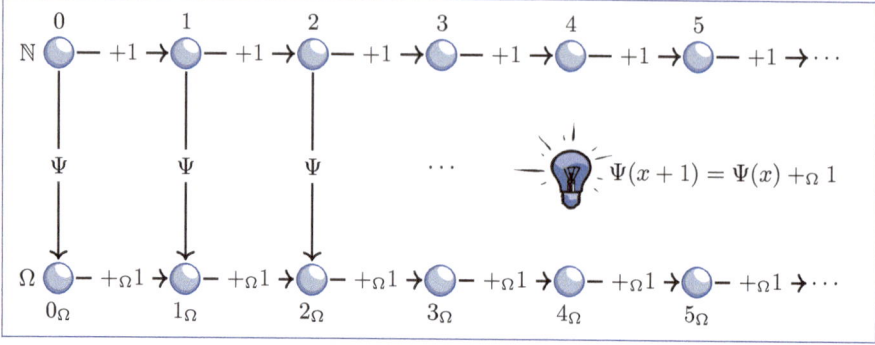

Figure 2.14: Two structures are isomorphic if their elements can be bijectively mapped to each other such that it is irrelevant whether an operation is performed on the elements of one structure or the corresponding elements of the other. If such a mapping exists, both structures are structurally identical and may only differ in the naming of their elements.

2.3 Principia Mathematica

The previous sections have emphasized how important the work of Frege and Peano was for the foundation of mathematics. Frege's predicate calculus and Peano's accurate notation are the key ingredients of modern mathematical logic, but neither of the two scientists succeeded in combining them in a synergistic way. This brilliant achievement was reserved for our third protagonist, whose life and work will be discussed more closely in this section.

Bertrand Arthur William Russell was born on May 18, 1872, as the third child of a British aristocratic family. When he was two years old, his sister and mother died of diphtheria, and when his father passed away a few years later, Bertrand fell into the care of his grandparents. His grandfather was the two-time British Prime Minister Lord John Russell, who resided with his family in Pembroke Lodge, a substantial stately residence in Richmond Park, near London. When the grandfather died in 1878, his grandmother took sole responsibility. Russell grew up in a wealthy, sheltered environment and never attended public school. Instead, he was taught by private tutors at Pembroke Lodge.

As a child, Russell discovered his passion for mathematics, and above all else, he was captivated by Euclid's *Elements*. The enormous fascination he felt as a child was to last into adulthood.

> "At the age of eleven, I began Euclid, with my brother as my tutor. This was one of the great events of my life, as dazzling as first love. I had not imagined that there was anything so delicious in the world. [...] From that moment until Whitehead and I finished Principia Mathematica, when I was thirty-eight, mathematics was my chief interest, and my chief source of happiness."

2.3 Principia Mathematica

Figure 2.15

BERTRAND RUSSELL
1872 – 1970

Bertrand Russell [93]

What Russell adds to this quote is equally remarkable. It reveals that he was already beginning to cultivate a profound philosophical perspective on mathematical concepts and methods at an exceptionally young age. He started to develop a logicism way of thinking, which became a recurring theme throughout his later work:

> "Like all happiness, however, it was not unalloyed. I had been told that Euclid proved things, and was much disappointed that he started with axioms. After first I refused to accept them unless my brother could offer me some reason for doing so, but he said: 'If you don't accept them we cannot go on', and as I wished to go on, I reluctantly admitted them pro ten. The doubt as to the premisses of mathematics which I felt at that moment remained with me, and determined the course of my subsequent work."

Bertrand Russell [93]

In addition to many positive days of his youth, Russell also recounts some negative ones in his autobiography. He grappled with loneliness in Pembroke Lodge and the social pressures of the Victorian era, which part of the reason he regarded his admission to the prestigious Trinity College in 1890 as a liberation. In Cambridge, he thrived in the company of his peers, and the university provided the ideal environment to nurture and further develop his exceptional intellect.

Russell's exceptional talent for mathematics and philosophy, which had already been noticed by his private tutors at Pembroke Lodge, was quickly recognized at Trinity College. One of the professors was the British philosopher and mathematician Alfred North Whitehead. He advocated for Russell from the beginning, and shortly after, a close friendship was to develop.

A pivotal event in his later life occurred in his fourth year of study when Russell, almost by accident, encountered the works of Frege and Cantor:

> "During my fourth year I read most of the great philosophers as well as masses of books on the philosophy of mathematics. James Ward was always giving me fresh books on this subject, and each time I returned them, saying that they were very bad books. [...] In the end, but after I had become a Fellow, I got from him two small books, neither of which he had read or supposed of any value. They were Georg Cantor's *Mannigfaltigkeitslehre*, and Frege's *Begriffsschrift*. These two books at last gave me the gist of what I wanted, but in the case of Frege I possessed the book for years before I could make out what it meant. Indeed, I did not understand it until I had myself independently discovered most of what it contained."
>
> Bertrand Russell [93]

After graduating in 1894, Russell seized the opportunity to conduct research at Cambridge without teaching obligations. As part of these activities, he traveled to Paris in July 1900 to attend the Second International Congress of Mathematicians. In hindsight, this congress ranks among the most important in the history of science. David Hilbert gave a groundbreaking speech that provided a glimpse of the century to come by presenting a list of problems that were still unsolved at the time but were of utmost importance for the future development of mathematics. Besides others, he addressed a question directly related to Gödel's work: The consistency of the arithmetic axioms. The question was answered many years later by the second incompleteness theorem in an utterly unexpected way:

> "But above all I wish to designate the following as the most important among the numerous questions which can be asked with regard to the axioms: To prove that they are not contradictory, that is, that a finite number of logical steps based upon them can never lead to contradictory results."

> "Vor allem aber möchte ich unter den zahlreichen Fragen, welche hinsichtlich der Axiome gestellt werden können, dies als das wichtigste Problem bezeichnen, zu beweisen, dass dieselben untereinander widerspruchslos sind, d. h., dass man aufgrund derselben mittelst einer endlichen Anzahl von logischen Schlüssen niemals zu Resultaten gelangen kann, die miteinander in Widerspruch stehen."
>
> David Hilbert [50, 47]

2.3 Principia Mathematica

The arithmetic axioms Hilbert refers to are the five Peano axioms discussed in Section 2.2. Giuseppe Peano himself participated in the congress, so it happened that the paths of our protagonists crossed. For Russell, the encounter with Peano was a lasting experience that would significantly influence his scientific future:

> "The Congress was a turning point in my intellectual life, because I there met Peano. I already knew him by name and had seen some of his work, but had not taken the trouble to master his notation. In discussions at the Congress I observed that he was always more precise than anyone else, and that he invariably got the better of any argument upon which he embarked. As the days went by, I decided that this must be owing to his mathematical logic. [...] It became clear to me that his notation afforded an instrument of logical analysis such as I had been seeking for years, and that by studying him I was acquiring a new and powerful technique for the work that I had long wanted to do."
>
> Bertrand Russell [93]

In the following months, Russell intensively studied Peano's work. Mentally, he was in line with Frege, but he could not consistently develop Frege's ideas due to the lack of a precise notion. The meeting with Peano was to bring about a lasting change. Suddenly, Russell had a symbolic language at hand, allowing him to express his ideas more precisely than ever before. Notations and concepts previously vague and fuzzy now appeared precise and clear. After all, the Congress of Mathematicians in Paris had become a turning point in Russell's life. It marked the beginning of a period of intellectual creativity, which he remembered himself with the following words:

> "The Whiteheads stayed with us at Fernhurst, and I explained my new ideas to him. Every evening the discussion ended with some difficulty, and every morning I found that the difficulty of the previous evening had solved itself while I slept. The time was one of intellectual intoxication. My sensations resembled those one has after climbing a mountain in a mist, when, on reaching the summit, the mist suddenly clears, and the country becomes visible for forty miles in every direction. For years, I have been endeavoring to analyze the fundamental notions of mathematics, such as order and cardinal numbers. Suddenly, in the space of a few weeks, I discovered what appeared to be definitive answers to the problems which had baffled me for years. And in the course of discovering these answers, I was introducing a new mathematical technique, by which regions formerly abandoned to the vaguenesses of philosophers were conquered for the precision of exact formulae. Intellectually, the month of September 1900 was the highest point of my life."
>
> Bertrand Russell [93]

Russell planned to publish his insights in the book *The Principles of Mathematics* and began working on the manuscript in September 1900. Initially, the pages flowed so quickly from his pen that he had already completed four out of seven parts by the beginning of winter. In May 1901, Russell suddenly ran into trouble:

> "I thought the work was nearly finished, but in the month of May I had an intellectual set-back almost as severe as the emotional set-back which I had had in February. Cantor had a proof that there is no greatest number, and it seemed to me that the number of all the things in the world ought to be the greatest possible. Accordingly, I examined his proof with some minuteness, and endeavored to apply it to the class of all the things there are. This led me to consider those classes which are not members of themselves, and to ask whether the class of such classes is or is not a member of itself. I found that either answer implies its contradictory."
>
> Bertrand Russell [93]

2.3.1 Cantor's Theorem

Russell's difficulties were caused by an antinomy, which is now referred to as *Cantor's Antinomy* in honor of its discoverer. It arises when the concept of the actual infinite is handled carelessly, for example, by treating the set of all sets as a totality.

Georg Cantor discovered the contradictory character of *the set of all sets* towards the end of the nineteenth century. He noticed that the assumption that a set L has the same *cardinality* as the set

$$M := \{f \mid f : L \to \{0, 1\}\}$$

leads to a contradiction. In colloquial language, M is the set of all functions mapping L into the range $\{0, 1\}$.

Let's take a closer look at Cantor's line of argument and assume the existence of a bijection β between the sets L and M. Then M could

> "be thought of in the form of a unique function of the two variables x and z: $\varphi(x, z)$, so that by each specialization of z an element $f(x) = \varphi(x, z)$ of M is obtained and conversely every element $f(x)$ of M arises from $\varphi(x, z)$ through a single specific specialization of z."

> "in der Form einer eindeutigen Funktion der beiden Veränderlichen x und z: $\varphi(x, z)$ gedacht werden, so dass durch jede Spezialisierung von z ein Element $f(x) = \varphi(x, z)$ von M erhalten wird und auch umgekehrt jedes

2.3 Principia Mathematica

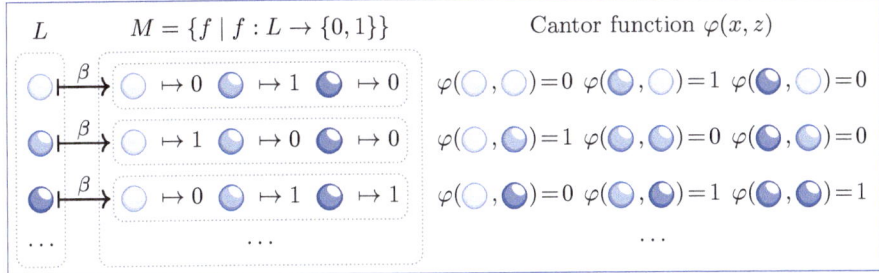

Figure 2.16: Visualization of the Cantor function $\varphi(x, z)$

Element $f(x)$ von M aus $\varphi(x, z)$ durch eine einzige bestimmte Spezialisierung von z hervorgeht."

Georg Cantor [7]

Cantor's words correspond to the following definition (Figure 2.16):

$$\varphi(x, z) := (\beta(z))(x) \qquad (x, z \in L)$$

Next, Cantor used the principle of *diagonalization* to turn $\varphi(x, z)$ into a function with a single variable:

> "If we understand by $g(x)$ that unique function of x, which only takes the values 0 or 1 and is different from $\varphi(x, x)$ for every value of x, [...]"

> "Denn versteht man unter $g(x)$ diejenige eindeutige Funktion von x, welche nur die Werte 0 oder 1 annimmt und für jeden Wert von x von $\varphi(x, x)$ verschieden ist, [...]"

Georg Cantor [7]

In modern notation, $g(x)$ is the function depicted in Figure 2.17:

$$g(x) := \begin{cases} 0 & \text{if } \varphi(x, x) = 1 \\ 1 & \text{if } \varphi(x, x) = 0 \end{cases}$$

The elements $\varphi(x, x)$ are the *diagonal elements* of φ.

The definition of $g(x)$ leads to a contradiction, as Cantor mentions after the previous quote:

> "[...] on the one hand, $g(x)$ is an element of M, on the other hand, $g(x)$ cannot result from any specialization $z = z_0$ from $\varphi(x, z)$, because $\varphi(z_0, z_0)$ is different from $g(z_0)$."

> "[...] so ist einerseits $g(x)$ ein Element von M, andererseits kann $g(x)$

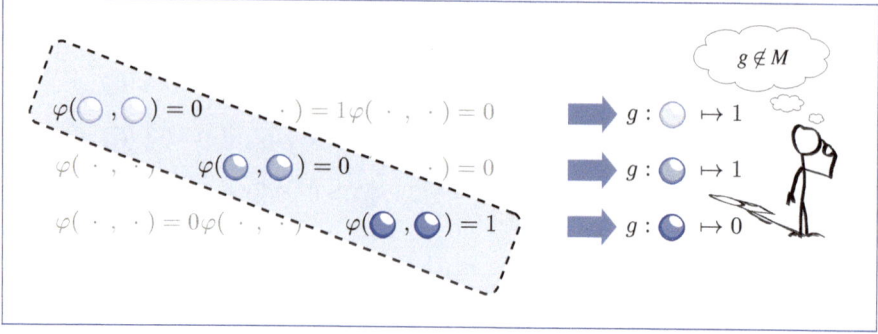

Figure 2.17: The principle of diagonalization. From the diagonal elements $\varphi(x,x)$ a new function $g(x)$ is derived, which differs from each function in M at the diagonal position.

> durch keine Spezialisierung $z = z_0$ aus $\varphi(x, z)$ hervorgehen, weil $\varphi(z_0, z_0)$ von $g(z_0)$ verschieden ist."
>
> Georg Cantor [7]

Consequently, we have to give up the assumption about the existence of a bijection between L and M.

We can strengthen this result even further. Since L can be mapped into M injectively by assigning each $y \in L$ the function mapping y to 1 and all other elements to 0, M is at least as large as L. As a result, we get:

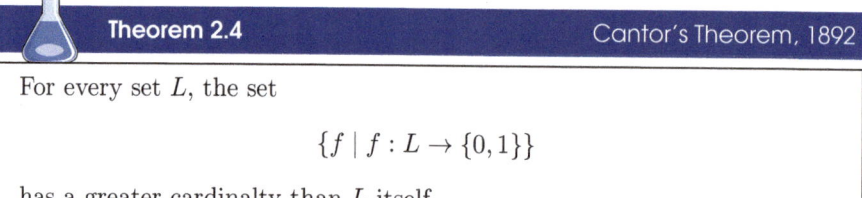

Theorem 2.4 — Cantor's Theorem, 1892

For every set L, the set

$$\{f \mid f : L \to \{0,1\}\}$$

has a greater cardinalty than L itself.

Georg Cantor proved this theorem in 1892. Contemporary textbooks typically present Cantor's theorem in a modernized form, which identifies each function of the form $f : L \to \{0,1\}$ with a subset of L. This is easily achievable one-to-one by including in the subset precisely those elements $x \in L$ satisfying $f(x) = 1$:

$$f \mapsto \{x \in L \mid f(x) = 1\}$$

This way, M can be identified with the set of all subsets of L. The set of all subsets of L is the *power set* of L, denoted by 2^L, enabling us to rewrite Theorem 2.4 in its modern form:

2.3 Principia Mathematica

Theorem 2.5 — Cantor's Theorem, modern formulation

For every set L, the power set 2^L has a greater cardinality than L itself.

The theorem is usually proved by assuming L and 2^L have the same cardinality, which implies the existence of a bijective mapping $\beta : L \to 2^L$. For each element $x \in L$, one of two cases must hold: Either x belongs to the image element ($x \in \beta(x)$), or it does not ($x \notin \beta(x)$). Consequently, the image set 2^L can be divided into two subsets

$$\{\beta(x) \mid x \in \beta(x)\} \tag{2.5}$$
$$\{\beta(x) \mid x \notin \beta(x)\} \tag{2.6}$$

and for all x, $\beta(x)$ must be contained in one or the other. Next, let us consider the set

$$G := \{x \in L \mid x \notin \beta(x)\}, \tag{2.7}$$

which includes exactly those elements of L that map into the set (2.6). Because β is a bijection, L must contain an element g with the property:

$$\beta(g) = G \tag{2.8}$$

As for all elements from L, either $g \in G$ or $g \notin G$ must hold. However, both cases immediately lead to a contradiction:

$$g \in G \stackrel{(2.7)}{\Rightarrow} g \notin \beta(g) \stackrel{(2.8)}{\Rightarrow} g \notin G$$
$$g \notin G \stackrel{(2.7)}{\Rightarrow} g \in \beta(g) \stackrel{(2.8)}{\Rightarrow} g \in G$$

Consequently, no bijective function $\beta : L \to 2^L$ can exist.

At first glance, the argument seems quite distinct from the original proof, but upon closer examination, it becomes apparent that we are offered old wine in new bottles. To see why, let us begin by rewriting the definition of the Cantor function φ as follows:

$$\varphi(x, z) = \begin{cases} 1 & \text{if } x \in \beta(z) \\ 0 & \text{if } x \notin \beta(z) \end{cases}$$

Considering only the diagonal elements, we get:

$$\varphi(x, x) = \begin{cases} 1 & \text{if } x \in \beta(x) \\ 0 & \text{if } x \notin \beta(x) \end{cases}$$

Thus, Cantor's function $g(x)$ can be denoted as follows:

$$g(x) = \begin{cases} 0 & \text{if } x \in \beta(x) \\ 1 & \text{if } x \notin \beta(x) \end{cases}$$

Colloquially speaking, g characterizes the subset of L that contains precisely those elements that do not appear in the subsets assigned to them. This set is identical to the set G used above in the revised proof. Thus, the modern variant also utilizes Cantor's diagonalization principle, although not as clearly as the original proof of 1892.

Overall, Cantor's theorem reveals a remarkable property of power sets: It exposes that the power set operation always creates a set with a greater cardinality than the original set itself. Consequently, there can be no *maximum infinity*.

Cantor was aware of this fact long before 1892. In fact, he had already proved the result as early as 1883 but with a more complex method than his newly developed diagonalization scheme.

> "I have already shown in the 'Foundations of a General Theory of Manifolds' (Leipzig 1883; Math. Annalen Vol. 21) by completely different means that the cardinalities have no maximum; there it was even proven that the totality of all cardinalities, if we think of them ordered by their size, forms a 'well-ordered set', so that there is a next larger one in nature for every cardinality, but also a next larger one follows any endlessly increasing set of cardinalities."

> "Ich habe bereits in den 'Grundlagen einer allgemeinen Mannigfaltigkeitslehre' (Leipzig 1883; Math. Annalen Bd. 21) durch ganz andere Hilfsmittel gezeigt, dass die Mächtigkeiten kein Maximum haben; dort wurde sogar bewiesen, dass der Inbegriff aller Mächtigkeiten, wenn wir letztere ihrer Grösse nach geordnet denken, eine 'wohlgeordnete Menge' bildet, so dass es in der Natur zu jeder Mächtigkeit eine nächst grössere gibt, aber auch auf jede ohne Ende steigende Menge von Mächtigkeiten eine nächst grössere folgt."

Georg Cantor [7]

Since no maximum infinity exists, we must never regard the *set of all sets* as a totality. If there were such a set, let's call it V, we would obtain the relation $|V| < |2^V|$ due to theorem 2.5. By definition, however, all elements of 2^V are contained in V, implying $|2^V| \leq |V|$. The resulting contradiction $|V| < |V|$ is called *Cantor's antinomy*. It clearly shows that the union of all sets is not a set itself.

Related mental constructs lead to akin contradictions. As early as 1897, the Italian mathematician Cesare Burali-Forti noted that if the set Ω of all ordinal

2.3 Principia Mathematica

numbers were to exist, it would be an ordinal number itself. Like all ordinal numbers, Ω would not only have a direct successor $\Omega+1$ but would also satisfy the relation $\Omega < \Omega+1$. However, by definition, the ordinal number $\Omega+1$ would be an element of Ω, making it smaller than Ω. The resulting contradiction $\Omega < \Omega$ is known as the *Burali-Forti paradox*. It shows that the union of all ordinal numbers is also too large to exist as a self-contained whole.

With a similar argument, it can be proved that the set of all *cardinal numbers* cannot exist either. The above quotation shows that Cantor was unaware of this in 1892, for he still called the *totality of all cardinalities* ("Inbegriff aller Mächtigkeiten") a well-ordered set.

For Cantor, however, the potential danger of forming sets that could not be considered closed entities was not a pressing problem. He had faith in mathematical intuition and believed he could determine, on a case-by-case basis, whether a set's definition was legitimate or not.

2.3.2 Russell's Antinomy

Paradoxes unveil their dark power when mathematics is carried out within a formal system where proving a statement is akin to a mechanical process. In such a system, there is no place for mathematical intuition, as a rigid set of rules decides whether a logical conclusion is valid. This explains why Russell could not ignore the discovered antinomy. He was trying to replicate all notions and concepts of classical mathematics in a formal system, so any contradiction formally derivable within the system would render his project meaningless.

On June 16, 1902, Russell wrote a letter to Frege reporting his discovery. The original letter was written in German and translated into English years later. Figure 2.18 shows a reprint of the English translation taken from [94]. After a friendly introduction, Russell quickly gets to the point and describes the discovered antinomy, first in a logical and then in a set-theoretical formulation. In the ensuing two sections, we will look at both variants.

2.3.2.1 Logical Formulation of the Antinomy

To bring the logical variant of Russell's antinomy to life, we must utilize *second-order predicates*, available in *higher-order logics*. They are similar to ordinary predicates, but expect a predicate in at least one argument position.

Defining second-order predicates is straightforward. Almost any formula can be considered the definition of such a predicate simply by treating all predicates occurring in it as free (predicate) variables. For example, **T** and **F** with

$$\mathbf{T}(P) := \exists x\, P(x)$$

> Friday's Hill.
> Haslemere.
> 16 June 1902
>
> Dear colleague,
>
> For a year and a half I have been acquainted with your "Grundgesetze der Arithmetik", but it is only now that I have been able to find the time for the thorough study I intended to make of your work. I find myself in complete agreement with you in all essentials, particularly when you reject any psychological element in logic and when you place a high value upon an ideography for the foundations of mathematics and of formal logic, which, incidentally, can hardly be distinguished. With regard to many particular questions, I find in your work discussions, distinctions, and definitions that one seeks in vain in the works of other logicians. Especially so far as function is concerned (§9 of your Begriffsschrift), I have been led on my own to views that are the same even in details. There is just one point where I have encountered a difficulty. You state that a function, too, can act as the indeterminate element. This I formerly believed, but now this view seems doubtful to me because of the following contradiction. Let w be the predicate: to be a predicate that cannot be predicated of itself. Can w be predicated of itself? From each answer its opposite follows. Therefore we must conclude that w is not a predicate. Likewise there is no class (as a totality) of those classes which, each taken as a totality, do not belong to themselves. From this I conclude that under certain circumstances a definable collection does not form a totality.
>
> I am on the point of finishing a book on the principles of mathematics and in it I should like to discuss your work very thoroughly. I already have your books or shall buy them soon, but I would be very grateful to you if you could send me reprints of your articles in various periodicals. In case this should be impossible, however, I will obtain them from a library.
>
> The exact treatment of logic in fundamental questions, where symbols fail, has remained very much behind; in your works I find the best I know of our time, and therefore I have permitted myself to express my deep respect to you. It is very regrettable that you have not come to publish the second volume of your Grundgesetze; I hope that this will still be done.
>
> Very respectfully yours,
> Bertrand Russell.
>
> The above contradiction, when expressed in Peano's ideography, reads as follows:
>
> $$w = \mathrm{cls} \cap x \ni (x \sim \varepsilon\, x) . \supset : w\, \varepsilon\, w . = . w \sim \varepsilon\, w.$$
>
> I have written to Peano about this, but he still owes me an answer.

Figure 2.18: Letter from Russell to Frege dated June 16, 1902 [94]

2.3 Principia Mathematica

$$\mathbf{F}(P) := \neg \exists x\, P(x)$$

are two unary second-order predicates with the argument P.

We stay in safe waters as long as we use first-order predicates as arguments in $\mathbf{T}$ and $\mathbf{F}$. When second-order predicates are permitted, more care must be taken (cf. [54]). In this case, the *self-predicating* expressions $\mathbf{T}(\mathbf{T})$ and $\mathbf{F}(\mathbf{F})$ are well-defined, both of them calling for a closer examination:

- The formula $\mathbf{T}(\mathbf{T})$ corresponds to

$$\exists x\, \mathbf{T}(x)$$

and claims the existence of an object x with the property $\mathbf{T}$. Since $\mathbf{T}$ expects a predicate as argument, this may be rephrased as: There is a predicate P with the property $\mathbf{T}$:

$$\exists P\, \mathbf{T}(P)$$

Resolving the definition of $\mathbf{T}$ a second time results in:

$$\exists P\, \exists x\, P(x)$$

This formula can be phrased as follows: There exists a predicate P and an individual x such that $P(x)$ is true. Or, more concisely: There is a *satisfiable* predicate. This is a true statement.

- The formula $\mathbf{F}(\mathbf{F})$ corresponds to

$$\neg \exists x\, \mathbf{F}(x)$$

and denies the existence of an object x with the property $\mathbf{F}$. Similar to the first case, the statement can be rephrased as: There is no predicate P with the property $\mathbf{F}$:

$$\neg \exists P\, \mathbf{F}(P)$$

Again, this formula can be rewritten by replacing $\mathbf{F}$ with its definition:

$$\neg \exists P\, \neg \exists x\, P(x)$$

This formula states that there is no predicate P which is true for no x. Or, more concisely: There are no *unsatisfiable* predicates. This is a false statement.

The predicates defined so far do not cause any harm: $\mathbf{T}(\mathbf{T})$ is a true statement, and $\mathbf{F}(\mathbf{F})$ is a false statement. However, we get into stormy waters when turning our attention to Russell's predicate w:

$$w(P) := \neg P(P)$$

Colloquially speaking, w applies exactly to those predicates P for which P(P) is false. Thus, w is *"the predicate of being a predicate that cannot be predicated of itself"*. According to what has been said above, $w(\mathbf{F})$ is true and $w(\mathbf{T})$ is false.

To evoke Russell's antinomy, let us ask whether w *"can be predicated of itself"* ($w(w)$) or not ($\neg w(w)$). In fact, *"from each answer its opposite follows"*:

$$w(w) \Rightarrow \neg w(w)$$
$$\neg w(w) \Rightarrow \neg\neg w(w) \Rightarrow w(w)$$

Or, in short:

$$w(w) \Leftrightarrow \neg w(w)$$

This is the logical formulation of Russell's antinomy.

2.3.2.2 Set-Theoretic Formulation of the Antinomy

The set-theoretic formulation of the antinomy is even easier to understand. In precise form, Russell describes it at the end of his letter in Peano's symbolic language:

$$w = \text{cls} \cap x \ni (x \sim \varepsilon\, x) \mathbin{.} \supset : w\,\varepsilon\,w \mathbin{.} = \mathbin{.} w \sim \varepsilon\, w.$$

The acronym 'cls' is an abbreviation for *class*. Thus, in modern terms,

$$w = \text{cls}$$

expresses that w is a set. The expression

$$x \sim \varepsilon\, x$$

means $x \notin x$, which allows the left-hand side of the implication to be rewritten as:

$$w = \{x \mid x \notin x\} \tag{2.9}$$

In words, w is the set of all sets that do not contain themselves. The definition of w immediately leads to a contradiction when asking whether this set contains itself or not:

$$w \in w \Rightarrow w \notin w$$
$$w \notin w \Rightarrow w \in w$$

Or, more concisely:

$$w \in w \Leftrightarrow w \notin w$$

2.3 Principia Mathematica

This is the set-theoretic formulation of Russell's antinomy. Russell's discovery makes unmistakably clear *"that under certain circumstances a definable collection does not form a totality."*

2.3.2.3 In Ruins

Frege quickly replied. In his letter of June 22, reprinted in Figure 2.19 in an English translation, he first thanked Russell for his interest in his work. Frege then acknowledged the fundamental correctness of Russell's observation. Towards the end of the letter, he politely pointed out that the logical formulation of Russell's antinomy is not reproducible with his concepts. Frege's logic explicitly prohibits the application of a predicate to itself, implying the expression P(P) is not well-defined.

Nevertheless, Frege admitted that his logic was not immune to the set-theoretical formulation of the antinomy. The set of all sets that do not contain themselves could be defined, albeit not as simply as initially presumed. For Frege, this was a disaster. Russell's antinomy had laid bare a fundamental flaw in his logic that cosmetic corrections could not remedy. The timing of this discovery couldn't have been worse either. At the end of his reply, Frege hinted that he had been just about to finish the second volume of the *Basic Laws of Arithmetic*.

Despite the devastating news, Frege's reply was candid and composed, making it likely he had yet to fully realize that his work, which had taken up many years of his life, lay in ruins. Notably, the afterword of the second volume of his *Basic Laws of Arithmetic* already reflected a much more pessimistic tone than his hastily penned letter:

> *"Hardly anything more unfortunate can befall a scientific writer than to have one of the foundations of his edifice shaken after the work is finished. This was the position I was placed in by a letter of Mr. Bertrand Russell, just when the printing of this volume was nearing its completion."*
>
> *"Einem wissenschaftlichen Schriftsteller kann kaum etwas Unerwünschteres begegnen, als daß ihm nach Vollendung einer Arbeit eine der Grundlagen seines Baues erschüttert wird. In diese Lage wurde ich durch einen Brief des Herrn Bertrand Russell versetzt, als der Druck dieses Bandes sich seinem Ende näherte."*
>
> Gottlob Frege [29, 24]

At first glance, it may not be immediately apparent how Russell's antinomy can be derived from Frege's basic law V, so let's delve deeper. First of all, we need to find a way to express the property of self-inclusion within Frege's logic. To understand how this can be achieved, let's briefly examine the implications of the relationships $x \in x$ and $x \notin x$:

Jena, 22 June 1902

Dear colleague,

Many thanks for your interesting letter of 16 June. I am pleased that you agree with me on many points and that you intend to discuss my work thoroughly. In response to your request I am sending you the following publications: [...]

I received an empty envelope that seems to be addressed by your hand. I surmise that you meant to send me something that has been lost by accident. If this is the case, I thank you for your kind intention. I am enclosing the front of the envelope.

When I now read my "Begriffsschrift" again, I find that I have changed my views on many points, as you will see if you compare it with my "Grundgesetze der Arithmetik". I ask you to delete the paragraph beginning "Nicht minder erkennt man" on page 7 of my "Begriffsschrift", since it is incorrect; incidentally, this had no detrimental effects on the rest of the booklet's contents.

Your discovery of the contradiction caused me the greatest surprise and, I would almost say, consternation, since it has shaken the basis on which I intended to build arithmetic. It seems, then, that transforming the generalization of an equality into an equality of courses-of-values (§9 of my Grundgesetze) is not always permitted, that my Rule V (§20. S. 36) is false, and that my explanations in §31 are not sufficient to ensure that my combinations of sings have a meaning in all cases. I must reflect further on that matter. It is all the more serious since, with the loss of my Rule V, not only the foundations of my arithmetic, but also the sole possible foundations of arithmetic, seem to vanish. Yet, I should think, it must be possible to set up conditions for the transformation of the generalization of an equality into an equality of courses-of-values such that the essentials of my proofs remain intact. In any case your discovery is very remarkable and will perhaps result in a great advance in logic, unwelcome as it may seem at first glance.

Incidentally, it seems to me that the expression "a predicate is predicted of itself" is not exact. A predicate is as a rule a first-level function, and this function requires an object as argument and cannot have itself as argument (subject). Therefore I would prefer to say "a concept is predicted of its own extension". If the function $\phi(\epsilon)$ is a concept, I denote its extension (or the corresponding class) by $\grave{\epsilon}\phi(\epsilon)$ (to be sure, the justification for this has now become questionable to me). In $\phi(\grave{\epsilon}\phi(\epsilon))$ or $\grave{\epsilon}\phi(\epsilon) \cap \grave{\epsilon}\phi(\epsilon)$ we then have a case in which the concept $\phi(\epsilon)$ is predicated of its own extension.

The second volume of my Grundgesetze is to appear shortly, I shall no doubt have to add an appendix in which your discovery is taken into account. If only I already had the right point of view for that!

Very respectfully yours,
G. Frege.

Figure 2.19: Frege's answer to Russell dated June 22, 1902 [30]

2.3 Principia Mathematica

- **Case 1:** $x \in x$

  In this case, there can be no predicate that is true for the elements in x but not for x itself. Therefore, the following formula must be a true statement:

  $$\neg \exists P\ (x = \{y \mid P(y)\} \land \neg P(x))$$

- **Case 2:** $x \notin x$

  In this case, there is a predicate that is true for all elements in x, but not for x itself. Therefore, the following formula must be a true statement:

  $$\exists P\ (x = \{y \mid P(y)\} \land \neg P(x)) \tag{2.10}$$

Putting both cases together, formula (2.10) is precisely true if x is a set that does not contain itself. It can be rephrased in the following logically equivalent form:

$$\neg \forall P\ (x = \{y \mid P(y)\} \to P(x)) \tag{2.11}$$

In Frege's logic, the set $\{y \mid P(y)\}$ is denoted by $\grave{\epsilon}P(\epsilon)$. Thus, formula (2.11) can be expressed in the following form, which we will abbreviate as W(x):

$$W(x) := \neg \forall P\ (x = \grave{\epsilon}P(\epsilon) \to P(x))$$

The predicate W(x) applies to those sets that do not contain themselves. Consequently, $\grave{\epsilon}W(\epsilon)$ is Russell's set w. It includes precisely those sets that do not contain themselves.

At this juncture, let's take another look at Frege's basic law V depicted in Figure 2.4. By substituting $\mathfrak{a}$ with $\grave{\epsilon}W(\epsilon)$, the law allows us to conclude the following:

$$\grave{\epsilon}W(\epsilon) = \grave{\epsilon}Q(\epsilon) \to (W(\grave{\epsilon}W(\epsilon)) \leftrightarrow Q(\grave{\epsilon}W(\epsilon)))$$

This formula can be weakened into

$$\grave{\epsilon}W(\epsilon) = \grave{\epsilon}Q(\epsilon) \to (W(\grave{\epsilon}W(\epsilon)) \to Q(\grave{\epsilon}W(\epsilon)))$$

which is logically equivalent to:

$$W(\grave{\epsilon}W(\epsilon)) \to (\grave{\epsilon}W(\epsilon) = \grave{\epsilon}Q(\epsilon) \to Q(\grave{\epsilon}W(\epsilon)))$$

Since Q can represent any predicate, Frege's logic allows us to draw the following conclusion:

$$W(\grave{\epsilon}W(\epsilon)) \to \forall P\ (\grave{\epsilon}W(\epsilon) = \grave{\epsilon}P(\epsilon) \to P(\grave{\epsilon}W(\epsilon))) \tag{2.12}$$

A closer look at this formula reveals that we have slipped the predicate W into the formula for a second time, as the right-hand side of the implication now matches the expression $\neg W(\grave{\epsilon}W(\epsilon))$. Therefore, formula (2.12) is the same as

$$W(\grave{\epsilon}W(\epsilon)) \to \neg W(\grave{\epsilon}W(\epsilon)) \tag{2.13}$$

On the other hand, the definition of the universal quantifier implies:

$$\forall P\ (\grave{\epsilon}W(\epsilon) = \grave{\epsilon}P(\epsilon) \to P(\grave{\epsilon}W(\epsilon))) \to (\grave{\epsilon}W(\epsilon) = \grave{\epsilon}W(\epsilon) \to W(\grave{\epsilon}W(\epsilon)))$$

$\grave{\epsilon}W(\epsilon) = \grave{\epsilon}W(\epsilon)$ is always true, simplifying the formula to:

$$\forall P\ (\grave{\epsilon}W(\epsilon) = \grave{\epsilon}P(\epsilon) \to P(\grave{\epsilon}W(\epsilon))) \to W(\grave{\epsilon}W(\epsilon))$$

The left-hand side corresponds to $\neg W(\grave{\epsilon}W(\epsilon))$, allowing for the following reduction:

$$\neg W(\grave{\epsilon}W(\epsilon)) \to W(\grave{\epsilon}W(\epsilon)) \tag{2.14}$$

Together, (2.13) and (2.14) yield the contradiction we were looking for:

$$\neg W(\grave{\epsilon}W(\epsilon)) \leftrightarrow W(\grave{\epsilon}W(\epsilon)) \tag{2.15}$$

According to this formula, the set of all sets that do not contain themselves is contained in itself exactly when it is not. An untenable situation! As harmless as the fifth basic law may seem, it opens a gateway for Russell's antinomy, unmasking Frege's logic as contradictory.

Unlike Cantor, who had encountered similar paradoxes years earlier, Frege immediately recognized their explosive power. To him, it was clear that we cannot rely on intuition to determine whether a mathematical definition describes an existing set. Well-defined rules had to be established, but after the discovery of the antinomy, Frege had yet to learn what they should look like. As the years passed, his confidence gradually turned into resignation. When his wife also passed away, he fell into a deep depression from which he never recovered.

Initially, Russell could not solve the problem either and reluctantly decided to publish *The Principles of Mathematics* without a satisfactory answer [90]:

> "Trivial or not, the matter was a challenge. Throughout the latter half of 1901 I supposed the solution would be easy, but by the end of that time I had concluded that it was a big job. I therefore decided to finish *The Principles of Mathematics*, leaving the solution in abeyance."
>
> Bertrand Russell [93]

In contrast to Frege, Russell firmly believed in finding a way to keep the set-theoretic antinomies at bay. Unmistakably, the contradictions arose from the composition of sets that are, in a sense, too large to be considered a totality.

Russell knew that eliminating the antinomies required a delicate balance. On the one hand, the underlying logic had to be constrained to prevent the formulation of self-referential statements like Russell's antinomy. On the other hand, the restricted logic needed to remain expressive enough to encompass all the concepts and conclusions of classical mathematics. Russell reflected on the events of 1903 and 1904 as follows:

2.3 Principia Mathematica

> "I was trying hard to solve the contradictions mentioned above. Every morning I would sit down before a blank sheet of paper. Throughout the day, with a brief interval for lunch, I would stare at the blank sheet. Often when evening came it was still empty. [...] the two summers of 1903 and 1904 remain in my mind as a period of complete intellectual deadlock. It was clear to me that I could not get on without solving the contradictions, and I was determined that no difficulty should turn me aside from the completion of *Principia Mathematica*, but it seemed quite likely that the whole of the rest of my life might be consumed in looking at that blank sheet of paper."
>
> Bertrand Russell [93]

2.3.3 Type Theory

Russell made his breakthrough in 1906. That year, he developed *ramified type theory* as an effective bulwark against the paradoxes. Russell first described the theory in his 1908 paper *Mathematical Logic as Based on the Theory of Types*. Nevertheless, it gained the most recognition for its application in the *Principia Mathematica*. Russell writes:

> "The following theory of symbolic logic recommended itself to me in the first instance by its ability to solve certain contradictions, of which the one best known to mathematicians is Burali-Forti's concerning the greatest ordinal."
>
> Bertrand Russell [91]

Russell continues with a detailed analysis of the discovered antinomies. He explains that all antinomies, as different as they may seem, share a crucial characteristic: *self-reference*.

> "In all the above contradictions (which are merely selections from an indefinite number) there is a common characteristic, which we may describe as self-reference or reflexiveness. [...] In each contradiction something is said about all cases of some kind, and from what is said a new case seems to be generated, which both is and is not of the same kind as the cases of which all were concerned in what was said"
>
> Bertrand Russell [91]

In the third section, he reveals the basic idea of type theory:

> "A *type* is defined as the range of significance of a propositional function, i. e., as the collection of arguments for which the said function has values."
>
> Bertrand Russell [91]

This colloquial description aligns well with the modern meaning of types. In fact, in its most general form, the notion of a type was not a new idea at the time. Even Euclid made a strict distinction between points and lines, thereby establishing a rudimentary type system.

Russell utilized types to invalidate certain formulas that were permissible under the old syntax. He designed his type system to act as a filter that eliminated potentially paradoxical formulas at the syntactic level. After thoroughly analyzing the antinomies, he decided to banish all formulas from logic that somehow refer to themselves. In [91], Russell formulated this basic idea in the form of a guideline, which he referred to as the *vicious circle principle*:

> *"The division of objects into types is necessitated by the reflexive fallacies which otherwise arise. These fallacies, as we saw, are to be avoided by what may be called the 'vicious-circle principle'; i. e., 'no totality can contain members defined in terms of itself'. This principle, in our technical language, becomes: 'Whatever contains an apparent variable must not be a possible value of that variable.' Thus whatever contains an apparent variable must be of a different type from the possible values of that variable; we will say that it is of higher type. Thus the apparent variables contained in an expression are what determines its type. This is the guiding principle in what follows."*
>
> Bertrand Russell [91]

To keep self-referential formulas out, Russell formulated two hierarchical orders. In combination, they form the ramified type theory of the *Principia Mathematica* (cf. [62]).

The first hierarchy requires that the arguments of a predicate must always be of a lower type than the predicate itself. At the lowest level are the *individual objects*. They constitute the *domain* or *universe* of an interpretation, embodying the conceptual entities upon which logical formulas make statements at the semantic level. In Gödel's system P, for instance, the individual objects are the natural numbers.

The next higher hierarchy level is formed by the *individual predicates* (*first-order predicates*), whose arguments are individual elements. Above them are *second-order predicates*, which expect at least one individual predicate among their arguments, and this hierarchy extends ad infinitum.

The type system of the *Principia Mathematica* is defined in colloquial language and sometimes remains ambiguous. Nowadays, formal definitions exist for the concept of type, such as the following:

2.3 Principia Mathematica

>  **Definition 2.6** — Simple types
>
> The set of (*simple*) *types* is recursively defined:
>
> - i and () are types.
> - If $\tau_1, \ldots, \tau_n$ are types, then $(\tau_1, \ldots, \tau_n)$ is also a type.

i and () are the *base types*, which can be combined recursively into complex structures. By successively applying the syntax rules given above, the following types, among others, can be derived:

$$i;\ ();\ (i);\ (i,i);\ (i,i,i);\ (());\ ((i,i));\ ((i),(i));\ ((i,i),(i),i);\ \ldots$$

In this type system, i represents the type of individual objects and () the type of propositional logic variables, which may be formally considered as zero-place individual predicates. The definition extends to expressions like (i), (i, i), and (i, i, i), which are the types of unary, binary, and ternary individual predicates, respectively. The meaning of the other types is analogous. For instance, ((i), i) represents the type of a second-order predicate that expects a unary individual predicate as its first argument and an individual element as its second.

By arranging the n free variables of a formula φ in a specified order, φ can be assigned a type in a straightforward manner. If the first free variable corresponds to an object of type τ_1, the second to an object of type τ_2, and so forth, then $(\tau_1, \ldots, \tau_n)$ is the type of φ.

The following examples provide clarity:

$$\varphi_1(\underbrace{x}_{i}) := (x = x) \qquad \text{Type}(\varphi_1) = (i)$$

$$\varphi_2(\underbrace{P}_{()}) := (P \vee \neg P) \qquad \text{Type}(\varphi_2) = (())$$

$$\varphi_3(\underbrace{x}_{i}) := \forall P\ (P(x) \vee \neg P(x)) \qquad \text{Typ}(\varphi_3) = (i)$$

$$\varphi_4(\underbrace{P}_{(i,i)}) := \forall x\, \exists y\ P(x,y) \qquad \text{Type}(\varphi_4) = ((i,i))$$

$$\varphi_5(\underbrace{P}_{(i)}, \underbrace{Q}_{(i)}) := \forall x\ (P(x) \to Q(x)) \qquad \text{Type}(\varphi_5) = ((i),(i))$$

$$\varphi_6(\underbrace{P}_{(i,i)}, \underbrace{Q}_{(i)}, \underbrace{y}_{i}) := \exists x\ (P(x,y) \vee Q(x)) \qquad \text{Type}(\varphi_6) = ((i,i),(i),i)$$

Restricting the syntactically valid formulas to well-typed expressions successfully banishes self-referential formulas like P(P).

Back then, Russell was unsure whether this kind of typing would suffice to keep all antinomies at bay and thus decided to implement another layer of safety. To understand his motives, let us examine the third example formula mentioned above:

$$\varphi_3(x) := \forall P\ (P(x) \vee \neg P(x)) \tag{2.16}$$

The type of φ_3 is (i), as the only free variable x is an individual variable. Additionally, the formula contains a universal quantifier that makes a statement about all unary single predicates of type (i). This leads to a situation where the formula quantifies over the same objects it belongs to, thereby violating the *vicious circle principle*.

To prevent self-references of this kind, Russell introduced a second hierarchy that distinguished formulas based on their syntactic structure, resulting in the *ramified type theory*. The second hierarchy emerged by assigning each formula an ordinal number whose value was independent of the first hierarchy level. In addition, he imposed restrictions on the expressiveness of all quantifiers. Henceforth, they were no longer permitted to quantify over arbitrary objects but only over objects of a particular order. As an example, consider the following formula:

$$\forall P^3\ (P(x) \vee \neg P(x)) \tag{2.17}$$

The universal quantifier no longer refers to all individual predicates but only those of order 3. Each formula is assigned an ordinal number that exceeds by one the ordinal numbers of all the formulas it references. In our example, this ordinal number is the number 4. Since, per definition, the ordinal number of a formula is greater than all ordinal numbers the quantifiers refer to, a self-reference, as seen in (2.16), can no longer arise.

Ramified type theory solved the problem of self-reference by strong means – and harsh consequences. In particular, the constraints imposed by the second hierarchy had restricted the expressiveness so much that it became difficult or even impossible to formulate many mathematical facts. Russell was aware of the dilemma and sought to weaken some restrictions by introducing a *reducibility axiom*. The outcome was a theory that achieved its intended purpose but appeared unnatural. Little remained of the beauty and elegance that mathematicians have always striven for.

In the preface of the second edition of the *Principia Mathematica*, Russell and Whitehead justified the presence of the reducibility axiom as follows:

> "This axiom has a purely pragmatic justification: it leads to the desired results, and to no others. But clearly it is not the sort of axiom with which we can rest content."
>
> Bertrand Russell, Alfred North Whitehead [100]

Later, Russell's ramified type theory was significantly simplified by Frank Plumpton Ramsey. The British mathematician had shown that the second

2.3 Principia Mathematica

hierarchy only contributed to the avoidance of so-called *semantic paradoxes* that arose from mixing the object level with the meta-level [83]. If a formal language, as it is common today, is strictly separated from its associated metalanguage, the semantic paradoxes disappear. Thus, Ramsey had shown that the second hierarchy level is unnecessary to avoid the antinomies, which is also why Gödel does not mention Russell's ramified type theory anywhere in his work. In Chapter 4, you will find that Gödel's system P is built upon a more elementary type system, easily understandable with the knowledge acquired in this section.

Although type theory in its original form plays no significant role today, its historical significance has to be acknowledged. Russell laid the foundation for a new field of research that would yield numerous insights over the years. Type theory received stimulating impetus, especially in the second half of the twentieth century, from the burgeoning field of computer science. In this area, various type systems emerged, playing crucial roles in the theory of programming languages, in compiler construction, as well as in hardware and software verification [58]. Readers who want to delve deeper into the modern branches of this field of research may enjoy the detailed account given in [62].

2.3.4 The Logic of the *Principia Mathematica*

Russell's ramified type theory paved the way for one of the paramount writings in mathematical world literature: the *Principia Mathematica* (Figure 2.20).

> *"After [I discovered the theory of types] it only remained to write the book out. Whitehead's teaching work left him not enough leisure for this mechanical job. I worked at it from ten to twelve hours a day for about eight month in the year, from 1907 to 1910."*
>
> Bertrand Russell [93]

Over several years of work, Russell and Whitehead assembled an axiomatic foundation for essential branches of classical mathematics. Upon opening, it becomes immediately apparent that the *Principia Mathematica* is not an ordinary book. This monumental work teems with thousands upon thousands of symbolic definitions and derivations, occasionally punctuated by explanatory text passages. The sample page in Figure 2.21 showcases the extraordinary style in which this work is composed.

Financially, however, the publication was a disaster, as Russell noted in his biography:

> *"The University Press estimated that there would be a loss of £600 on the book, and while the syndics were willing to bear a loss of £300, they did not feel that they could go above this figure. The Royal Society very*

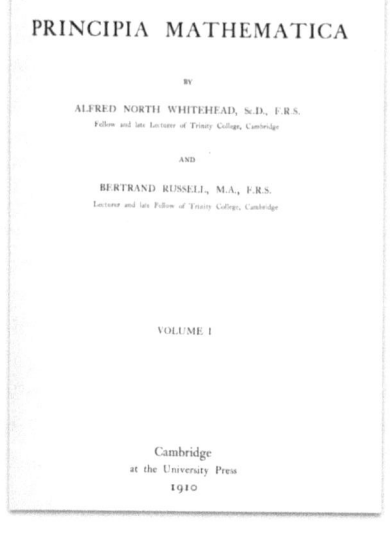

Figure 2.20: Title page of the *Pricipia Mathematica* from 1910

> generously contributed £200, and the remaining £100 we had to find ourselves. We thus earned minus £50 each by ten years' work."
>
> Bertrand Russell [93]

The *Principia Mathematica* comprises more than 1800 pages, divided into three volumes. The first volume appeared in 1910, and the second and third followed in 1912 and 1913. Volume I commences with a detailed explanation of the goals and methodologies of the *Principia Mathematica*. It continues with an idiomatic introduction to type theory, followed by Part I, which is the most significant in the context of this book. It consists of a comprehensive account of mathematical logic, divided into five sections labeled A to E. In Section A, Russell and Whitehead start by developing their calculus's propositional component, subsequently extending it to a typed predicate logic in Section B.

The *Principia Mathematica* refers to axioms as *primitive propositions*, abbreviated as Pp. The propositional axioms are listed in Figure 2.22. Little effort is needed to translate them into modern notation, except for the first axiom:

$$\text{Taut}: (p \vee p) \rightarrow p \tag{PM.2}$$
$$\text{Add}: q \rightarrow (p \vee q) \tag{PM.3}$$
$$\text{Perm}: (p \vee q) \rightarrow (q \vee p) \tag{PM.4}$$
$$\text{Assoc}: p \vee (q \vee r) \rightarrow q \vee (p \vee r) \tag{PM.5}$$
$$\text{Sum}: (q \rightarrow r) \rightarrow (p \vee q \rightarrow p \vee r) \tag{PM.6}$$

Unlike Frege, who had chosen negation '¬' and implication '→' as the basic connectives, Russell and Whitehead relied on disjunction '∨' and implication '→' in their axioms. The plaintext names *Taut, Add, Perm, Assoc,* and *Sum*

2.3 Principia Mathematica

> SECTION A]   CARDINAL COUPLES   379
>
> *54·42. $\vdash :: \alpha \in 2 . \supset :. \beta \subset \alpha . \exists ! \beta . \beta \neq \alpha . \equiv . \beta \in \iota``\alpha$
>
> Dem.
>
> $\vdash . *54·4 . \quad \supset \vdash :: \alpha = \iota'x \cup \iota'y . \supset :.$
>
> $\qquad \beta \subset \alpha . \exists ! \beta . \equiv : \beta = \Lambda . \lor . \beta = \iota'x . \lor . \beta = \iota'y . \lor . \beta = \alpha : \exists ! \beta :$
>
> [*24·53·56.*51·161] $\equiv : \beta = \iota'x . \lor . \beta = \iota'y . \lor . \beta = \alpha$ (1)
>
> $\vdash . *54·25 . \text{Transp} . *52·22 . \supset \vdash : x \neq y . \supset . \iota'x \cup \iota'y \neq \iota'x . \iota'x \cup \iota'y \neq \iota'y :$
>
> [*13·12] $\supset \vdash : \alpha = \iota'x \cup \iota'y . x \neq y . \supset . \alpha \neq \iota'x . \alpha \neq \iota'y$ (2)
>
> $\vdash . (1) . (2) . \supset \vdash :: \alpha = \iota'x \cup \iota'y . x \neq y . \supset :.$
>
> $\qquad \beta \subset \alpha . \exists ! \beta . \beta \neq \alpha . \equiv : \beta = \iota'x . \lor . \beta = \iota'y :$
>
> [*51·235] $\equiv : (\exists z) . z \in \alpha . \beta = \iota'z :$
>
> [*37·6] $\equiv : \beta \in \iota``\alpha$ (3)
>
> $\vdash . (3) . *11·11·35 . *54·101 . \supset \vdash . \text{Prop}$
>
> *54·43. $\vdash :. \alpha, \beta \in 1 . \supset : \alpha \cap \beta = \Lambda . \equiv . \alpha \cup \beta \in 2$
>
> Dem.
>
> $\vdash . *54·26 . \supset \vdash :. \alpha = \iota'x . \beta = \iota'y . \supset : \alpha \cup \beta \in 2 . \equiv . x \neq y .$
>
> [*51·231] $\equiv . \iota'x \cap \iota'y = \Lambda .$
>
> [*13·12] $\equiv . \alpha \cap \beta = \Lambda$ (1)
>
> $\vdash . (1) . *11·11·35 . \supset$
>
> $\quad \vdash :. (\exists x, y) . \alpha = \iota'x . \beta = \iota'y . \supset : \alpha \cup \beta \in 2 . \equiv . \alpha \cap \beta = \Lambda$ (2)
>
> $\vdash . (2) . *11·54 . *52·1 . \supset \vdash . \text{Prop}$
>
> From this proposition it will follow, when arithmetical addition has been defined, that $1 + 1 = 2$.

Figure 2.21: Probably the most famous page of the *Principia Mathematica*. It shows how the formally developed theory of ordinal numbers can be utilized to derive the arithmetic relationship $1 + 1 = 2$.

are also taken from the *Principia*. Most of the time Russell and Whitehead referenced axioms by their names rather than by their numbers.

The first axiom was formulated in colloquial language and did not sound very meaningful at first glance: *"Anything implied by a true premiss is true"*. At second glance, it becomes clear that this formulation hides a well-known inference rule: The modus ponens. Russell and Whitehead express themselves more precisely on page 99, where they formulate the inference rule in the context of predicate logic:

> *1·11. When ϕx can be asserted, where x is a real variable, and $\phi x \supset \psi x$ can be asserted, where x is a real variable, then ψx can be asserted, where x is a real variable. Pp.

> The following are the primitive propositions employed in the calculus of propositions. The letters "Pp" stand for "primitive proposition."
>
> (1) Anything implied by a true premiss is true Pp.
> This is the rule which justifies inference.
>
> (2) $\vdash : p \vee p . \supset . p$ Pp,
> i.e. if p or p is true, then p is true.
>
> (3) $\vdash : q . \supset . p \vee q$ Pp,
> i.e. if q is true, then p or q is true.
>
> (4) $\vdash : p \vee q . \supset . q \vee p$ Pp,
> i.e. if p or q is true, then q or p is true.
>
> (5) $\vdash : p \vee (q \vee r) . \supset . q \vee (p \vee r)$ Pp,
> i.e. if either p is true or "q or r" is true, then either q is true or "p or r" is true.
>
> (6) $\vdash :. q \supset r . \supset : p \vee q . \supset . p \vee r$ Pp,
> i.e. if q implies r, then "p or q" implies "p or r."

Figure 2.22: Propositional axioms of the *Principia Mathematica* [99]

Next, we want to see the logic of the *Principia Mathematica* brought to action. For this purpose, we will look at a few selected proofs from the 1st volume.

■ *Principia Mathematica*, Volume I, Page 104:

$*2{\cdot}05. \quad \vdash :. q \supset r . \supset : p \supset q . \supset . p \supset r$

Dem.

$\left[\text{Sum } \dfrac{\sim p}{p} \right] \quad \vdash :. q \supset r . \supset : \sim p \vee q . \supset . \sim p \vee r \quad (1)$

$[(1).(*1{\cdot}01)] \quad \vdash :. q \supset r . \supset : p \supset q . \supset . p \supset r$

The proof of Theorem *2.05 consists of two derivation steps. In the first step, an instance of the 6th axiom (Sum) is formed by substituting the placeholder p with $\sim p$. In the second step, the disjunction is replaced by the implication operator according to the following definition:

$*1{\cdot}01. \quad p \supset q . = . \sim p \vee q \quad \text{Df.}$

After freeing the formulas from the dust of their old-fashioned notation, a trivial proof comes to light:

$(q \to r) \to ((p \to q) \to (p \to r))$	*2.05
1. $\vdash \quad (q \to r) \to ((\neg p \vee q) \to (\neg p \vee r))$	(PM.6)
2. $\vdash \quad (q \to r) \to ((p \to q) \to (p \to r))$	(Def)

2.3 *Principia Mathematica*

The next two proofs are just as easy to understand. With **2.1**, Russell and Whitehead yield a prominent propositional theorem: the *law of excluded middle* (lat. *tertium non datur*).

■ *Principia Mathematica*, Volume I, Page 105:

∗2·08. ⊦ . p ⊃ p

Dem.

$\left[*2{\cdot}05 \dfrac{p \vee p, p}{q, \ r} \right]$ ⊦ :: p ∨ p . ⊃ . p : ⊃ :. p . ⊃ . p ∨ p : ⊃ . p ⊃ p   (1)

[Taut] ⊦ : p ∨ p . ⊃ . p   (2)

[(1).(2).∗1·11] ⊦ :. p . ⊃ . p ∨ p : ⊃ . p ⊃ p   (3)

[2·07] ⊦ : p . ⊃ . p ∨ p   (4)

[(3).(4).∗1·11] ⊦ . p ⊃ p

∗2·1. ⊦ . ∼p ∨ p [Id . (∗1·01)]

After replacing the dot symbols with parenthesis using Peano's rules and swapping the symbols '∼' and '⊃' for '¬' and '→', respectively, the derivation sequences appear in this guise:

p → p	∗2.08
1. ⊦ ((p ∨ p) → p) → ((p → (p ∨ p)) → (p → p))	(∗2.05)
2. ⊦ (p ∨ p) → p	(PM.2)
3. ⊦ (p → (p ∨ p)) → (p → p)	(MP, 2,1)
4. ⊦ p → (p ∨ p)	(PM.3)
5. ⊦ p → p	(MP, 4,3)

¬p ∨ p	∗2.1
1. ⊦ p → p	(∗2.08)
2. ⊦ ¬p ∨ p	(Def)

In Section B of the *Principia Mathematica*, Russell and Whitehead introduced the *universal quantifier* '∀' and the *existential quantifier* '∃'. In addition, they supplemented the axioms with a series of elementary propositions of predicate logic. Among the numerous definitions given at the beginning of this section were these:

- *Principia Mathematica*, Volume 1, Page 135:

*9·03. $(x) . \phi x . \lor . p : = . (x) . \phi x \lor p$ Df
*9·04. $p . \lor . (x) . \phi x : = . (x) . p \lor \phi x$ Df
*9·05. $(\exists x) . \phi x . \lor . p : = . (\exists x) . \phi x \lor p$ Df
*9·06. $p . \lor . (\exists x) . \phi x : = . (\exists x) . p \lor \phi x$ Df
*9·07. $(x) . \phi x . \lor . (\exists y) . \psi y : = : (x) : (\exists y) . \phi x \lor \psi y$ Df
*9·08. $(\exists y) . \psi y . \lor . (x) . \phi x : = : (x) : (\exists y) . \psi y \lor \phi x$ Df

The first four definitions allow the scope of a quantifier in a disjunctively connected formula to be limited to those parts in which the quantified variable occurs *freely*. Modern predicate logic also formalizes this fact, but not in the form of a definition, but in the form of an axiom:

$$\forall \xi \, (\varphi \lor \psi) \rightarrow (\varphi \lor \forall \xi \, \psi) \quad \text{(for all } \varphi \text{ with } \xi \notin \varphi) \qquad (2.18)$$

The notation $\xi \notin \varphi$ expresses that the variable ξ does not occur freely in φ, that is, each occurrence of ξ, if there is any, falls within the scope of a quantifier. In modern formulations of predicate logic, this axiom is sufficient to derive the other variants of quantifier shifting.

On the next page, further axioms follow:

- *Principia Mathematica*, Volume 1, Page 136:

*9·1. $\vdash : \phi x . \supset . (\exists z) . \phi z$ Pp
*9·11. $\vdash : \phi x \lor \phi y . \supset . (\exists z) . \phi z$ Pp

Proposition *9.1 is of high relevance. Below, we will show that it is equivalent to a well-known axiom of modern predicate logic. Equally important is this one:

- *Principia Mathematica*, Volume 1, Page 137:

$\vdash : [\phi y] . \supset . (x) . \phi x$ Pp.

The square brackets play a unique role, which Russell and Whitehead explain on page 137:

> "[...] *if we put*
>
> $\vdash : \phi y . \supset . (x) . \phi x$'
>
> *that means:* 'However y may be chosen, ϕy implies $(x).\phi x$' *which is in general false. What we mean is:* 'If ϕy is true however y may be chosen, then $(x).\phi x$ is true.' *But we have not supplied a symbol for the mere hypothesis of what is asserted in* '$\vdash .\phi y$', *where y is a real variable, and it is not worth while to supply a symbol, because it would be very rarely required. If for the moment, we use the symbol $[\phi y]$ to express this*

2.3 Principia Mathematica

hypothesis, then our primitive proposition is

$$\vdash :[\phi y] . \supset .(x).\phi x \quad \text{Pp.} \qquad \text{"}$$

Immediately after introducing this axiom, Russell and Whitehead point out its true purpose: it is meant solely to derive new theorems:

> "In practice, this proposition is only used for *inference*, not for implication. [...] This process will be called 'turning a real variable into an apparent variable.'"

When expressed in the form of an inference rule, the axiom takes on the following form:

$$\frac{\varphi}{\forall x\, \varphi} \qquad (G)$$

(G) is referred to as the *generalization rule*. In addition to the modus ponens, it is the second important rule of inference in modern predicate logic.

The last proof we will look at in this section is this one:

■ *Principia Mathematica*, Volume 1, Page 138:

*9·2. $\vdash : (x) . \phi x . \supset . \phi y$

The above proposition states the principle of deduction from the general to the particular, *i.e.* "what holds in all cases, holds in any one case."

Dem.

$$\vdash . *2\cdot 1 . \supset \vdash . \sim\phi y \vee \phi y \qquad (1)$$
$$\vdash . *9\cdot 1 . \supset \vdash :\sim\phi y \vee \phi y . \supset . (\exists x) . \sim\phi x \vee \phi y \qquad (2)$$
$$\vdash . (1) . . (2) . *1\cdot 11 . \supset \vdash . (\exists x) . \sim\phi x \vee \phi y \qquad (3)$$
$$[(3).(*9\cdot 05)] \qquad \vdash : (\exists x) . \sim\phi x . \vee . \phi y \qquad (4)$$
$$[(4).(*9\cdot 01. *1\cdot 01)] \quad \vdash : (x) . \phi x . \supset . \phi y$$

Again, this sequence can be translated easily into a modern-looking proof:

$\forall x\, \phi(x) \to \phi(y)$	*9.2
1. $\vdash\ \neg\phi(y) \vee \phi(y)$	(*2.1)
2. $\vdash\ \neg\phi(y) \vee \phi(y) \to \exists x\, (\neg\phi(x) \vee \phi(y))$	(*9.1)
3. $\vdash\ \exists x\, (\neg\phi(x) \vee \phi(y))$	(MP, 1,2)
4. $\vdash\ \exists x\, \neg\phi(x) \vee \phi(y)$	(*9.05, 3)
5. $\vdash\ \neg\forall x\, \phi(x) \vee \phi(y)$	(Def)
6. $\vdash\ \forall x\, \phi(x) \to \phi(y)$	(Def)

Table 2.1: A complete axiom system for first-order predicate logic

Axioms	
$(\varphi \vee \varphi) \rightarrow \varphi$	(PM.2)
$\psi \rightarrow (\varphi \vee \psi)$	(PM.3)
$(\varphi \vee \psi) \rightarrow (\psi \vee \varphi)$	(PM.4)
$(\psi \rightarrow \chi) \rightarrow (\varphi \vee \psi \rightarrow \varphi \vee \chi)$	(PM.6)
$\forall \xi \, \varphi \rightarrow \varphi[\xi \leftarrow \sigma]$ (for every collision-free substitution)	(2.19)
$\forall \xi \, (\varphi \vee \psi) \rightarrow (\varphi \vee \forall \xi \, \psi)$ (for all φ with $\xi \notin \varphi$)	(2.18)
Inference rules	
$\dfrac{\varphi, \varphi \rightarrow \psi}{\psi}$ (MP) $\dfrac{\varphi}{\forall \xi \, \varphi}$ (G)	

In present-day predicate logic, this theorem exists in a very similar form as an axiom:

$$\forall \xi \, \varphi \rightarrow \varphi[\xi \leftarrow \sigma] \quad \text{(for all collision-free substitutions)} \quad (2.19)$$

In (2.19), the substitution of ξ with σ happens through syntactic replacement, that is, every free occurrence of the variable ξ is textually replaced by the character sequence σ. This way, however, a variable that occurs freely in φ may end up in the scope of a quantifier. Such *collisions* must be prohibited as they can lead to substantively false formulas. In Section 4.1.2, we will provide an in-depth discussion of the concept of substitution and illustrate the proper application of this axiom.

The discussed axioms, definitions, theorems, and inference rules of the *Principia Mathematica* were not randomly chosen. Written side by side, as done in Table 2.1, they form an axiom system commonly used in modern textbooks for defining first-order predicate logic. The propositional axioms are sometimes chosen differently, though, as many contemporary authors prefer Łukasiewicz's axioms presented in Section 2.1.2.

It is not a mistake that Russell's fifth axiom (Assoc) is not listed, as in 1926, Hilbert's student Paul Bernays showed that the propositional logic axioms are not independent. In Section 4.3, we will reveal how to deduce the fifth axiom from the others.

Overall, the considerations in this section have shown that a familiar logic lies beneath the seemingly complex structure of Russell's type theory. All

2.3 Principia Mathematica

axioms and rules of inference employed in modern predicate logic have their counterparts in the *Principia Mathematica*, in either the same or a similar form. It is primarily the antiquated notation that prevents us from recognizing this connection right away.

In hindsight, it becomes evident that Russell and Whitehead played a significant role in popularizing formal mathematics. Their work was a compelling showcase for the precision and expressive power of formal systems in general and predicate logic in particular. Thus, one may wonder if students should work through the 1800 pages as part of their training. The honest answer is no. The logic presented in the *Principia Mathematica* is, in many ways, too immature to replace modern textbooks, and its intricate type theory is mainly obsolete in contemporary logic.

Nevertheless, the *Principia Mathematica* is still relevant today. Russell and Whitehead formalized classical mathematics to such an extent that the notion of proof reduces to the symbolic manipulation of symbol strings. In particular, it is noteworthy that the authors did not just demonstrate their method on selected examples but applied it in meticulous hard work to large parts of mathematics. In this sense, the over 1800 pages are empirical proof for the formalistic and logicistic belief that formal systems can model all concepts and inference methods of ordinary mathematics. The *Principia Mathematica* demonstrated that formal mathematics works and this is their true significance.

The intellectual endeavor required to compose such a monumental work is hard to quantify. What is evident, however, is that the ten years dedicated to the *Principia Mathematica* shaped Russell's mind:

> "[...] *I always found myself hoping that perhaps Principia Mathematica would be finished some day. Moreover the difficulties appeared to me in the nature of a challenge, which it would be pusillanimous not to meet and overcome. So I persisted, and in the end the work was finished, but my intellect never quite recovered from the strain. I have been ever since definitely less capable of dealing with difficult abstractions than I was before. This is part, though by no means the whole, of the reason for the change in the nature of my work.*"
>
> Bertrand Russell [93]

In the following years, Russell withdrew almost entirely from logic, redirecting his focus toward social and philosophical matters. He made significant contributions in these fields and was honored with the Nobel Prize for Literature in 1950. Russell became world famous, and many people nowadays associate his name exclusively with his philosophical work. Many do not know that the renowned philosopher Bertrand Russell was also one of the greatest mathematicians ever.

Figure 2.23

Ernst Zermelo
1871 – 1953

2.4 Axiomatic Set Theory

The next stop on our journey is Berlin, where Ernst Friedrich Ferdinand Zermelo was born on July 27, 1871. His father, Theodor Zermelo, was a high school teacher and had six children with his wife Maria Auguste. Ernst was the only son and grew up with an older and four younger sisters. Several tragic events punctuated Zermelo's life. At the age of seven, he had to cope with the loss of his mother shortly after the birth of his youngest sister. Then, in 1889, just before his high school graduation, he and his sisters tragically lost their father, leaving them as orphans. His parents had made financial provisions, but the guardianship court ordered that most of the assets had to be spent for the care of the younger sisters. As he grew up, Zermelo's life changed for the better. Due to his excellent academic performance, he was granted two scholarships, allowing him to enroll at the Berlin Friedrich Wilhelm University, now known as the Humboldt University.

Zermelo had a wide range of interests and attended lectures in mathematics, physics, and philosophy. As was customary at the time, he spent a few semesters at other universities. This led him to take part in a lecture at the University of Halle-Wittenberg by a man whose scientific work would play a key role in his life: Georg Cantor. At that time, however, set theory was neither one of Zermelo's fields of interest nor covered in the lecture mentioned above; Cantor was lecturing on elliptical functions.

After earning his doctorate in 1894, Zermelo initially worked as an assistant to Max Planck at the Berlin Institute for Theoretical Physics. In 1897, he relocated to Göttingen, rekindling his focus on mathematics. The tranquil university town, steeped in tradition, not only provided Zermelo with an exceptional working environment. With the appointment of David Hilbert, Göttingen rose

2.4 Axiomatic Set Theory

to prominence as the world's leading center for mathematics. In the summer of 1898, Zermelo got acquainted with set theory in a lecture by Arthur Schoenflies and in a seminar by Felix Klein [17]. In 1899, he habilitated in statistical mechanics and later worked as a private lecturer at the university. It was the influence of David Hilbert that gradually shifted his interest to the foundations of mathematics. In [112], Zermelo writes:

> "Already 30 years ago, when I was a private lecturer in Göttingen, I began to deal with the fundamental questions of mathematics under the influence of D. Hilbert, to whom I owe most of my scientific development, especially with the fundamental problems of Cantor's set theory, which only came to my full awareness in the then so fruitful cooperation of the Göttingen mathematicians. It was the time when the 'antinomies', the apparent 'contradictions' in set theory, attracted the most general attention and prompted both qualified and unqualified pens to the boldest as well as the most anxious attempts at solutions."

> "Schon vor 30 Jahren, als ich Privatdozent in Göttingen war, begann ich unter dem Einflusse D. Hilberts, dem ich überhaupt das meiste in meiner wissenschaftlichen Entwickelung zu verdanken habe, mich mit den Grundlagenfragen der Mathematik zu beschäftigen, insbesondere aber mit den grundlegenden Problemen der Cantorschen Mengenlehre, die mir in der damals so fruchtbaren Zusammenarbeit der Göttinger Mathematiker erst in ihrer vollen Bedeutung zum Bewußtsein kamen. Es war damals die Zeit, wo die 'Antinomien', die scheinbaren 'Widersprüche' in der Mengenlehre, die allgemeinste Aufmerksamkeit auf sich zogen und berufene wie unberufene Federn zu den kühnsten wie zu den ängstlichsten Lösungsversuchen veranlaßten."

Ernst Zermelo, 1930

Zermelo drew his attention to a problem that Hilbert had identified as one of mathematics's most compelling and fundamental questions. This question continues to captivate mathematicians today: the *continuum hypothesis*.

2.4.1 Continuum Hypothesis

The *continuum hypothesis* (CH), conjectured by Georg Cantor, postulates a relationship between the cardinality of a set M and the cardinality of its power set 2^M. To understand its exact formulation, let us recall *Cantor's Theorem*, proven on page 63. This theorem states that the power set 2^M has a greater cardinality than M itself. For instance, Cantor's theorem implies the following hierarchy:

$$|\mathbb{N}| < |2^{\mathbb{N}}| < |2^{2^{\mathbb{N}}}| < |2^{2^{2^{\mathbb{N}}}}| < |2^{2^{2^{2^{\mathbb{N}}}}}| < \cdots$$

Cantor employed the Hebrew letter Aleph (ℵ) to label the cardinalities of infinite sets. The smallest infinity corresponds to the cardinality of the natural numbers and is denoted by the *cardinal number* $\aleph_0$. No smaller infinities than $|\mathbb{N}|$ exist, as all infinite subsets of $\mathbb{N}$ can be bijectively mapped onto $\mathbb{N}$. The cardinal number $\aleph_1$ denotes the next larger infinity, and so on. If a set M has the cardinality $\aleph_n$, then $2^{\aleph_n}$ denotes the cardinality of the power set 2^M.

It is straightforward to demonstrate that the set of real numbers, denoted as $\mathbb{R}$, has the same cardinality as the power set of natural numbers. Symbolically, this equivalence is expressed as follows:

$$|\mathbb{R}| = |2^{\mathbb{N}}| = 2^{\aleph_0} \tag{2.20}$$

Hence, the set of real numbers has a greater cardinality than the set of natural numbers. Cantor was captivated by the question of whether additional infinities were hiding between the sets $\mathbb{N}$ and $\mathbb{R}$. In a letter to Richard Dedekind dated July 11, 1877, he conjectured this was not the case.

In his letter, Cantor formulated his hypothesis in a rather intricate manner. If he were to phrase it today, Cantor would probably use a formulation similar to this one:

> "Every infinite subset of the real numbers has the cardinality of the natural numbers or the cardinality of the real numbers."

Or, more concisely:

> "There is no set M with $|\mathbb{N}| < |M| < |\mathbb{R}|$."

Assuming the continuum hypothesis were true, the real numbers would be second in the infinitely long list of infinities. Symbolically, this conjecture is expressed in the following form:

$$|\mathbb{R}| \stackrel{?}{=} \aleph_1 \tag{2.21}$$

According to (2.20), this can be rewritten as

$$2^{\aleph_0} \stackrel{?}{=} \aleph_1.$$

Extrapolating this equation to

$$2^{\aleph_n} \stackrel{?}{=} \aleph_{n+1} \tag{2.22}$$

leads to the *generalized continuum hypothesis* (GCH). In colloquial terms, this hypothesis states that the power set operation moves seamlessly from one infinity to the next (Figure 2.24).

2.4 Axiomatic Set Theory

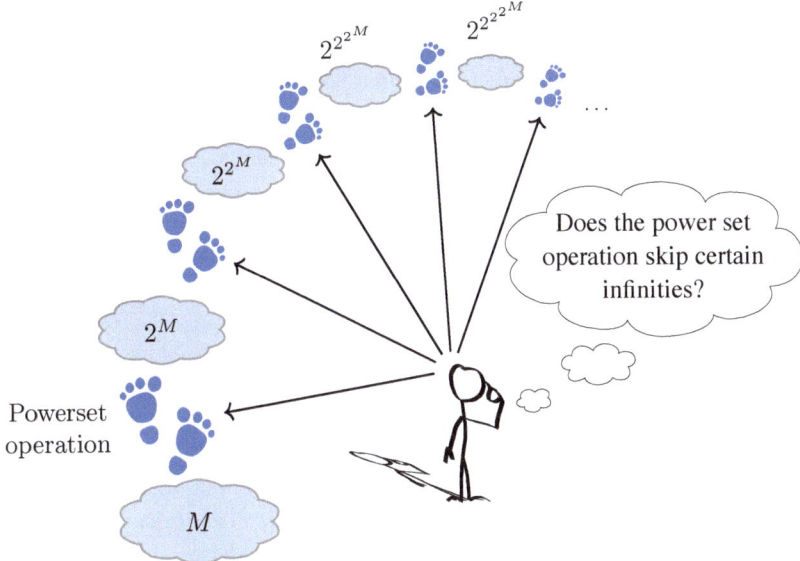

Figure 2.24: The generalized continuum hypothesis conjectures that the powerset operation does not skip any infinities while traversing from one infinity to the next.

Cantor continued to work on the continuum hypothesis until the end of his life. Several times, he believed he had found a proof. Other times, Cantor thought he had successfully refuted the hypothesis. But errors kept cropping up, constantly shattering his perceived successes. No matter how hard he tried, he was denied solving this great riddle of set theory during his lifetime.

Hilbert regarded the resolution of the continuum hypothesis as a matter of utmost urgency, emphasizing it in his speech at the second International Congress of Mathematicians in Paris. He placed it prominently at the top of his renowned list of unsolved problems:

> *"Two systems, i.e., two assemblages of ordinary real numbers or points, are said to be (according to Cantor) equivalent or of equal cardinal number, if they can be brought into a relation to one another such that to every number of the one assemblage corresponds one and only one definite number of the other. The investigations of Cantor on such assemblages of points suggest a very plausible theorem, which nevertheless, in spite of the most strenuous efforts, no one has succeeded in proving. This is the theorem:*
> *Every system of infinitely many real numbers, i.e., every assemblage of numbers (or points), is either equivalent to the assemblage of natural integers, 1, 2, 3,... or to the assemblage of all real numbers and therefore to the continuum, that is, to the points of a line; as regards equivalence there are, therefore, only two assemblages of numbers, the countable assemblage and the continuum.*

From this theorem it would follow at once that the continuum has the next cardinal number beyond that of the countable assemblage; the proof of this theorem would, therefore, form a new bridge between the countable assemblage and the continuum."

"Zwei Systeme, d. h. zwei Mengen von gewöhnlichen reellen Zahlen (oder Punkten) heißen nach Cantor äquivalent oder von gleicher Mächtigkeit, wenn sie zu einander in eine derartige Beziehung gebracht werden können, daß einer jeden Zahl der einen Menge eine und nur eine bestimmte Zahl der anderen Menge entspricht. Die Untersuchungen von Cantor über solche Punktmengen machen einen Satz sehr wahrscheinlich, dessen Beweis jedoch trotz eifrigster Bemühungen bisher noch Niemandem gelungen ist; dieser Satz lautet: 'Jedes System von unendlich vielen reellen Zahlen d. h. jede unendliche Zahlen- (oder Punkt)menge ist entweder der Menge der ganzen natürlichen Zahlen 1, 2, 3, ... oder der Menge sämmtlicher reellen Zahlen und mithin dem Continuum, d. h. etwa den Punkten einer Strecke aequivalent; im Sinne der Aequivalenz giebt es hiernach nur zwei Zahlenmengen, die abzählbare Menge und das Continuum.' Aus diesem Satz würde zugleich folgen, daß das Continuum die nächste Mächtigkeit über die Mächtigkeit der abzählbaren Mengen hinaus bildet; der Beweis dieses Satzes würde mithin eine neue Brücke schlagen zwischen der abzählbaren Menge und dem Continuum."

David Hilbert, Paris, 1900 [50, 47]

2.4.2 Well-Ordering Theorem

Equally important are the words that follow. Let's listen again:

"Let me mention another very remarkable statement of Cantor's which stands in the closest connection with the theorem mentioned and which, perhaps, offers the key to its proof. Any system of real numbers is said to be ordered, if for every two numbers of the system it is determined which one is the earlier and which the later, and if at the same time this determination is of such a kind that, if a is before b and b is before c, then a always comes before c. The natural arrangement of numbers of a system is defined to be that in which the smaller precedes the larger. But there are, as is easily seen infinitely many other ways in which the numbers of a system may be arranged."

"Es sei noch eine andere sehr merkwürdige Behauptung Cantors erwähnt, die mit dem genannten Satze in engstem Zusammenhange steht und die vielleicht den Schlüssel zum Beweise dieses Satzes liefert. Irgend ein System von reellen Zahlen heißt geordnet, wenn von irgend zwei Zahlen des Systems festgesetzt ist, welches die frühere und welches die spätere sein soll, und dabei diese Festsetzung eine derartige ist, daß, wenn eine Zahl a früher als die Zahl b und b früher als c ist, so auch stets a früher

2.4 Axiomatic Set Theory

als c erscheint. Die natürliche Anordnung der Zahlen eines Systems heiße diejenige, bei der die kleinere als die frühere, die größere als die spätere festgesetzt wird. Es giebt aber, wie leicht zu sehen ist, noch unendlich viele andere Arten, wie man die Zahlen eines Systems ordnen kann."

David Hilbert, Paris, 1900 [50, 47]

Hilbert's description aligns with the classical definition of a totally ordered set:

Definition 2.7   Partial order, total order

Let M be a set and '$<$' a binary relation on M.

- '$<$' is a *partial order* on M, if it is
  - *irreflexive*,   ☞ $x \not< x$
  - *asymmetric*, and   ☞ From $x < y$ follows $y \not< x$
  - *transitive*.   ☞ From $x < y$ and $y < z$ follows $x < z$

- '$<$' is a *linear order* or *total order* on M, if
  - '$<$' is a partial order, and
  - all elements are strongly connected.
    ☞ For all x and y with $x \neq y$, either $x < y$ or $y < x$

Ordered sets are abundant in mathematics. For instance, the well-known number sets $\mathbb{N}$, $\mathbb{Z}$, $\mathbb{Q}$, and $\mathbb{R}$ are totally ordered by the standard less-than relation '$<$'. Hilbert continues:

" If we think of a definite arrangement of numbers and select from them a particular system of these numbers, a so-called partial system or assemblage, this partial system will also prove to be ordered. Now Cantor considers a particular kind of ordered assemblage which he designates as a well ordered assemblage and which is characterized in this way, that not only in the assemblage itself but also in every partial assemblage there exists a first number. The system of integers 1, 2, 3, ... in their natural order is evidently a well ordered assemblage. On the other hand the system of all real numbers, i.e., the continuum in its natural order, is evidently not well ordered. For, if we think of the points of a segment of a straight line, with its initial point excluded, as our partial assemblage, it will have no first element."

"Wenn wir eine bestimmte Ordnung der Zahlen ins Auge fassen und aus denselben irgend ein besonderes System dieser Zahlen, ein sogenanntes Teilsystem oder eine Teilmenge, herausgreifen, so erscheint diese Teilmenge ebenfalls geordnet. Cantor betrachtet nun eine besondere Art

von geordneten Mengen, die er als wohlgeordnete Mengen bezeichnet und die dadurch charakterisirt sind, daß nicht nur in der Menge selbst, sondern auch in jeder Teilmenge eine früheste Zahl existirt. Das System der ganzen Zahlen 1, 2, 3, ... in dieser seiner natürlichen Ordnung ist offenbar eine wohlgeordnete Menge. Dagegen ist das System aller reellen Zahlen, d. h. das Continuum in seiner natürlichen Ordnung offenbar nicht wohlgeordnet. Denn, wenn wir als Teilmenge die Punkte einer endlichen Strecke mit Ausnahme des Anfangspunktes der Strecke ins Auge fassen, so besitzt diese Teilmenge jedenfalls kein frühestes Element."

David Hilbert, Paris, 1900 [50, 47]

Thus, Cantor was not interested in arbitrary orders but only in those that lead to what is known as a *well-ordering*.

Definition 2.8 — Well-Ordering

Let M be a set and '<' a binary relation on M.

- '<' is a *well-ordering* on M, if
  - '<' is a total order on M, and
  - every non-empty subset $N \subseteq M$ has a smallest element.
    - ☞ There exists an $x \in N$ with $x < y$ for all $y \in N$ with $y \neq x$

Among the already mentioned sets $\mathbb{N}$, $\mathbb{Z}$, $\mathbb{Q}$, and $\mathbb{R}$, only $\mathbb{N}$ is well-ordered by the standard less-than relation '<'. In this set, every non-empty subset contains a minimal element, that is, an element that is smaller than all other elements within that subset.

We reach a pivotal point in Hilbert's speech:

> "The question now arises whether the totality of all numbers may not be arranged in another manner so that every partial assemblage may have a first element, i. e., whether the continuum cannot be considered as a well ordered assemblage – a question which Cantor thinks must be answered in the affirmative. It appears to me most desirable to obtain a direct proof of this remarkable statement of Cantor's, perhaps by actually giving an arrangement of numbers such that in every partial system a first number can be pointed out."

> "Es erhebt sich nun die Frage, ob sich die Gesamtheit aller Zahlen nicht in anderer Weise so ordnen läßt, daß jede Teilmenge ein frühestes Element hat, d. h. ob das Continuum auch als wohlgeordnete Menge aufgefaßt werden kann, was Cantor bejahen zu müssen glaubt. Es erscheint mir höchst wünschenswert, einen direkten Beweis dieser merkwürdigen Behauptung von Cantor zu gewinnen, etwa durch wirkliche Angabe einer

2.4 Axiomatic Set Theory

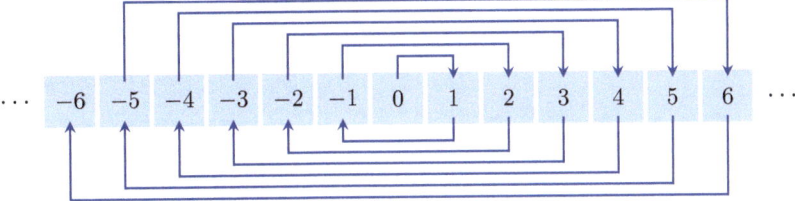

Figure 2.25: One of infinitely many well-orderings of the integers.

solchen Ordnung der Zahlen, bei welcher in jedem Teilsystem eine früheste Zahl aufgewiesen werden kann."

David Hilbert, Paris, 1900 [50, 47]

Apparently, Cantor considered it possible to find an arrangement of the real numbers that fulfills the definition of a well-ordering.

Well-ordering the integers rather than the real numbers does not impose any problems. To ensure that every non-empty subset of $\mathbb{Z}$ contains a minimal element, it suffices to arrange the integers according to the following scheme (Figure 2.25):

$$0 < 1 < -1 < 2 < -2 < 3 < -3 < 4 < -4 < 5 < -5 < 6 < -6 < \ldots$$

With a similar scheme, the set of rational numbers $\mathbb{Q}$ can be well-ordered, too (Figure 2.26):

$$0 < \tfrac{1}{1} < -\tfrac{1}{1} < \tfrac{2}{1} < \tfrac{1}{2} < -\tfrac{1}{2} < -\tfrac{2}{1} < \tfrac{3}{1} < \tfrac{1}{3} < -\tfrac{1}{3} < -\tfrac{3}{1} < \tfrac{4}{1} < \tfrac{3}{2} < \tfrac{2}{3} < \tfrac{1}{4} < \ldots$$

By definition, 0 is the smallest number in this sequence, followed by the fractions whose absolute sum of numerator and denominator equals 2. Then come the fractions whose absolute sum equals 3, and so on. The chosen order ensures that each non-empty subset contains a minimal element.

Hilbert and his colleagues puzzled whether the real numbers could also be well-ordered. But unlike the integers and the rational numbers, the continuum resisted all attempts to construct such an order explicitly. For Hilbert, deciding the well-ordering hypothesis was a pressing problem. Around 1900, he suspected it might hold the key to resolving the even more significant continuum hypothesis.

Eventually, Zermelo turned his attention to the mysteries of well-ordered sets, leading to a breakthrough in 1904. That year, he presented a proof that decided the well-ordering hypothesis positively [111]:

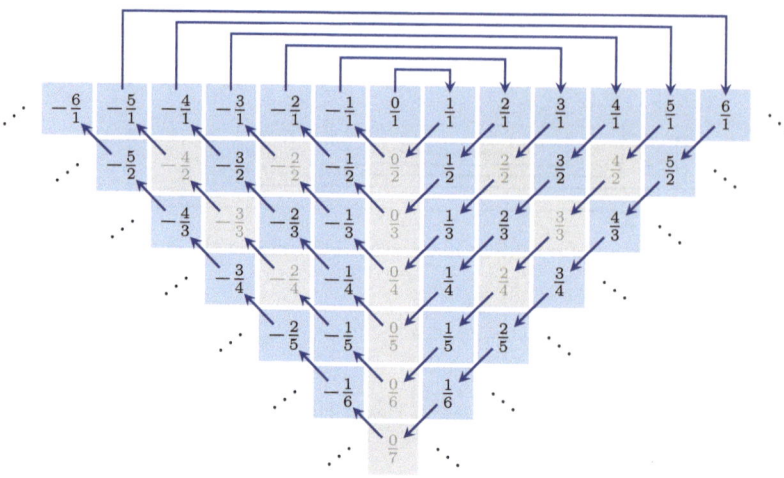

Figure 2.26: One of infinitely many well-orderings of the rational numbers. The gray-shaded fractions are unreduced and only listed to emphasize the enumeration scheme.

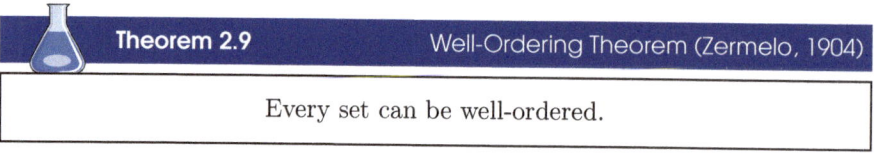

Every set can be well-ordered.

Zermelo's proof marked the beginning of a development that eventually gave rise to axiomatic set theory. However, at the time of publication, the proof was regarded with suspicion and triggered many adverse reactions. Understanding the historical events requires a closer look at Zermelo's line of reasoning.

The proof that every set M can be well-ordered unfolds in four steps:

1. Zermelo introduces the notion of γ-sets, which are well-ordered subsets of M with certain additional properties.

2. He demonstrates that the elements within γ-sets are consistently arranged. For any given set of γ-sets, there exists a unique relation, denoted as '$\prec$', that well-orders the elements in each γ-set.

3. Zermelo defines the set L_γ as the union of all γ-sets and proves that the relation '$\prec$' also establishes a well-ordering on L_γ.

4. Finally, Zermelo shows that L_γ is a γ-set itself, containing all elements of M. This implies that '$\prec$' well-orders the entire set M.

Having roughly outlined the proof, let us fill some of the gaps:

2.4 Axiomatic Set Theory

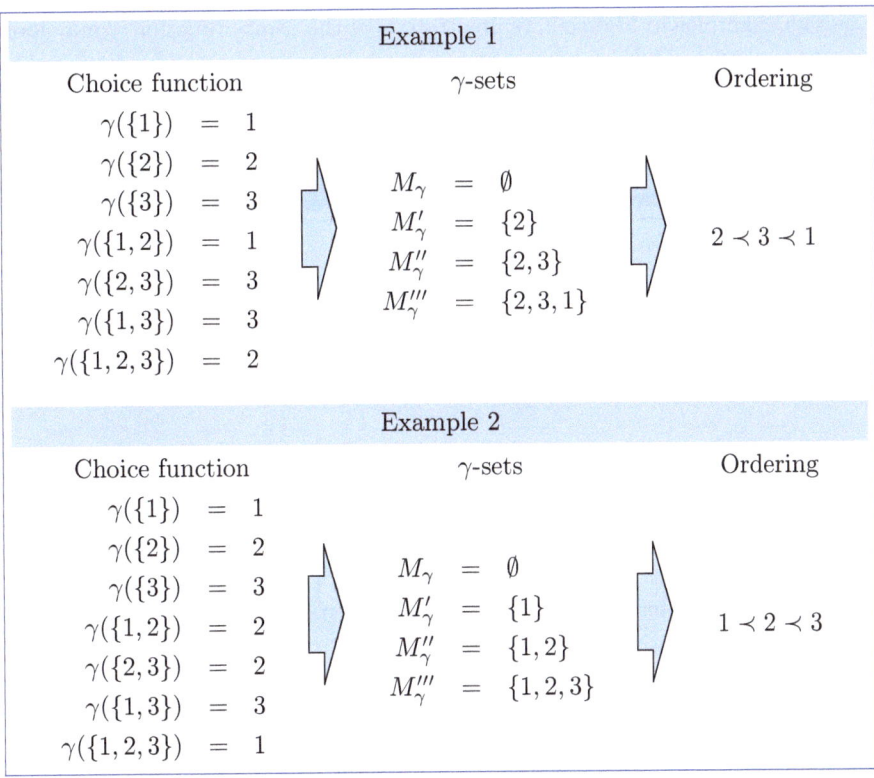

Figure 2.27: Each choice function γ leads to a well-ordering, here demonstrated on the finite set $\{1, 2, 3\}$. This property constitutes the core of Zermelo's proof of the well-ordering theorem.

Step 1

The definition of γ-sets relies on a *choice function* γ, mapping every non-empty subset $M' \subset M$ to an arbitrary element within M'. To put it in Zermelo's words:

> "Imagine that with every subset M' there is associated an arbitrary element m'_1 that occurs in M' itself: let m'_1 be called the 'distinguished' element of M'. This yields a 'covering' γ of the set M by certain elements of the set M."

> "Jeder Teilmenge M' denke man sich ein beliebiges Element m' zugeordnet, das in M' selbst vorkommt und das 'ausgezeichnete' Element von M' genannt werden möge. So entsteht eine 'Belegung' γ der Menge M mit Elementen der Menge M von besonderer Art."

Ernst Zermelo [111, 104]

The two examples in Figure 2.27 illustrate how the choice function γ may look like for the set $\{1, 2, 3\}$.

The selection function is utilized in the next definition:

 Definition 2.10 γ-Set

A subset M_γ of M is a γ-set, if

- M_γ is well-ordered and
- $\gamma(M \backslash A_a) = a$ for all $a \in M_\gamma$

A_a is the *initial segment* of M_γ, containing all elements smaller than a.

According to this definition, a γ-set is a well-ordered set M_γ with the additional property that the distinguished element of $M \backslash A_a$ is the minimal element of $M_\gamma \backslash A_a$. This formulation already suggests that there is little freedom in constructing γ-sets.

As an example, let us consider the set $M = \{1, 2, 3\}$ and the first function from Figure 2.27. The simplest γ-set is the empty set. The next-simplest γ-sets are those with a single element. For every γ-set of the form $\{a_1\}$, the initial segment A_{a_1} is empty. Therefore, we have:

$$a_1 = \gamma(M \backslash \emptyset) = \gamma(\{1, 2, 3\}) = 2$$

Consequently, $\{2\}$ is the only γ-set with a single element.

A similar argument applies for γ-sets $\{a_1, a_2\}$ with two elements. If a_1 is the smaller element, then, as before, $a_1 = \gamma(\{1, 2, 3\}) = 2$. As a result, the larger element a_2 is also uniquely determined:

$$a_2 = \gamma(M \backslash \{a_1\}) = \gamma(\{1, 3\}) = 3$$

Thus, $\{2, 3\}$ is the only γ-set with two elements, and within this set, the ordering relation $2 \prec 3$ holds.

In the same way, the uniquely determined γ-set $\{a_1, a_2, a_3\}$ with three elements can be obtained. The smallest element is equal to 2, the next larger element is equal to 3, and the largest element a_3 must satisfy:

$$a_3 = \gamma(M \backslash \{a_1, a_2\}) = \gamma(\{1\}) = 1$$

Thus, $\{1, 2, 3\}$ is the only γ-set with three elements, and within this set, the ordering relation $2 \prec 3 \prec 1$ holds.

2.4 Axiomatic Set Theory

Step 2

All four calculated γ-sets $\emptyset$, $\{2\}$, $\{2,3\}$ and $\{2,3,1\}$ share a crucial property: When considered as ordered sequences, one set is always the initial segment of the other:

$\emptyset$ is the initial segment of 2 ...
... is the initial segment of $2 \prec 3$ ...
... is the initial segment of $2 \prec 3 \prec 1$.

The example has demonstrated why this has to be the case, at least for finite sets. The γ-sets are created deterministically by successively adding a uniquely determined element. With a generalized argument, Zermelo established this property not only for finite but for arbitrary γ-sets. In his original proof, this consideration is the subject of point (5):

> "(5) Whenever M'_γ and M''_γ are any two distinct γ-sets (associated, however, with the same covering γ chosen once for all!), one of the two is identical with a segment of the other."
>
> "5) Sind M'_γ und M''_γ irgend zwei verschiedene γ-Mengen (die aber zu derselben ein für allemal gewählten Belegung γ gehören!), so ist immer eine von beiden identisch mit einem Abschnitte der anderen."
>
> Ernst Zermelo [111, 104]

This interim result leads to interesting conclusions. If two elements a and b are contained in two different γ-sets, the ordering relationship between a and b must be the same in both sets. This property is part of point 6 in Zermelo's proof:

> "6) Consequences. If two γ-sets have an element a in common, they also have the segment A of the preceding elements in common. If they have two elements a, b in common, then either in both sets $a \prec b$ or in both sets $b \prec a$."
>
> "6) Folgerungen. Haben zwei γ-Mengen ein Element a gemeinsam, so haben sie auch den Abschnitt A der vorangehenden Elemente gemein. Haben sie zwei Elemente a, b gemein, so ist in beiden Mengen entweder $a \prec b$ oder $b \prec a$."
>
> Ernst Zermelo [111, 104]

Step 3

Zermelo refers to an element that occurs in any γ-set as a γ-element and defines L_γ as the set of all γ-elements. For the examples in Figure 2.27, L_γ is the set $\{1,2,3\}$, that is, the set M itself.

The elements of L_γ can be ordered by applying the order in which they occur in the individual γ-sets. This is justified by the prior results. For two γ-elements a and b, property (5) guarantees the existence of a γ-set containing both, and property (6) ensures that the order relations are consistent across all γ-sets. Hence, the larger γ-set can be utilized to determine the smaller element among a and b. Zermelo expresses this idea as follows:

> "If a and b are two arbitrary γ-elements and if M'_γ and M''_γ are any two γ-sets to which they respectively belong, then according to (5), the larger of the two γ-sets contains both elements and determines whether the order relation is $a \prec b$ oder $b \prec a$. According to (6) this order relation is independent of the γ-sets selected."

> "Sind a, b zwei beliebige γ-Elemente und M'_γ und M''_γ irgend zwei γ-Mengen, denen sie angehören, so enthält nach 5) die größere der beiden γ-Mengen beide Elemente und bestimmt die Ordnungsbeziehung $a \prec b$ oder $b \prec a$. Diese Ordnungsbeziehung ist nach 6) unabhängig von der Wahl der verwendeten γ-Menge."

Ernst Zermelo [111, 104]

The two examples in Figure 2.27 yield the orderings $2 \prec 3 \prec 1$ and $1 \prec 2 \prec 3$, respectively.

Step 4

Next, Zermelo proves two central properties of L_γ:

> "(7) If we call any element of M that occurs in some γ-set as 'γ-element', the following theorem holds: The totality L_γ of all γ-elements can be so ordered that it will itself be a γ-set, and it contains all elements of the original set M. M itself is thereby well-ordered."

> "7) Bezeichnet man als 'γ-Element' jedes Element von M, das in irgendeiner γ-Menge vorkommt, so gilt der Satz: die Gesamtheit L_γ aller γ-Elemente läßt sich so ordnen, daß sie selbst eine γ-Menge darstellt, und umfaßt alle Elemente der ursprünglichen Menge M. Die letztere ist damit selbst wohlgeordnet."

Ernst Zermelo [104, 111]

This quote is followed by paragraphs (I) to (V), proving the stated properties. With the proof of (7), he had crossed the finish line, as it was now established: L_γ is itself a γ-set and thus well-ordered. Since the set L_γ contains all elements of M, the relation '$\prec$' must be a well-ordering on M.

2.4 Axiomatic Set Theory

The proof of the well-ordering theorem caused a great stir among mathematicians and boosted Zermelo's academic career. In 1905, he was appointed professor at the University of Göttingen.

2.4.3 Criticism of Zermelo's Proof

2.4.3.1 The Axiom of Choice

Hilbert's dream of deciding the continuum hypothesis through the well-ordering hypothesis did not come true. Zermelo's proof neither revealed how a corresponding order on the real numbers could be constructed, nor did it provide any clue on how the continuum hypothesis could be decided.

The proof is non-constructive for a simple reason: the utilization of the choice function γ. Zermelo had no need to care about the details of this function because his proof only required its mere existence for any given set M. For finite sets, a choice function can always be constructed in the demonstrated manner, thus affirming the question of its existence. However, the situation becomes significantly more challenging for infinite sets, as the elements cannot be listed one after the other. Zermelo believed this was only a matter of representation, not affecting the existence of such a function. Even if it is infeasible to explicitly write down the definition of a choice function for a complex set, he thought it should still be possible to prove its existence. However, despite all efforts, such a proof was never found.

Today, we know the proof Zermelo had been looking for does not exist, with far-reaching consequences: We are compelled to replace certainty with belief by simply postulating the existence of a choice function. This is precisely the purpose of the *axiom of choice* (AC), which, in its simplest form, reads as follows:

> "Every set of non-empty sets has a choice function."

An equivalent formulation is this one (Figure 2.28):

> "Let M be a set of non-empty sets. Then one element can be taken from each set of M and the chosen elements be combined into a new set."

Another equivalent formulation is the well-ordering theorem itself. That is because the well-ordering theorem not only follows from the axiom of choice, but the axiom of choice also follows from the well-ordering theorem. The proof builds upon the following train of thought: Suppose M is a set of non-empty sets. Now, the well-ordering theorem states that the union of all sets contained in M has a well-ordering, and from this, it follows that each set from M contains

Figure 2.28: Illustration of the axiom of choice

a minimal element. Combining the minimal elements into a new set then yields a choice set for M.

For short, the axiom of choice and the well-ordering theorem are equivalent. Consequently, we can postulate the well-ordering theorem as an axiom just as well as the axiom of choice, which puts us in a precarious situation. The axiom of choice looks like a trivial statement. At the same time, it is equivalent to a theorem whose substantive meaning is not entirely convincing, even at a second glance. Overall, this raises the question of whether we can fully trust our intuition and whether the axiom of choice indeed describes a trivial truth.

Based on considerations of this kind, numerous mathematicians had turned against Zermelo, including notable figures such as Giuseppe Peano, Émile Borel, and Henri Poincaré [10, 84]. Borel regarded Zermelo's proof merely as a contradiction of the axiom of choice, as it implied the absurd assertion that all sets could be well-ordered. Peano saw no way to deduce the axiom from elementary principles, thus doubting its correctness. Whether it described an intuitively convincing fact was irrelevant to him.

The debate over the axiom of choice divided the scientific community, and both camps occasionally lapsed into polemics. For instance, Zermelo pointed out in [105] that Russell's antinomy could be easily written down in Peano's logic with a few pen strokes:

> "Of course, there remains to Peano a simple way of proving the theorems in question, as well as many others, from his own principles. He need only use Russell's antinomy, lately much discussed, since, as is well known, everything can be proved from contradictory premisses."

> "Freilich hätte Herr Peano noch ein einfaches Mittel, die in Frage stehenden Sätze wie noch viele andere aus seinen eigenen Prinzipien zu beweisen. Er brauchte nur von der neuerdings viel erörterten 'Russellschen Antinomie' Gebrauch zu machen, da sich aus widersprechenden Prämissen bekanntlich alles beweisen lässt."

2.4 Axiomatic Set Theory

Ernst Zermelo [110, 105]

Zermelo had no proof for the axiom of choice, and we have already foreshadowed that he had no chance of finding one. The astonishing fact that AC is neither provable nor disprovable within classical mathematics was established in two stages. In 1938, Kurt Gödel showed that adding the axiom does not generate contradictions in classical mathematics [33]. The proof was later completed in 1963 by Paul Cohen. The American mathematician succeeded in proving that the negation of AC does not lead to contradictions either [11].

The independence of AC implies that we are free to choose between mathematics with the axiom of choice and mathematics without; neither bears the risk of creating contradictions. However, the question remains: Which mathematics should we adopt? Which one feels right?

- **Option 1:** Accepting AC

  If the axiom of choice is accepted, there is no need to worry about choice functions, as they exist by definition. With the help of AC, many theorems known in classical mathematics can be proved, such as the theorem that every vector space has a basis. On the other hand, theorems are provable that put our intuition to the test. A well-known example is the Banach-Tarski paradox. It states that a sphere can be split into a finite number of parts in a way that makes it possible to reassemble them into two new spheres, each having the same size as the original sphere [2].

- **Option 2:** Rejecting AC

  Rejecting the axiom of choice renders several established mathematical theorems false. The well-ordering theorem becomes a false statement within this mathematical framework, and not every vector space has a basis. Interestingly, some counterintuitive theorems would also cease to hold, including the mentioned theorem by Banach and Tarski.

The intellectual wounds inflicted by the heated debate surrounding the axiom of choice have mainly healed by now. Most contemporary mathematicians consider the axiom of choice legitimate, and hardly anyone still demands its banishment today.

2.4.3.2 Non-Predicative Definitions

The French mathematician Henri Poincaré attacked Zermelo's proof from a different angle. He directed his criticism towards the definition of the set L_γ, which consists of all elements occurring in some γ-set. L_γ is, as Zermelo has shown, also a γ-set and thus referenced in its definition. Poincaré commented on this matter as follows:

> "The union of all M_γ can only mean the union of all those M_γ in whose definition the set Γ does not occur. Consequently, that M_γ, which consists of Γ and the distinguished element of $E - \Gamma$, must be excluded. Although I am quite inclined to accept Zermelo's axiom [axiom of choice], I reject his proof."
>
> "La somme logique de tous M_γ, cela doit vouloir dire la somme logique de tous les M_γ dans la définition desquels ne figure pas la notion de Γ; et alors le M_γ, formé par Γ et l'élément distingué de $E - \Gamma$ doit être exclu. Aussi, quoique je sois plutôt disposé à admettre l'axiom de Zermelo, je rejette sa démonstration."
>
> Henri Poincaré [81]

Poincaré used E and Γ to denote Zermelo's sets M and L_γ, respectively.

Does Poincaré's criticism remind you of the *Principia Mathematica*? If so, you have already developed a good sense of the sentiments of mathematicians in the first half of the twentieth century. The same point of criticism led Russell to formulate his vicious circle principle, discussed in Section 2.3.3.

2.4.4 Zermelo's Axioms

After the criticism had surfaced, Zermelo initially insisted on the correctness of his arguments. Four years later, however, he presented a revised proof that addressed the issue of using a non-predicative definition. This revised proof appeared in 1908 in an article titled *A new proof of the possibility of a well-ordering* [110, 105]. In this work, Zermelo introduced four axioms that postulated all the properties a set must satisfy for the proof to succeed. These axioms were part of a more elaborate axiom system described by Zermelo in the same journal, in an article titled *Investigations on the Foundations of Set Theory I* [109, 106]. In hindsight, this work was of utmost importance for modern mathematics, as it laid the foundation for axiomatic set theory.

For Zermelo, the definition of sets was a tightrope walk. If his work was to have practical use, the definition had to be broad enough to include all sets that did not cause any harm. At the same time, he had to be careful not to open a gateway for antinomies. Zermelo opted for a constructive approach by establishing several rules that allowed sets to be constructed from other sets in specific ways. Here are four of his fundamental rules:

- The empty set $\emptyset$ is a set.   ☞ *Axiom of empty set*
- If M and N are sets, then $\{M, N\}$ is a set.   ☞ *Axiom of pairing*
- If M is a set, then $\bigcup M$ is a set.   ☞ *Axiom of union*
- If M is a set, then 2^M is a set.   ☞ *Axiom of power set*

2.4 Axiomatic Set Theory

Herein, $\bigcup M$ is a shorthand notion for:

$$\bigcup_{x \in M} x$$

Zermelo supplemented these laws with the *axiom of definiteness*, stating that two sets are equal if and only if they contain the same elements.

Let us examine the full extent of these axioms. If they sufficed to establish the existence of all ordinary sets, they should, at the very least, entail the existence of the union, the intersection, and the difference of two sets M and N, denoted by $M \cup N$, $M \cap N$, and $M \setminus N$, respectively. The existence of $M \cup N$ is easy to prove. The pairing axiom guarantees the existence of $\{M, N\}$, and the axiom of union ensures the formation of the set $\bigcup \{M, N\} = M \cup N$. However, as things stand, the axioms are too weak to form intersections or complements. To ascertain the existence of

$$M \cap N := \{x \mid x \in M \wedge x \in N\}$$
$$M \setminus N := \{x \mid x \in M \wedge x \notin N\}$$

the so-called *comprehension schema* is needed, which, in its general form, reads as follows:

"*If φ is a formula, then $\{x \mid \varphi(x)\}$ is a set.*"

However, Zermelo could not include the comprehension schema in this general form, as the choice $\varphi = x \notin x$ brings forth Russell's antinomy. Therefore, he opted for the weaker *separation schema*:

■ If M is a set and φ is a formula, then $\{x \mid x \in M \wedge \varphi(x)\}$ is a set.

☞ Axiom schema of separation

Like in the general comprehension schema, φ can be substituted with any formula. However, the axiom schema can be safely employed, as φ operates solely on the elements of another set. The axiom does not permit the aggregation of arbitrary elements but only to separate elements from an already established set.

With the separation schema in hand, it is straightforward to define the intersection and the difference of two sets:

$$M \cap N := \{x \mid x \in (M \cup N) \wedge \varphi(x)\} \quad \text{with } \varphi(x) := x \in M \wedge x \in N$$
$$M \setminus N := \{x \mid x \in M \wedge \varphi(x)\} \quad \text{with } \varphi(x) := x \notin N$$

Can we now construct all the sets we are familiar with? The answer is no, as our current methods only allow us to generate finite sets. To delve into the realm of the infinite, the so-called *transfinite*, the existence of infinite sets has to be backed up axiomatically.

- $\{\emptyset, \{\emptyset\}, \{\{\emptyset\}\}, \{\{\{\emptyset\}\}\}, \ldots\}$ is a set.   ☞ *Axiom of infinity*

Below, we will refer to the postulated set as *Zermelo's number sequence*, abbreviated as Z_0.

Still, the discussed axioms do not suffice to generate all sets that are generally considered harmless. Among the first to notice this gap was the German-Israeli mathematician Abraham Fraenkel. In his 1922 publication titled *Zu den Grundlagen der Cantor-Zermelo'schen Mengenlehre*, he gives the following example:

> "The seven Zermelo axioms are not sufficient to establish set theory. To prove this assertion, consider the following simple example: Let Z_0 be [Zermelo's number sequence] [...]; the power set $\mathfrak{U}Z_0$ (set of all subsets of Z_0) is denoted by Z_1, $\mathfrak{U}Z_1$ by Z_2, etc. Then the axioms, as their examination easily shows, do not allow the formation of the set $\{Z_0, Z_1, \ldots\}$, and therefore also not the formation of the union set."
>
> "Die sieben Zermeloschen Axiome reichen nicht aus zur Begründung der Mengenlehre. Zum Nachweis dieser Behauptung diene etwa das folgende einfache Beispiel: Es sei Z_0 die [Zermelo'sche Zahlenreihe] [...]; die Potenzmenge $\mathfrak{U}Z_0$ (Menge aller Untermengen von Z_0) werde mit Z_1, $\mathfrak{U}Z_1$ mit Z_2 bezeichnet usw. Dann gestatten die Axiome, wie deren Durchmusterung leicht zeigt, nicht die Bildung der Menge $\{Z_0, Z_1, \ldots\}$, also auch nicht die Bildung der Vereinigungsmenge."
>
> Abraham Fraenkel, 1922 [20]

To permit the formation of the set

$$\{Z_0, Z_1, \ldots\} = \{Z_0, 2^{Z_0}, 2^{2^{Z_0}}, 2^{2^{2^{Z_0}}}, 2^{2^{2^{2^{Z_0}}}}, \ldots\} \tag{2.23}$$

Fraenkel introduced the so-called *replacement schema*:

> "If M is a set and each element of M is replaced by a 'thing of the domain $\mathfrak{B}$', then M again becomes a set."
>
> "Ist M eine Menge und wird jedes Element von M durch ein 'Ding des Bereiches $\mathfrak{B}$' ersetzt, so geht M wiederum in eine Menge über."
>
> Abraham Fraenkel, 1922 [20]

This schema guarantees the following:

- If M is a set and f is a function that assigns a set to each element of M, then

$$\{f(x) \mid x \in M\}$$

is a set.   ☞ *Axiom schema of replacement*

2.4 Axiomatic Set Theory

The replacement schema permits the derivation of Fraenkel's example set $\{Z_0, Z_1, \ldots\}$ from Zermelo's number sequence using the following function:

$$f(x) := \begin{cases} Z_0 & \text{for } x = \emptyset \\ 2^{Z_0} & \text{for } x = \{\emptyset\} \\ 2^{2^{Z_0}} & \text{for } x = \{\{\emptyset\}\} \\ 2^{2^{2^{Z_0}}} & \text{for } x = \{\{\{\emptyset\}\}\} \\ \ldots & \ldots \end{cases}$$

As a side effect, adding the replacement schema allows us to eliminate the separation schema and the pairing axiom, as both become derivable as theorems.

However, one last problem remains: The axioms formulated so far allow us to construct certain sets explicitly, but they do not exclude the existence of others. To keep such uninvited guests at bay, Fraenkel formulated the *Axiom of Limitation*:

> "[It] is clear that the axiom system does not have a 'categorical character', i.e., it does not completely determine the totality of sets. [...], so here the indicated evils can be remedied by a 'limitation axiom' to be set up as the ninth and last axiom, which imposes on the concept of set or more conveniently on the domain $\mathfrak{B}$ the smallest extent compatible with the other axioms."

> "[Es] geht hervor, dass das Axiomensystem keinen 'kategorischen Charakter' besitzt, nämlich die Gesamtheit der Mengen nicht vollständig festlegt. [...], so kann hier den angegebenen Übelständen durch ein als neuntes und letztes Axiom aufzustellendes 'Beschränktheitsaxiom' abgeholfen werden, das dem Mengenbegriff oder zweckmäßiger dem Bereich $\mathfrak{B}$ den geringsten mit den übrigen Axiomen verträglichen Umfang auferlegt."

Abraham Fraenkel, 1922 [20]

John von Neumann was dissatisfied with the formulation, as the axiom explicitly referred to the others. In his own set theory, developed in 1925, he replaced it with an equivalent axiom that prohibited infinitely descending inclusion chains. In 1930, Zermelo adopted this revised axiom in the following form:

> "Every (regressive) chain of elements, in which each link is an element of the preceding one, terminates with a finite index at a primal element."

> "Jede (rückschreitende) Kette von Elementen, in welcher jedes Glied Element des vorangehenden ist, bricht mit endlichem Index ab bei einem Urelement."

Zermelo, 1930 [107]

This is Zermelo's *axiom of foundation*. It denies the existence of infinitely descending inclusion chains of the form $M_0 \ni M_1 \ni M_2 \ni \ldots$.

Together, the presented axioms constitute the *Zermelo-Fraenkel set theory* (ZF). Adding the axiom of choice (AC) yields ZFC (*Zermelo-Fraenkel set theory with Choice*), which most contemporary mathematicians widely accept as the formal foundation of mathematics. This also clarifies the precise meaning of the above formulation: the axiom of choice or its negation "does not generate contradictions in classical mathematics". It expresses that the sets ZF $\cup$ {AC} and ZF $\cup$ {¬AC} are consistent, provided no contradictions are derivable within ZF itself.

Over the course of time, other axiom systems have been proposed. Examples are the set theory of Wilhelm Ackermann [1] and the lesser-known *Morse-Kelley set theory* [63, 71]. The most famous descendant is the *Neumann-Bernays-Gödel set theory*, NBG in short. This theory distinguishes between *sets* and *classes*. For instance, in ZF and ZFC, the set of all sets is not permitted to exist, but in NBG, it exists as a class. This approach avoids antinomies not by excluding contentious objects but by categorizing them as classes. Classes, however, are subject to significant limitations, such as not being allowed as elements of sets or other classes. Another notable property of NBG set theory is its finite axiomatizability, which means it is definable without resorting to axiom schemata.

Developing an axiom system for set theory was a tremendous achievement, yet Zermelo was denied one challenge for the rest of his life: proving the consistency of his axioms. Zermelo tried hard to preempt the looming criticism his set theory might lead to antinomies, but the proof was unexpectedly elusive. Ultimately, he had no option but to publish his axiom system without formal assurance.

In 1910, Zermelo accepted an offer from the University of Zurich, but due to declining health, he had to relinquish his chair just six years later. From that point on, he lived in Breisgau, Germany, and worked as an honorary professor at the University of Freiburg starting in 1926.

After Gödel published the incompleteness theorems in 1931, it became evident why Zermelo failed to prove the consistency of his axiom system. The expressive power of ZF and ZFC is equivalent to classical mathematics, and according to Gödel's second incompleteness theorem, no system of such expressiveness can prove its own consistency. In short, the consistency of ZF or ZFC is unprovable with the methods of classical mathematics.

Zermelo never accepted the proofs of the incompleteness theorems and became one of Gödel's most vigorous opponents. When both met in September 1931 at the meeting of the German Mathematicians Association in Bad Elster, Zermelo left no doubt about what he thought of the young man's absurd results. Initially, he refused any conversation, but a personal discussion even-

2.4 Axiomatic Set Theory

tually took place, remaining surprisingly civil. Just six days later, however, Zermelo claimed to have discovered an error in the proof, followed by a correspondence in which Gödel tried hard to elucidate his derivation. Zermelo was not swayed by the provided arguments and finally made his criticism public in 1932 [108]. Gödel was no man of confrontation and made no further attempts to settle the dispute.

Zermelo taught at Freiburg until 1935. That year, the National Socialists revoked his permission to teach, and he was not allowed to return to the university until 1946. By that time, however, the health of the now seventy-five-year-old was already so impaired that he could no longer lecture. Ernst Zermelo died on May 21, 1953, in Freiburg at the age of 81.

This is where our journey into the history of mathematical logic comes to an end. It is time to set the stage for the true subject of our interest: Gödel's proof of the incompleteness theorems. So let it be: Curtain up!

3 Proof Sketch

> *"One of Crete's own prophets has said it: 'Cretans are always liars, evil brutes, idle bellies'"*
>
> Epistle of Paul to Titus, 1:12–14

Having acquired the necessary knowledge in Chapter 2 to comprehend both the formal details and the philosophical dimensions of the incompleteness theorems, the time has come to return to Gödel's work. In the few passages quoted so far, Gödel had taken stock of the mathematics of the early twentieth century. With the *Principia Mathematica* and the Zermelo-Fraenkel set theory, he mentioned two of the predominant axiomatic systems of that era. While these terms were still empty shells towards the end of Chapter 1, they appear in bright colors before our mind's eye after our historical excursion. We are now well-equipped to master the rest of Gödel's work. So let us listen again!

> geführt sind. Es liegt daher die Vermutung nahe, daß diese Axiome und Schlußregeln dazu ausreichen, alle mathematischen Fragen, die sich in den betreffenden Systemen überhaupt formal ausdrücken lassen, auch zu entscheiden. Im folgenden wird gezeigt, daß dies nicht der Fall ist, sondern daß es in den beiden angeführten

> It is reasonable therefore to make the conjecture that these axioms and rules of inference are also sufficient to decide all mathematical questions which can be formally expressed in the given systems. In what follows it will be shown that this is not the case,

The *Principia Mathematica* and Zermelo-Fraenkel set theory are two axiomatic systems expressive enough to formalize the concepts and conclusions of classical mathematics, implying that any classical proof written down with pencil and paper is reproducible in these systems. As a result, the *Principia Mathematica* and Zermelo-Fraenkel set theory would be complete in the sense of Definition 1.5 if the concepts and conclusions of mathematics were sufficient to either prove or disprove any mathematical statement, and this was an unspoken assumption for thousands of years.

This explains why completeness was subordinate to the scientific discussion for so long. Many considered completeness a fundamental trait of mathematics, which they never seriously questioned. It is also worth noting that Russell, Whitehead, and Zermelo established their systems when set-theoretical antinomies were ubiquitous. In the face of the antinomies, consistency became a central concern, attracting the interest of most mathematicians like no other.

After being freed from the antinomies at the beginning of the twentieth century, many mathematicians believed that the *Principia Mathematica* and the Zermelo-Fraenkel set theory would provide what they had been after for so long: a correct and complete formal system for ordinary mathematics. In the passage quoted above, Gödel announces his intention to refute this very belief: The systems he mentioned (and not only these!) contain *undecidable* statements, that is, neither the statement itself nor its negation can be derived from the axioms.

3.1 Arithmetic Formulas

Right at the outset of his work, Gödel reveals a noteworthy aspect of the first incompleteness theorem:

> nicht der Fall ist, sondern daß es in den beiden angeführten Systemen sogar relativ einfache Probleme aus der Theorie der gewöhnlichen ganzen Zahlen gibt⁴), die sich aus den Axiomen nicht
>
> ⁴) D. h. genauer, es gibt unentscheidbare Sätze, in denen außer den logischen Konstanten: — (nicht), ∨ (oder), (x) (für alle), = (identisch mit) keine anderen Begriffe vorkommen als + (Addition), . (Multiplikation), beide bezogen auf natürliche Zahlen, wobei auch die Präfixe (x) sich nur auf natürliche Zahlen beziehen dürfen.
>
> 174   Kurt Gödel,
>
> entscheiden lassen. Dieser Umstand liegt nicht etwa an der speziellen

> but rather that, in both of the cited systems, there exist relatively simple problems of the theory of ordinary whole numbers which cannot be decided on the
>
> ⁴) More precisely, there exist undecidable sentences in which, other than the logical constants: — (not), ∨ (or), (x) (for all), = (identical with), the only concepts occurring are + (addition), . (multiplication) (of natural numbers), and where the prefix (x) refers only to natural numbers.
>
> 174   Kurt Gödel,
>
> basis of the axioms⁴).

3.1 Arithmetic Formulas

In order to find undecidable statements, there is no need to resort to exotic branches of mathematics; we are going to find them at the heart of this science in the well-known theory of ordinary integers.

In footnote 4, Gödel points out what he means by arithmetic statements. Syntactically, these are formulas in which, besides the propositional connectives and the predicate quantifiers, only the arithmetic operators '+' and '×', as well as the equality operator, may occur. Semantically, the formulas are interpreted over the range of the natural numbers. Today we refer to such formulas as *formulas of Peano arithmetic* or, more briefly, *PA formulas*.

Let's look at an example:

$$\neg \forall n \, (\neg(n + n = n) \lor$$
$$\forall n' \, (n' = n \lor \neg(n' \times n' = n') \lor$$
$$\forall n'' \, (\neg(n'' = n' + n') \lor$$
$$\neg \forall x \, (\forall y \, \neg(x = n'' \times y) \lor \forall y \, \neg(x = n' + n' + n' + y) \lor \neg($$
$$\forall p \, (p = n' \lor \neg \forall y \, (\forall z \, \neg(y \times z = p) \lor (y = n' \lor y = p)) \lor$$
$$\forall q \, (q = n' \lor \neg \forall y \, (\forall z \, \neg(y \times z = q) \lor (y = n' \lor y = q)) \lor$$
$$\neg(x = p + q))))))))$$

The formula can be written down more concisely with the help of the existential quantifier '∃', the conjunction operator '∧', and the implication operator '→'. As usual, these operators serve as syntactic abbreviations, namely:

$$\exists \xi \, \varphi := \neg \forall \xi \, \neg \varphi$$
$$\varphi \land \psi := \neg(\neg \varphi \lor \neg \psi)$$
$$\varphi \rightarrow \psi := \neg \varphi \lor \psi$$

This allows us to rewrite the example formula as such:

$$\exists n \, (n + n = n \land$$
$$\exists n' \, (\neg(n' = n) \land n' \times n' = n' \land$$
$$\exists n'' \, (n'' = n' + n' \land$$
$$\forall x \, (\exists y \, x = n'' \times y \land \exists y \, x = n' + n' + n' + y \rightarrow ($$
$$\exists p \, (\neg(p = n') \land \forall y \, (\exists z \, (y \times z = p) \rightarrow (y = n' \lor y = p)) \land$$
$$\exists q \, (\neg(q = n') \land \forall y \, (\exists z \, (y \times z = q) \rightarrow (y = n' \lor y = q)) \land$$
$$x = p + q))))))$$

To uncover its substantive meaning, let us inspect the individual components:

- $n + n = n$

  This formula is true if and only if n is interpreted as a natural number n fulfilling the condition $n + n = n$. Since only 0 meets this requirement, we may mentally substitute each occurrence of n with that number.

- $\neg(n' = n) \wedge n' \times n' = n'$

  This formula is true if and only if n' is interpreted as a natural number n' that is different from 0 and simultaneously fulfills the condition $n' \cdot n' = n'$. Only 1 satisfies this requirement, so we may mentally substitute each occurrence of n' with that number.

- $n'' = n' + n'$

  From what has been said so far, the meaning of this formula is easy to grasp. After forcing variable n' to be interpreted as the number 1, this formula is true if and only if n'' is interpreted as the number 2.

These early examples demonstrate how natural numbers can be referenced within an arithmetic formula. To simplify things further, we allow the natural numbers to occur as constant symbols. Adopting the syntax of the formal system outlined in Section 1.2, we will employ the symbol $\bar{n}$ to denote the numerical value n. Additionally writing $\xi \neq \zeta$ for $\neg(\xi = \zeta)$ transforms the original formula into:

$$\forall x \, (\exists y \, x = \bar{2} \times y \wedge \exists y \, x = \bar{3} + y \rightarrow (\\ \exists p \, (p \neq \bar{1} \wedge \forall y \, (\exists z \, (y \times z = p) \rightarrow (y = \bar{1} \vee y = p)) \wedge \\ \exists q \, (q \neq \bar{1} \wedge \forall y \, (\exists z \, (y \times z = q) \rightarrow (y = \bar{1} \vee y = q)) \wedge \\ x = p + q))))$$

Now, the remaining subformulas can be understood with little effort:

- $\exists y \, x = \bar{2} \times y$

  This formula is true if and only if x is interpreted as an even number. For clarity, we abbreviate subformulas of this kind by the expression even(x).

- $\exists y \, z \times y = x$

  This formula is similar in structure to the previous one. It is true if and only if z and x are interpreted such that the number assigned to z divides the number assigned to x. In the following, we express this property by the ordinary mathematical notation $z \mid x$.

- $\exists y \, x = \bar{3} + y$

  This formula is just as easy to understand. It is true if and only if the variable x is interpreted as a natural number greater than 2. In the following, we express this fact by $x > \bar{2}$, or equivalently, by $x \geq \bar{3}$.

3.1 Arithmetic Formulas

With the proposed simplifications, we can represent the example formula in a much cleaner and more concise way:

$$\forall x \, (\text{even}(x) \wedge x > \overline{2} \rightarrow (\\
\exists p \, (p \neq \overline{1} \wedge \forall y \, (y \,|\, p \rightarrow (y = \overline{1} \vee y = p)) \wedge \\
\exists q \, (q \neq \overline{1} \wedge \forall y \, (y \,|\, q \rightarrow (y = \overline{1} \vee y = q)) \wedge \\
x = p + q))))$$

- $p \neq \overline{1} \wedge \forall y \, (y \,|\, p \rightarrow (y = \overline{1} \vee y = p))$

  This formula is true if and only if p is interpreted as a natural number $\neq 1$ that is divisible only by 1 and itself. This is equivalent to the statement "*p is a prime number*", which we will shortly express as prime(p).

Applying the new abbreviation lets the initial formula shrink to:

$$\forall x \, (\text{even}(x) \wedge x > \overline{2} \rightarrow \exists p \, (\text{prime}(p) \wedge \exists q \, (\text{prime}(q) \wedge x = p + q)))$$

In this form, the formula unveils its true face. It encompasses *Goldbach's strong conjecture*, which remains one of the most renowned open questions in number theory to this day (Figure 3.1):

> "*Every even integer n > 2 can be written as the sum of two primes.*"
>
> Goldbach's (strong) conjecture

The conjecture is named after Christian Goldbach. In 1742, the German mathematician postulated in a letter to Leonhard Euler that any natural number greater than 2 is expressable as a sum of three prime numbers. His hypothesis hides in a hastily formulated marginal note, depicted in Figure 3.2. Be aware that Goldbach considered the number 1 to be a prime number. Therefore, in contemporary terms, his original conjecture must be reformulated to state that any natural number greater than 2 is expressable as a sum of three numbers, each being either prime or equal to 1.

Today, the historical formulation is referred to as *Goldbach's weak conjecture*, as it is deducable from the strong variant.

Whether Goldbach was right, we do not know. Despite mounting evidence, formal proof is still pending. Is Goldbach's conjecture perhaps a statement in the Gödelian sense, which cannot be proven in the system of classical mathematics? The fact that neither a proof nor a counterexample has been found for such a long time may nourish this suspicion, but it does not offer certainty. Let us recall that the famous conjecture of Pierre de Fermat that the equation $a^n + b^n = c^n$ has no solutions in the positive integers for $n > 2$ resisted all proof attempts for over three hundred years. It was not until 1995 that the British mathematician Andrew Wiles was able to present a flawless derivation of the

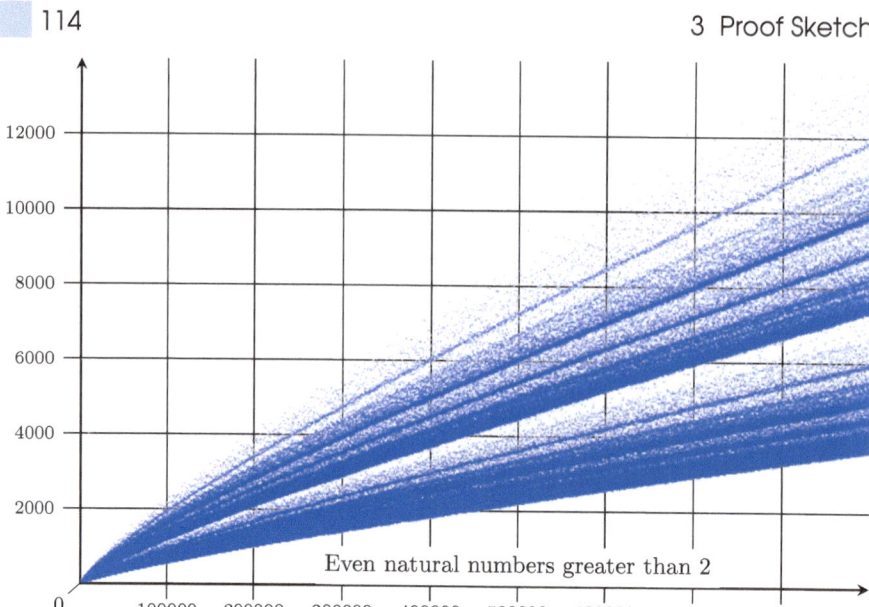

Figure 3.1: According to Goldbach's strong conjecture, all even numbers $n > 2$ can be written as the sum of two prime numbers. In the diagram above, the even numbers are plotted on the x-axis, and the data points indicate the number of possible decompositions. Goldbach's conjecture is true exactly when the x-axis is free of data points.

"Nachdem ich dieses wieder durchgelesen, finde ich, daß sich die conjecture in summo rigore demonstrieren lässet in casu $n+1$, si successerit in casu n, et $n+1$ dividi possit in duos numeros primos. Die Demonstration ist sehr leicht. Es scheinet wenigstens, daß eine jede Zahl, die größer ist als 2, ein aggregatum trium numerorum primorum sey."	"Having read this again, I find that the conjecture can be demonstrated in summo rigore in the case of $n+1$, if it succeeds in the case of n, and $n+1$ can be divided into two prime numbers. The demonstration is very easy. At least it seems that every number greater than 2 is an aggregate of three prime numbers."

Figure 3.2: Original formulation of Goldbach's (weak) conjecture

Taniyama-Shimura conjecture, from which Fermat's Last Theorem emerged as a corollary [102, 95].

Gödel's undecidable statements bear significant similarities to the conjectures of Goldbach and Fermat. All three originate in number theory and can be expressed in simple terms with the additive and multiplicative properties of the natural numbers. However, their complexities couldn't be more different.

3.1 Arithmetic Formulas

Figure 3.3

PIERRE DE FERMAT
1607 – 1665

Unlike Goldbach's and Fermat's conjectures, which can be written down succinctly with a few pen strokes, Gödel's formulas are true monstrosities, far too vast to be conveyed in plain text.

Gödel had already stressed the connection between his undecidable formulas and the conjectures of Goldbach and Fermat in Königsberg when he first publicly formulated his incompleteness theorem. We've previously come across his words on page 27, although a portion of the quote was replaced by ellipses for didactic reasons. With our current knowledge, we are now able to comprehend the full quote:

> "One can – assuming the consistency of classical mathematics – even give examples of sentences (namely those of Goldbach's or Fermat's type) that are indeed correct in content, but unprovable in the formal system of classical mathematics."

> "Man kann – unter Voraussetzung der Widerspruchsfreiheit der klassischen Mathematik – sogar Beispiele für Sätze (und zwar solche von der Art des Goldbach'schen oder Fermat'schen) angeben, die zwar inhaltlich richtig, aber im formalen System der klassischen Mathematik unbeweisbar sind."

Kurt Gödel [9]

> entscheiden lassen. Dieser Umstand liegt nicht etwa an der speziellen Natur der aufgestellten Systeme, sondern gilt für eine sehr weite Klasse formaler Systeme, zu denen insbesondere alle gehören, die aus den beiden angeführten durch Hinzufügung endlich vieler Axiome entstehen[5]), vorausgesetzt, daß durch die hinzugefügten Axiome keine

> ⁵⁾ Dabei werden in PM nur solche Axiome als verschieden gezählt, die aus einander nicht bloß durch Typenwechsel entstehen.

> This situation does not depend upon the special nature of the constructed systems, but rather holds for a very wide class of formal systems, among which are included, in particular, all those which arise from the given systems by addition of finitely many axioms ⁵⁾,
>
> ⁵⁾ In PM only those axioms are considered distinct which do not arise from each other by a change of types.

In this passage, Gödel provides a glimpse of the far-reaching implications of his results. He is about to demonstrate the incompleteness of a specific formal system, which he refers to as system P. This system is a basic variation of the *Principia Mathematica* but also expressable in the terminology of Zermelo-Fraenkel set theory. However, this doesn't imply that Gödel has only demonstrated the incompleteness of these two systems. A significant part of his findings is that his proof applies to all formal systems that are expressive enough to make statements about the additive and multiplicative properties of natural numbers. Consequently, formal systems such as the *Principia Mathematica* and Zermelo-Fraenkel set theory cannot be *completed*, that is, it is impossible to escape the incompleteness theorems by adding a finite number of axioms. Each additional axiom tears a hole at another location and exposes new undecidable statements.

Let us accept these preliminary remarks, vague as they may be. In Section 6.1.2, Gödel will precisely state when a formal system gets into the pull of the first incompleteness theorem and why every attempt for its completion is doomed to fail.

To be entirely accurate, we need to revise some of what has just been said.

> entstehen ⁵⁾, vorausgesetzt, daß durch die hinzugefügten Axiome keine falschen Sätze von der in Fußnote ⁴⁾ angegebenen Art beweisbar werden.

> assuming that no false sentences of the kind given in footnote 4 become provable by means of the additional axioms.

With this additional remark, Gödel highlights a fundamental property common to all formal systems that involve the ordinary propositional logic apparatus.

3.2 The Arithmetization of Syntax

Such a system can only be incomplete if it is consistent. The reason for this is simple: If a contradictory pair of formulas φ and $\neg\varphi$ can be derived from the axioms, then every other formula is derivable, too. This implies that all substantively true formulas are provable, making the formal system, useless as it may be, complete in the sense of Definition 1.5. This topic will be revisited on page 177, where it will be demonstrated step by step how arbitrary statements can be proven from a contradictory pair of formulas.

> Wir skizzieren, bevor wir auf Details eingehen, zunächst den Hauptgedanken des Beweises, natürlich ohne auf Exaktheit Anspruch zu erheben. Die Formeln eines formalen Systems (wir beschränken

> Before we go into details, let us first sketch the main ideas of the proof, naturally without making any claim to rigor.

In this passage, Gödel announces his agenda: He will start by sketching the general line of reasoning without claiming to be exact. Due to the complexity of the subject and the disturbing result, this decision was not only a wise move but also fits well with the structure of this book. Staying aligned with the original work, we can intuitively approach Gödel's first incompleteness theorem without having to introduce a tangle of definitions, propositions, and derivations. There is no need to worry, though, as Gödel will supply all the details later.

3.2 The Arithmetization of Syntax

> zu erheben. Die Formeln eines formalen Systems (wir beschränken uns hier auf das System PM) sind äußerlich betrachtet endliche Reihen der Grundzeichen (Variable, logische Konstante und Klammern bzw. Trennungspunkte) und man kann leicht genau präzisieren, welche Reihen von Grundzeichen sinnvolle Formeln sind und welche nicht[6]). Analog sind Beweise vom formalen Standpunkt nichts anderes als endliche Reihen von Formeln (mit bestimmten angebbaren Eigenschaften). Für metamathematische Betrachtungen

---

[6]) Wir verstehen hier und im folgenden unter „Formel aus PM" immer eine ohne Abkürzungen (d. h. ohne Verwendung von Definitionen) geschriebene Formel. Definitionen dienen ja nur der kürzeren Schreibweise und sind daher prinzipiell überflüssig.

> The formulas of a formal system (we limit ourselves here to the system PM) are, considered from the outside, finite sequences of primitive symbols (variables, logical constants, and parentheses or dots) and one can easily make completely precise which sequences of primitive symbols are meaningful formulas and which are not [6]). Analogously, from the formal standpoint, proofs are nothing but finite sequences of formulas (with certain specifiable properties).
>
> ---
>
> [6]) By a "formula of PM", we always understood here and in the sequal a formula written without abbreviations (i.e. without use of definitions). Definitions serve only to make writing briefer and are therefore theoretically superflous.

The content of this passage sounds familiar to us. In Chapter 2, we have learned that the formulas of a formal system are finite series of elementary symbols, and by adhering to precisely defined syntax rules, it becomes straightforward to determine which series of symbols are meaningful formulas and which are not. In addition, we have seen that proofs, when viewed formally, are nothing but a finite series of formulas with certain specifiable properties.

> angebbaren Eigenschaften). Für metamathematische Betrachtungen ist es natürlich gleichgültig, welche Gegenstände man als Grundzeichen nimmt, und wir entschließen uns dazu, natürliche Zahlen [7]) als solche zu verwenden. Dementsprechend ist dann eine Formel eine endliche Folge natürlicher Zahlen [8]) und eine Beweisfigur eine endliche Folge von endlichen Folgen natürlicher Zahlen. Die meta-
>
> ---
>
> [7]) D. h. wir bilden die Grundzeichen in eineindeutiger Weise auf natürliche Zahlen ab. (Vgl. die Durchführung auf S. 179.)
>
> [8]) D. h. eine Belegung eines Abschnittes der Zahlenreihe mit natürlichen Zahlen. (Zahlen können ja nicht in räumliche Anordnung gebracht werden.)

> Naturally, for metamathematical considerations, it makes no difference which objects one takes as primitive symbols, and we decide to use natural numbers [7]) for that purpose. Accordingly, a formula is a finite sequence of natural numbers [8]) and a proof-figure is a finite sequence of finite sequences of natural numbers.
>
> ---
>
> [7]) That is, we map the primitive symbols in one-to-one fashion onto the natural numbers. (Cf. page 179) to see how this is done.)
>
> [8]) That is, a mapping of a segment of the natural number sequence into the natural numbers. (Numbers, of course, cannot be spatially ordered.)

In these few lines, Gödel describes a pivotal aspect of his proof. He emphasizes that the choice of the alphabet used to write down formulas is mostly irrelevant. For instance, it does not matter whether the logical implication is symbolized

3.2 The Arithmetization of Syntax

by '⊃', as Russell did, or by the more modern variant '→'. Just as well, and this is a key element in the proof of the first incompleteness theorem, formulas can be represented by natural numbers. The transition from symbols to numbers is called the *arithmetization of syntax* and the process of mapping a formula to a number as *Gödelization*. The numerical representation of a formula φ is called the *Gödel number* of φ, abbreviated by $\ulcorner \varphi \urcorner$.

The fact that every string of characters, thus every formula and every proof of a formal system, can be Gödelized is not particularly surprising in the information age. A sequence of characters can be arithmetized simply by typing it into a computer console and interpreting the memory image as a natural number.

Let us try the aforesaid on three formulas from the *Principia Mathematica*, namely those three that conclude the proof of Theorem **∗2.08** on page 81.

- Example 1: $\varphi_1 := (p \to (p \lor p)) \to (p \to p)$

```
echo "(p->(pvp))->(p->p)" | hexdump
```
```
28 70 2D 3E 28 70 76 70 29 29 2D 3E 28    (p->(pvp))->(
70 2D 3E 70 29 00 00 00 00 00 00 00 00    p->p)........
```

☞ $\ulcorner \varphi_1 \urcorner$ = $28702D3E2870767029292D3E28702D3E7029_{16}$ (Hexadecimal)
    = $3522663200367977117319339317059413685989417$ (Decimal)

- Example 2: $\varphi_2 := (p \to (p \lor p))$

```
echo "(p->(pvp))" | hexdump
```
```
28 70 2D 3E 28 70 76 70 29 29 00 00 00    (p->(pvp))...
```

☞ $\ulcorner \varphi_2 \urcorner$ = $28702D3E287076702929_{16}$ (Hexadecimal)
    = 190963954738685029656873 (Decimal)

- Example 3: $(p \to p)$

```
echo "(p->p)" | hexdump
```
```
28 70 2D 3E 70 29 00 00 00 00 00 00 00    (p->p).......
```

☞ $\ulcorner \varphi_3 \urcorner$ = $28702D3E7029_{16}$ (Hexadecimal)
    = 44462260514857 (Decimal)

The fact that φ_3 was obtained by applying the modus ponens from φ_1 and φ_2 can now be formulated arithmetically by expressing the syntactic transformation as a numerical equation involving the Gödel numbers $\ulcorner\varphi_1\urcorner$, $\ulcorner\varphi_2\urcorner$ and $\ulcorner\varphi_3\urcorner$:

$$\ulcorner\varphi_1\urcorner = \ulcorner\varphi_2\urcorner \cdot 16^{16} + \underbrace{11582}_{\text{`->'}} \cdot 16^{12} + \ulcorner\varphi_3\urcorner$$

This approach generalizes to arbitrary formulas. A formula φ_3 is derived from two formulas, φ_1 and φ_2, by modus ponens if and only if the Gödel numbers $\ulcorner\varphi_1\urcorner$, $\ulcorner\varphi_2\urcorner$ and $\ulcorner\varphi_3\urcorner$ satisfy the relationship

$$\ulcorner\varphi_1\urcorner = \ulcorner\varphi_2\urcorner \cdot 16^{2 \cdot l(\varphi_3)+4} + 11582 \cdot 16^{2 \cdot l(\varphi_3)} + \ulcorner\varphi_3\urcorner$$

where $l(\varphi_3)$ denotes the number of characters in φ_3. Replacing $l(\varphi_3)$ with

$$\tfrac{1}{2}(\lfloor \log_{16} \ulcorner\varphi_3\urcorner \rfloor + 1)$$

results in

$$\ulcorner\varphi_1\urcorner = \ulcorner\varphi_2\urcorner \cdot 16^{\lfloor \log_{16}\ulcorner\varphi_3\urcorner \rfloor+5} + 11582 \cdot 16^{\lfloor \log_{16}\ulcorner\varphi_3\urcorner \rfloor+1} + \ulcorner\varphi_3\urcorner$$

which is an ordinary equation of number theory. This becomes even more apparent after replacing the placeholders $\ulcorner\varphi_1\urcorner$, $\ulcorner\varphi_2\urcorner$, and $\ulcorner\varphi_3\urcorner$ with ordinary variables x, y and z:

$$x = y \cdot 16^{\lfloor \log_{16} z \rfloor+5} + 11582 \cdot 16^{\lfloor \log_{16} z \rfloor+1} + z \qquad (3.1)$$

With (3.1), we have successfully constructed an equation conveying two substantively different meanings:

- The first meaning is arithmetical. Equation (3.1) establishes a numerical relationship between three natural numbers, x, y, and z, and corresponds to an ordinary statement from number theory.

- Apart from that, the equation has a metatheoretical meaning. If x, y, and z are substituted with the Gödel numbers of φ_1, φ_2, and φ_3, then (3.1) is true exactly when φ_3 can be derived from φ_2 and φ_1 using the modus ponens rule of inference.

In summary, this unveils a pivotal isomorphism between formal systems on one hand and ordinary arithmetic on the other. Every syntactic manipulation carried out within a formal system can be interpreted in an arithmetic context. Simultaneously, a wide range of number-theoretic formulas can be understood as metatheoretic statements about formal systems (Figure 3.4).

Please note that the chosen Gödelization method is one of many. As a matter of fact, the selected variant would be highly unsuitable for proving the first incompleteness theorem, as most syntactic operations are cumbersome to express, leading to complex arithmetic relations. This is why Gödel, as Section 4.5 will

3.2 The Arithmetization of Syntax

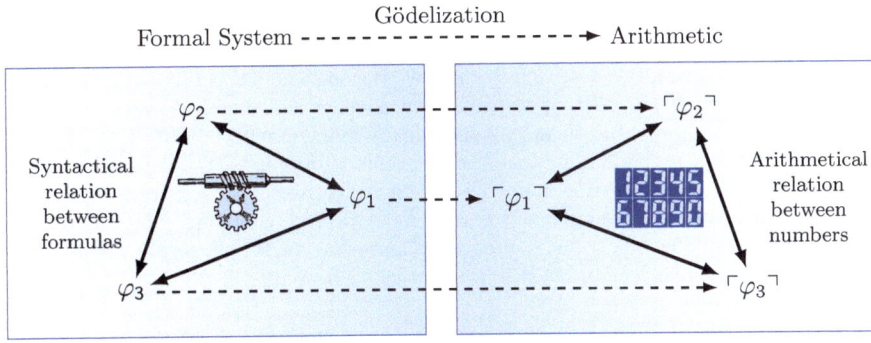

Figure 3.4: The procedure of Gödelization. Formulas and proofs are mapped to natural numbers, thereby creating an isomorphic image in the realm of arithmetic.

demonstrate, has chosen a completely different, mathematically more pleasant encoding.

Later in the article, a laborious part follows in which Gödel works through an extensive list of meta-statements, translating them into an equivalent list of arithmetic definitions. You will observe that a wide range of concepts and properties is expressable through arithmetic definitions, including simple ones, such as the property of a character being a variable, as well as more intricate ones, such as the property of a character string being a well-defined formula, a proof chain, or the end formula of such a chain.

The latter example holds particular significance and means the following: There exists an arithmetic formula $F(v)$, which becomes a substantively true statement if and only if v is interpreted as the Gödel number of a provable formula. In Gödel's words:

> endliche Folge von endlichen Folgen natürlicher Zahlen. Die metamathematischen Begriffe (Sätze) werden dadurch zu Begriffen (Sätzen) über natürliche Zahlen bzw. Folgen von solchen [9]) und daher (wenigstens teilweise) in den Symbolen des Systems PM selbst ausdrückbar. Insbesondere kann man zeigen, daß die Begriffe „Formel", „Beweisfigur", „beweisbare Formel" innerhalb des Systems PM definierbar sind, d. h. man kann z. B. eine Formel $F(v)$ aus PM mit einer freien Variablen v (vom Typus einer Zahlenfolge) angeben [10]), so daß $F(v)$ inhaltlich interpretiert besagt: v ist eine beweisbare Formel. Nun

---

[9]) m. a. W.: Das oben beschriebene Verfahren liefert ein isomorphes Bild des Systems PM im Bereich der Arithmetik und man kann alle metamathematischen Überlegungen ebenso gut an diesem isomorphen Bild vornehmen. Dies geschieht in der folgenden Beweisskizze, d. h. unter „Formel", „Satz", „Variable" etc. sind immer die entsprechenden Gegenstände des isomorphen Bildes zu verstehen.

[10]) Es wäre sehr leicht (nur etwas umständlich), diese Formel tatsächlich hinzuschreiben.

> Metamathematical concepts (assertions) thereby become concepts (assertions) about natural numbers or sequences of such,[9]) and therefore (at least partially) expressible in the symbolism of the sytem PM itself. It can be shown, in particular, that the concepts "formula", "proof-figure", "provable formula" are definable within the system PM, i.e. one can produce,[10]) for example, a formula $F(v)$ of PM with one free variable v (of the type of a sequence of numbers) such that $F(v)$, when intuitively interpreted, says: v is a provable formula.
>
> ---
>
> [9]) In other words: the process described above provides an isomorphic image of the system PM in the domain of arithmetic and one can just as well carry out all metamathematical arguments in this isomorphic image. This occurs in the following sketch of the proof, i.e. by "formula", "sentence", "variable", etc., one is always to understand the corresponding objects of the isomorphic image.
>
> [10]) It would be very easy (though somewhat tedious) actually to write this formula down

The construction of $F(v)$ is an essential building block in the proof of the first incompleteness theorem and will be addressed in detail later in this book. Remember that only the axioms and inference rules determine whether a formula is provable. As a result, $F(v)$ will be different for every formal system.

3.3 I Am Unprovable!

At this point, Gödel's work reaches its first climax. Gödel outlines the construction of a formula A that is undecidable within the logic of the *Principia Mathematica*. A and $\neg A$ are both unprovable; neither can be derived from the axioms by repeated application of inference rules.

> inhaltlich interpretiert besagt: v ist eine beweisbare Formel. Nun stellen wir einen unentscheidbaren Satz des Systems PM, d. h. einen Satz A, für den weder A noch *non-A* beweisbar ist, folgendermaßen her:
>
> Über formal unentscheidbare Sätze der Principia Mathematica etc. 175
>
> Eine Formel aus PM mit genau einer freien Variablen, u. zw. vom Typus der natürlichen Zahlen (Klasse von Klassen) wollen wir ein Klassenzeichen nennen. Die Klassenzeichen denken wir uns irgend-

> Now we obtain an undecidable proposition of the system PM, i.e. a proposition A for which neither A nor *non-A* is provable, as follows:

3.3 I Am Unprovable!

> On formally undecidable propositions of Principia Mathematica etc. 175
>
> A formula of PM with exactly one free variable, which is of the type of the natural numbers (class of classes), will be called a class-expression.

In the German original of this passage, Gödel coins the term *Klassenzeichen*, which plays a fundamental role throughout the rest of his work. This term sounds rather unusual even to native German speakers, as it is virtually non-existent in the German language. English-speaking writers usually translate it as *class expression* [13, 82], *class sign* [35], or *statement form* [101]. Gödel employs the term to describe a formula with a single free variable interpreted as a natural number.

> Klassenzeichen nennen. Die Klassenzeichen denken wir uns irgendwie in eine Folge geordnet[11]), bezeichnen das n-te mit $R(n)$ und
>
> ---
> [11]) Etwa nach steigender Gliedersumme und bei gleicher Summe lexikographisch.

> We think of the class-expressions ordered in a sequence in some manner[11]), we denote the n-th by $R(n)$,
>
> ---
> [11]) Say, according to increasing sum of the terms, and lexicographically for equal sums.

Since the class expressions are countable, we can think of them as being arranged in some sequence:

$$\varphi_0(\xi), \varphi_1(\xi), \varphi_2(\xi), \varphi_3(\xi), \varphi_4(\xi), \varphi_5(\xi), \ldots \qquad (3.2)$$

We call $\varphi_n(\xi)$ the n-th class expression. In his original paper, Gödel uses a slightly different terminology and denotes the n-th class expressions with $R(n)$. Thus,

$$R(n) = \varphi_n(\xi).$$

> wie in eine Folge geordnet[11]), bezeichnen das n-te mit $R(n)$ und bemerken, daß sich der Begriff „Klassenzeichen" sowie die ordnende Relation R im System PM definieren lassen. Sei α ein beliebiges

> and we note that the concept "class expression" as well as the ordering relation R can be defined in the system PM.

This sentence is of particular significance. It asserts that the concept of a class expression and the ordering relation mentioned above are both definable within the system of the *Principia Mathematica*. That is, there exists a formula $\psi_K(\xi)$ which is true exactly when the free variable ξ is interpreted as the Gödel number of a class expression, and another formula $\psi_R(\xi, \zeta)$ which is true exactly when ξ is interpreted as the number n and ζ as the Gödel number of the n-th class expression.

In most formal languages, natural numbers have a direct correspondence in the form of unique symbol strings. For example, recall Section 1.2, where we integrated such a notion into the example system E by treating the expression $\overline{n}$ as the following syntactical abbreviation:

$$\overline{n} \;:=\; \underbrace{\mathsf{s}(\mathsf{s}(\ldots \mathsf{s}(0)\ldots))}_{n \text{ times}}$$

By substituting the free variable ξ in a class expression $\alpha(\xi)$ with an expression of the form $\overline{n}$, a closed formula $\alpha(\overline{n})$ is obtained. In his proof sketch (and only there!), Gödel uses the somewhat peculiar notation $[\alpha; n]$ for this formula:

$$[\alpha; n] \;=\; \alpha(\overline{n})$$

Gödel again anticipates a result from the main part of his article and emphasizes that for any class expression $\alpha(\xi)$, the process of substituting the free variable ξ by the sign for the natural number n can also be characterized arithmetically and thus turns out to be definable within PM, the logic of the *Principia Mathematica*:

> Relation R im System PM definieren lassen. Sei α ein beliebiges Klassenzeichen; mit $[\alpha; n]$ bezeichnen wir diejenige Formel, welche aus dem Klassenzeichen α dadurch entsteht, daß man die freie Variable durch das Zeichen für die natürliche Zahl n ersetzt. Auch die Tripel-Relation $x = [y; z]$ erweist sich als innerhalb PM definierbar.

> Let α be an arbitrary class expression; by $[\alpha; n]$ we denote that formula which arises from the class-expression α by substitution of the symbol for the natural number n for the free variable. The ternary relation $x = [y; z]$ also turns out to be definable within PM.

3.3 I Am Unprovable!

	0	1	2	3	4	5		q	
$\varphi_0(\xi)$	$\varphi_0(\overline{0})$	$\varphi_0(\overline{1})$	$\varphi_0(\overline{2})$	$\varphi_0(\overline{3})$	$\varphi_0(\overline{4})$	$\varphi_0(\overline{5})$	...	$\varphi_0(\overline{q})$	...
$\varphi_1(\xi)$	$\varphi_1(\overline{0})$	$\varphi_1(\overline{1})$	$\varphi_1(\overline{2})$	$\varphi_1(\overline{3})$	$\varphi_1(\overline{4})$	$\varphi_1(\overline{5})$	...	$\varphi_1(\overline{q})$	...
$\varphi_2(\xi)$	$\varphi_2(\overline{0})$	$\varphi_2(\overline{1})$	$\varphi_2(\overline{2})$	$\varphi_2(\overline{3})$	$\varphi_2(\overline{4})$	$\varphi_2(\overline{5})$	...	$\varphi_2(\overline{q})$	...
$\varphi_3(\xi)$	$\varphi_3(\overline{0})$	$\varphi_3(\overline{1})$	$\varphi_3(\overline{2})$	$\varphi_3(\overline{3})$	$\varphi_3(\overline{4})$	$\varphi_3(\overline{5})$	...	$\varphi_3(\overline{q})$	...
$\varphi_4(\xi)$	$\varphi_4(\overline{0})$	$\varphi_4(\overline{1})$	$\varphi_4(\overline{2})$	$\varphi_4(\overline{3})$	$\varphi_4(\overline{4})$	$\varphi_4(\overline{5})$	...	$\varphi_4(\overline{q})$	...
$\varphi_5(\xi)$	$\varphi_5(\overline{0})$	$\varphi_5(\overline{1})$	$\varphi_5(\overline{2})$	$\varphi_5(\overline{3})$	$\varphi_5(\overline{4})$	$\varphi_5(\overline{5})$	...	$\varphi_5(\overline{q})$	...
⋮	⋮	⋮	⋮	⋮	⋮	⋮	⋱	⋮	
$\varphi_q(\xi)$	$\varphi_q(\overline{0})$	$\varphi_q(\overline{1})$	$\varphi_q(\overline{2})$	$\varphi_q(\overline{3})$	$\varphi_q(\overline{4})$	$\varphi_q(\overline{5})$	...	$\varphi_q(\overline{q})$	...
⋮	⋮	⋮	⋮	⋮	⋮	⋮	⋱	⋮	

Figure 3.5: Excerpt from the infinite table of class-expression instances. Later, Gödel will show that the main diagonal contains an undecidable formula $\varphi_q(\overline{q})$, that is, neither $\varphi_q(\overline{q})$ nor $\neg\varphi_q(\overline{q})$ is provable.

For a better understanding of the main argument that follows, it is helpful to imagine all class expressions being arranged in an infinite table, as shown in Figure 3.5. The table is structured such that the n-th class expression appears in the n-th row, and each column contains a specific instance of that formula. In particular, the instance in the n-th column is obtained by replacing the free variable of the class expression with the expression $\overline{n}$. The *diagonal elements* $\varphi_n(\overline{n})$, located on the main diagonal of the table, are of particular interest. They are created by instantiating the n-th class expression with the term representation of the natural number n.

Nun definieren wir eine Klasse K natürlicher Zahlen folgendermaßen:

$$n \, \varepsilon \, K \equiv \overline{Bew}\,[R(n);n]\,^{11\text{a}}) \qquad (1)$$

(wobei $Bew\,x$ bedeutet: x ist eine beweisbare Formel). Da die Begriffe, welche im Definiens vorkommen, sämtlich in PM definierbar sind, so auch der daraus zusammengesetzte Begriff K, d. h. es gibt ein Klassenzeichen S^{12}), so daß die Formel $[S;n]$ inhaltlich gedeutet besagt, daß die natürliche Zahl n zu K gehört. S ist als Klassen-

---
[11a]) Durch Überstreichen wird die Negation bezeichnet.
[12]) Es macht wieder nicht die geringsten Schwierigkeiten, die Formel S tatsächlich hinzuschreiben.

We now define a class K of natural numbers in the following way:

$$n \, \varepsilon \, K \equiv \overline{Bew}\,[R(n);n]\,^{11\text{a}}) \qquad (1)$$

> (where $Bew\ x$ means: x is a provable formula). Since the concepts occurring in the definiens are all definable in PM, so also is the concept K which is built up from them, i.e. there is a class-expression $S^{12})$ such that the formula $[S; n]$, intuitively interpreted, says that the natural number n belongs to K.
>
> ---
>
> $^{11a})$ The bar above denotes negation.
> $^{12})$ Again there is not the slightest difficulty in actually writing down the formula S.

In our notation, the definition of Gödel's set K reads as follows:

$$K := \{n \mid \nvdash \varphi_n(\overline{n})\}$$

Accordingly, K contains a natural number if and only if the n-th diagonal element is unprovable:

$$n \in K \Leftrightarrow \nvdash \varphi_n(\overline{n}) \tag{3.3}$$

Gödel points out that the membership relation for K is definable within the logic of the *Principia Mathematica* and any related system. More precisely, Gödel asserts the existence of a formula $S(\xi)$ that becomes a true statement if and only if the free variable ξ is replaced by the term representation $\overline{n}$ of a natural number n from the set K. In short:

$$\models S(\overline{n}) \Leftrightarrow n \in K \tag{3.4}$$

An important observation follows: $S(\xi)$ is a formula with exactly one free variable and thus a class expression itself!

> besagt, daß die natürliche Zahl n zu K gehört. S ist als Klassenzeichen mit einem bestimmten $R(q)$ identisch, d. h. es gilt
>
> $$S = R(q)$$
>
> für eine bestimmte natürliche Zahl q. Wir zeigen nun, daß der

> As a class expression, S is identical with some definite $R(q)$, i.e.
>
> $$S = R(q)$$
>
> holds for some definite natural number q.

The line of reasoning is compelling: If $S(\xi)$ itself is a class expression, it must appear in some row of the table of all class expressions. If we denote the row number, just as Gödel did, with q, then $S(\xi)$ is identical to the formula $\varphi_q(\xi)$:

$$S(\xi) = \varphi_q(\xi) \tag{3.5}$$

3.3 I Am Unprovable!

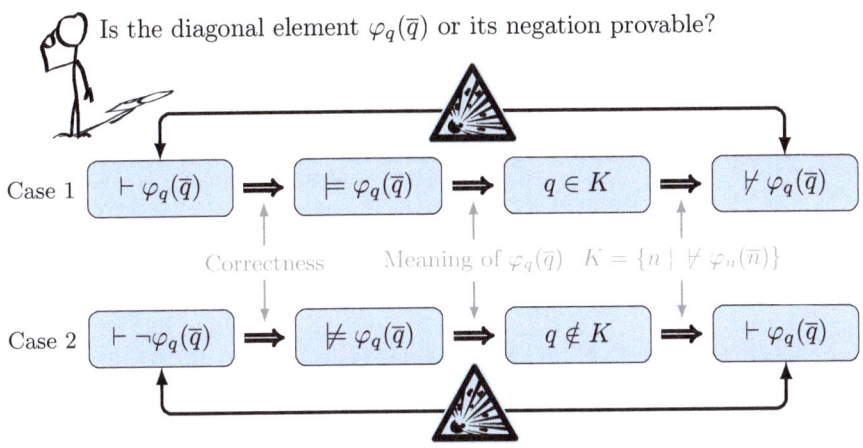

Figure 3.6: Gödel's line of reasoning. The assumption that the diagonal element $\varphi_q(\bar{q})$ is provable leads to a contradiction, just as the assumption that its negation $\neg\varphi_q(\bar{q})$ is provable.

Substituting ξ with the term $\bar{q}$ yields $\varphi_q(\bar{q})$, which is the formula we have been looking for: Neither the formula itself nor its negation is provable, assuming that the underlying formal system is correct. Let us see why (Figure 3.6):

- **Case 1: Suppose $\varphi_q(\bar{q})$ is provable.**

  Assuming the formal system is correct, every provable formula is substantively true ($\vdash \varphi_q(\bar{q})$ implies $\models \varphi_q(\bar{q})$). According to (3.5), $\varphi_q(\bar{q})$ is the same as $S(\bar{q})$, which means, according to (3.4), that q is a member of K. Now it follows from (3.3) that $\varphi_q(\bar{q})$ is unprovable.

- **Case 2: Suppose $\neg\varphi_q(\bar{q})$ is provable.**

  In this case, $\varphi_q(\bar{q})$ is substantively false, which means, according to (3.5) and (3.4), that K does not contain q. Then, according to (3.3), there is a proof for $\varphi_q(\bar{q})$, contradicting the property of correct formal systems that no formula is derivable together with its negation.

We have finally crossed the finish line: Assuming the formal system is correct, $\varphi_q(\bar{q})$ is an undecidable statement, that is, neither $\varphi_q(\bar{q})$ nor $\neg\varphi_q(\bar{q})$ is provable. In Gödel's words, the argument sounds like this:

> für eine bestimmte natürliche Zahl q. Wir zeigen nun, daß der Satz $[R(q); q]$ [13]) in PM unentscheidbar ist. Denn angenommen der Satz $[R(q); q]$ wäre beweisbar, dann wäre er auch richtig, d. h. aber nach dem obigen q würde zu K gehören, d. h. nach (1) es würde $\overline{Bew}\,[R(q); q]$ gelten, im Widerspruch mit der Annahme. Wäre dagegen die Negation von $[R(q); q]$ beweisbar, so würde $\overline{q\,\varepsilon\,K}$,

> d. h. $Bew\,[R(q);q]$ gelten. $[R(q);q]$ wäre also zugleich mit seiner Negation beweisbar, was wiederum unmöglich ist.
>
> ---
>
> [13]) Man beachte, daß „$[R(q);q]$" (oder was dasselbe bedeutet „$[S;q]$") bloß eine metamathematische Beschreibung des unentscheidbaren Satzes ist. Doch kann man, sobald man die Formel S ermittelt hat, natürlich auch die Zahl q bestimmen und damit den unentscheidbaren Satz selbst effektiv hinschreiben.

> We now show that the proposition $[R(q);q]$ [13]) is undecidable in PM. For, if the proposition $[R(q);q]$ were assumed to be provable, then it would be true, i.e. according to what was said above, q would belong to K, i.e. according to (1), $\overline{Bew}\,[R(q);q]$ would hold, contradicting our assumption. On the other hand, if the negation of $[R(q);q]$ were provable, then $\overline{q\,\varepsilon\,K}$ would hold, i.e. $Bew\,[R(q);q]$ would be true. Hence, $[R(q);q]$ together with its negation would be provable, which is again impossible.
>
> ---
>
> [13]) One should observe that "$[R(q);q]$" (or the synonymous "$[S;q]$") is merely a metamathematical description of the undecidable proposition. Nevertheless, as soon as one has obtained the formula S, one can, of course, also determine the number q, and thereby effectively write down the undecidable proposition itself.

Let us take a closer look at footnote 13. First, Gödel points out that the string $[S;q]$, in our notation, the string $\varphi_q(\bar{q})$, is only a metamathematical description of the undecidable proposition and not the proposition itself. However, this does not mean we could not write down the undecidable formula in plain text. It would indeed be possible since we could explicitly construct $S(\xi)$ and thus theoretically be able to determine its position in the sequence (3.2). This position is the number q. If we then replaced in $S(\xi)$ all occurrences of the free variable ξ with the term $\bar{q}$, we would obtain the undecidable formula $\varphi_q(\bar{q})$ in plain text.

In practice, however, we would quickly run into problems. The formula would soon become so monstrous that after just a few construction steps, we would have used up our earthly repertoire of ink and paper. Therefore, we better refrain from any attempt in this direction.

Next, let us interpret the content of $\varphi_q(\bar{q})$ substantively. First, (3.3) can be utilized to rewrite (3.4) as follows:

$$\models S(\bar{n}) \Leftrightarrow \not\vdash \varphi_n(\bar{n}) \tag{3.6}$$

$\varphi_q(\bar{q})$ is the formula $S(\bar{q})$. Thus (3.6) lets us derive:

$$\models \varphi_q(\bar{q}) \Leftrightarrow \not\vdash \varphi_q(\bar{q}) \tag{3.7}$$

3.4 Gödel, Richard and the Liar

Now, the substantive meaning of $\varphi_q(\bar{q})$ is right in front of us:

"I am unprovable!"

The above considerations have revealed the trickiness of Gödel's approach. Through the principle of diagonalization, he managed to construct a formula $\varphi_q(\bar{q})$ that makes a metatheoretical statement about itself. The formula postulates its unprovability and thus cannot be proven, assuming the underlying formal system is correct.

3.4 Gödel, Richard and the Liar

The self-reference described above is reminiscent of two well-known paradoxes that Gödel immediately addresses:

> Die Analogie dieses Schlusses mit der Antinomie Richard springt in die Augen; auch mit dem "Lügner" besteht eine nahe Verwandtschaft [14]), denn der unentscheidbare Satz $[R(q); q]$ besagt ja, daß q zu K gehört, d. h. nach (1), daß $[R(q); q]$ nicht beweisbar ist. Wir haben also einen Satz vor uns, der seine eigene Unbeweisbarkeit behauptet [15]). Die eben auseinandergesetzte Beweismethode
>
> ---
> [14]) Es läßt sich überhaupt jede epistemologische Antinomie zu einem derartigen Unentscheidbarkeitsbeweis verwenden.
> [15]) Ein solcher Satz hat entgegen dem Anschein nichts Zirkelhaftes an sich, denn er behauptet zunächst die Unbeweisbarkeit einer ganz bestimmten Formel (nämlich der q-ten in der lexikographischen Anordnung bei einer bestimmten Einsetzung), und erst nachträglich (gewissermaßen zufällig) stellt sich heraus, daß diese Formel gerade die ist, in der er selbst ausgedrückt wurde.

> The analogy of this result with Richard's antinomy is immediately evident; there is also a close relationship [14]) with the Liar Paradox, for the undecidable proposition $[R(q); q]$ says that q belongs to K, i.e. according to (1), that $[R(q); q]$ is not provable. Thus we have a proposition before us which asserts its own unprovability [15]).
>
> ---
> [14]) Every epistemological antinomy can be used for a similar proof of undecidability.
> [15]) Contrary to appeareances, such a proposition is not circular, for, to begin with, it asserts the unprovability of a quite definite formula (namely, the q-th in the lexicographical ordering, after a certain substitution) and only subsequently (accidentially, as it were) does it turn out that this formula itself is precisely the one whose unprovability is expressed.

Although knowing anything about the abovementioned paradoxes is unnecessary to execute Gödel's proof, they give valuable clues as to why undecidable

Figure 3.7: The Liar Paradox. Both the assumption that the liar is telling the truth and the assumption that he is lying lead to contradictions.

propositions must exist in any sufficiently expressive formal system. So, let's take a closer look!

3.4.1 The Liar Paradox

The liar's paradox arises whenever a statement substantively denies its truth. In its purest form, it emerges from the simple exclamation:

"I am lying!"

Figure 3.7 illustrates why asking whether this statement is true or false puts us in a precarious position. On the one hand, the statement cannot be true, as it claims to be false. On the other hand, it cannot be false either since this is precisely what it claims.

The Liar Paradox exists in various forms and can also arise even if a sentence does not directly refer to itself. The following example demonstrates that an indirect statement extending over several sentences can produce the same contradictory result:

Socrates: *"What Plato says is false!"*

Plato: *"What Socrates says is true!"*

When comparing the liar's paradox with the undecidable formula $\varphi_q(\overline{q})$, a striking similarity catches the eye: *self-reference*. Gödel's formula $\varphi_q(\overline{q})$ postulates its unprovability, thus making a statement about itself, just like the alleged liar does. However, there is a crucial difference: unlike the liar, the formula $\varphi_q(\overline{q})$ does not claim its falsehood but merely its unprovability. At first, the

3.4 Gödel, Richard and the Liar

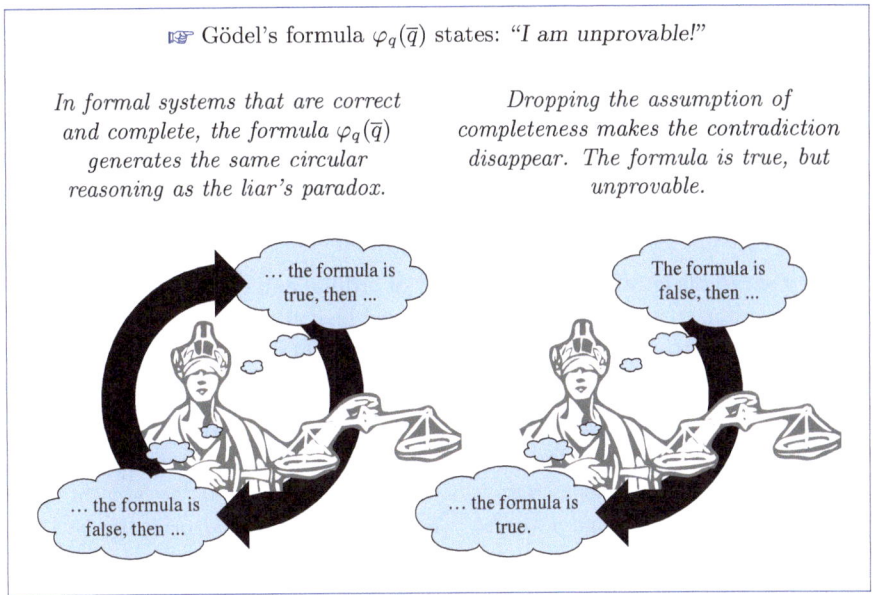

Figure 3.8: Gödel's main argument generates the same contradictory self-reference that underlies the liar's paradox. However, the contradiction disappears when the assumption about the existence of a correct and, at the same time, complete formal system is dropped.

difference sounds marginal because, in formal systems that are both correct and complete, the true statements and the provable statements are identical. In these systems, whether to ask about truth or provability is synonymous. Indeed, assuming the underlying formal system is both correct and complete makes it a victim of the same vicious circle in which the liar is trapped: On the one hand, if $\varphi_q(\overline{q})$ were true, it would be unprovable, contradicting completeness. On the other hand, if $\varphi_q(\overline{q})$ were false, it would be provable, which is impossible in a correct formal system (Figure 3.8 left).

Unlike in the case of the liar paradox, we can break the vicious circle by dropping the completeness assumption (Figure 3.8 right). In this case, only the assumption that $\varphi_q(\overline{q})$ is false leads to a contradiction, but not the assumption that it is true. In incomplete formal systems, and only in these, formulas can exist that are true but unprovable.

It is important to note that the pure existence of Gödel's undecidable formula does not create an antinomy, even if it may seem so from a distance. The antinomy arises only when a formula with the substantive meaning of $\varphi_q(\overline{q})$ is supposed to exist within a formal system that is both correct and complete. In an incomplete formal system, however, the notions of truth and provability do not coincide, which is the crucial property that makes the contradiction disappear.

3.4.2 Richard's Antinomy

Richard's antinomy was pronounced in 1905 by the French mathematics teacher Jules Richard in a letter to the editor of the journal *Revue générale des sciences pures et appliquées* and published in the June issue of the same year [85]. The contribution was reprinted in 1906 in the *Acta Mathematica* [86] and later translated into English. The quotes given below refer to the version in [87].

Richard wrote his letter when the theories of ordinal and cardinal numbers were floundering due to antinomies, and he believed he could provoke similar contradictions with much simpler means. Richard thought he had found antinomies in the realm of real numbers – the continuum:

> "It is not necessary to go so far as the theory of ordinal numbers to find such contradictions. Here is one that presents itself at the moment we study the continuum and to which some others could probably be reduced."
>
> Jules Richard [87]

At the outset, Richard observed that some real numbers have a colloquial description while others do not. This discovery led him to define a set, containing all real numbers for which such a description exists. His letter contained a detailed description of how he intended to define this set, which we are going to refer to as *Richard's set E*:

> "I am going to define a certain set of numbers, which I shall call the set E, through the following considerations. Let us write all permutations of the twenty-six letters of the French alphabet taken two at a time, putting these permutations in alphabetical order; then, after them, all permutations taken three at a time, in alphabetical order; then, after them, all permutations taken four at a time, and so forth. These permutations may contain the same letter repeated several times; they are permutations with repetitions. [...] The definition of a number being made up of words, and these words of letters, some of these permutations will be definitions of numbers. Let us cross out from our permutations all those that are not definitions of numbers."
>
> Jules Richard [87]

Following Richard's instructions, an infinitely long table is created, encompassing all finite sequences that can be crafted with the letters of our alphabet. While many sequences lack meaningful content, some will describe a number (Figure 3.9 left). The numbers definable in this manner constitute Richard's set E.

At this point, we are on the verge of giving rise to Richard's antinomy.

3.4 Gödel, Richard and the Liar

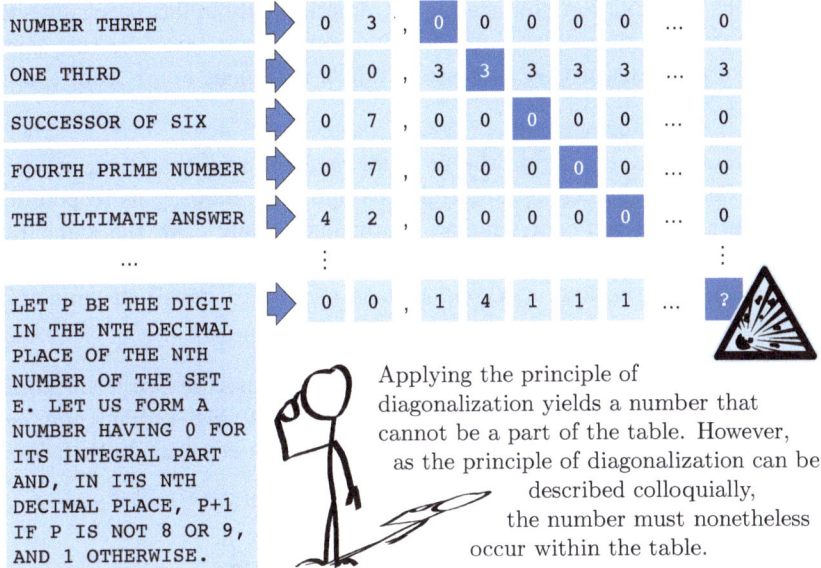

Figure 3.9: Richard's antinomy arises from the imprecision of colloquial language, implying the construction of a number which is demonstrably non-constructible.

> "Now here comes the contradiction. We can form a number not belonging to this set. 'Let p be the digit in the nth decimal place of the nth number of the set E; let us form a number having 0 for its integral part and, in its nth decimal place, $p + 1$ if p is not 8 or 9, and 1 otherwise.' This number N does not belong to the set E. If it were the nth number of the set E, the digit in its nth decimal place would be the same as the one in the nth decimal place of that number, which is not the case."
>
> Jules Richard [87]

Richard's argument sounds compelling. On the one hand, the principle of diagonalization allows us to construct a number x that cannot be contained in the table, implying $x \notin E$. On the other hand, a colloquial description of the diagonalization method must appear somewhere in the table, implying $x \in E$. Thus, we have found a number that simultaneously satisfies $x \in E$ and $x \notin E$. An untenable situation!

Richard's antinomy arises from a self-reference embedded in the construction of the diagonalization statement. Assuming the diagonal statement appears in the n-th row, the statement expresses that the n-th digit – which is its own diagonal digit – is different from itself. This assumption leads to a contradiction for any specific numerical value. The fact that the principle of diagonalization evokes Richard's antinomy puts it in direct proximity to the proof of the first incompleteness theorem: Gödel's undecidable formula appears on the diagonal of the table of all class expression instances. We cannot stress enough that Gödel's construction produces a contradiction only in formal systems that are both correct and complete. Abandoning completeness resolves this contradiction, a crucial distinction that sets Gödel's argument apart from classical antinomies.

Richard's antinomy highlights the need for caution when handling colloquial phrases. Natural language is powerful yet imprecise, allowing us to pen down self-referential and self-contradictory statements easily. At the same time, Richard's example illustrates that contradictions are not always as apparent as in the case of the liar; a second look is needed to unveil the underlying self-reference.

The Liar's Paradox and Richard's Antinomy are known as *semantic antinomies*, as they exploit the imprecision of colloquial language to blend object and metalanguage deliberately. In formal logics that strictly separate both language levels, these antinomies typically vanish.

3.4.3 When Is a Formal System Affected?

In the next section, Gödel outlines the properties that a formal system must exhibit for the argumentation of the proof sketch to be applicable:

> barkeit behauptet¹⁵). Die eben auseinandergesetzte Beweismethode
>
> 176   Kurt Gödel,
>
> läßt sich offenbar auf jedes formale System anwenden, das erstens inhaltlich gedeutet über genügend Ausdrucksmittel verfügt, um die in der obigen Überlegung vorkommenden Begriffe (insbesondere den Begriff "beweisbare Formel") zu definieren, und in dem zweitens jede beweisbare Formel auch inhaltlich richtig ist. Die nun folgende

> The method of proof which has
>
> 176   Kurt Gödel,
>
> just been explained can obviously be applied to every formal system which, first, possesses sufficient means of expression when interpreted according to its meaning to define the concepts (especially the concept "provable formula") occurring in the above argument; and, secondly, in which every provable formula is true.

Gödel states that the proof sketch applies to any formal system satisfying two properties. Firstly, the formal system must have sufficient means of expression

3.4 Gödel, Richard and the Liar

to define the term "provable formula" within its object language. Secondly, every provable formula must be substantively correct. As a result, we can only follow the line of argument in those formal systems that are correct in the sense of Definition 1.5.

From this point onward, Gödel begins to fill in the gaps in the proof sketch. In the following passages, he meticulously explains what it means to define a mathematical term or concept within a formal system. He also derives in detail under which conditions such a definition is possible or impossible.

> jede beweisbare Formel auch inhaltlich richtig ist. **Die nun folgende exakte Durchführung des obigen Beweises wird unter anderem die Aufgabe haben, die zweite der eben angeführten Voraussetzungen durch eine rein formale und weit schwächere zu ersetzen.**

> In the precise execution of the above proof, which now follows, we shall have the task (among others) of replacing the second of the assumptions just mentioned by a purely formal and much weaker assumption.

This sentence contains a compelling announcement. Gödel states that in the exact execution of the above proof, he will significantly weaken the second of the abovementioned conditions: the correctness of the underlying formal system. It was important to him not to base the proof of the incompleteness theorems on the semantic concept of truth, as he created his work at a time when set-theoretical paradoxes were still widely discussed, and many of his contemporaries were skeptical or even hostile towards the concept of truth. It was a time when, in Gödel's words,

> "a concept of objective mathematical truth [...] was viewed with the greatest suspicion and rejected as meaningless in wide circles."
>
> *"ein Konzept der objektiven mathematischen Wahrheit [...] mit größtem Misstrauen betrachtet und in weiten Kreisen als bedeutungsleer zurückgewiesen wurde."*
>
> Kurt Gödel [14]

This brings up an important matter: Gödel's first incompleteness theorem exists in several variants, with three of the most significant shown in Figure 3.10. At the very top is the semantic variant, which Gödel refers to in his proof sketch. It is the weakest of the three, making a statement about correct formal systems – those in which all provable statements are substantively true. At the very bottom is the syntactic variant. It dispenses with the substantive

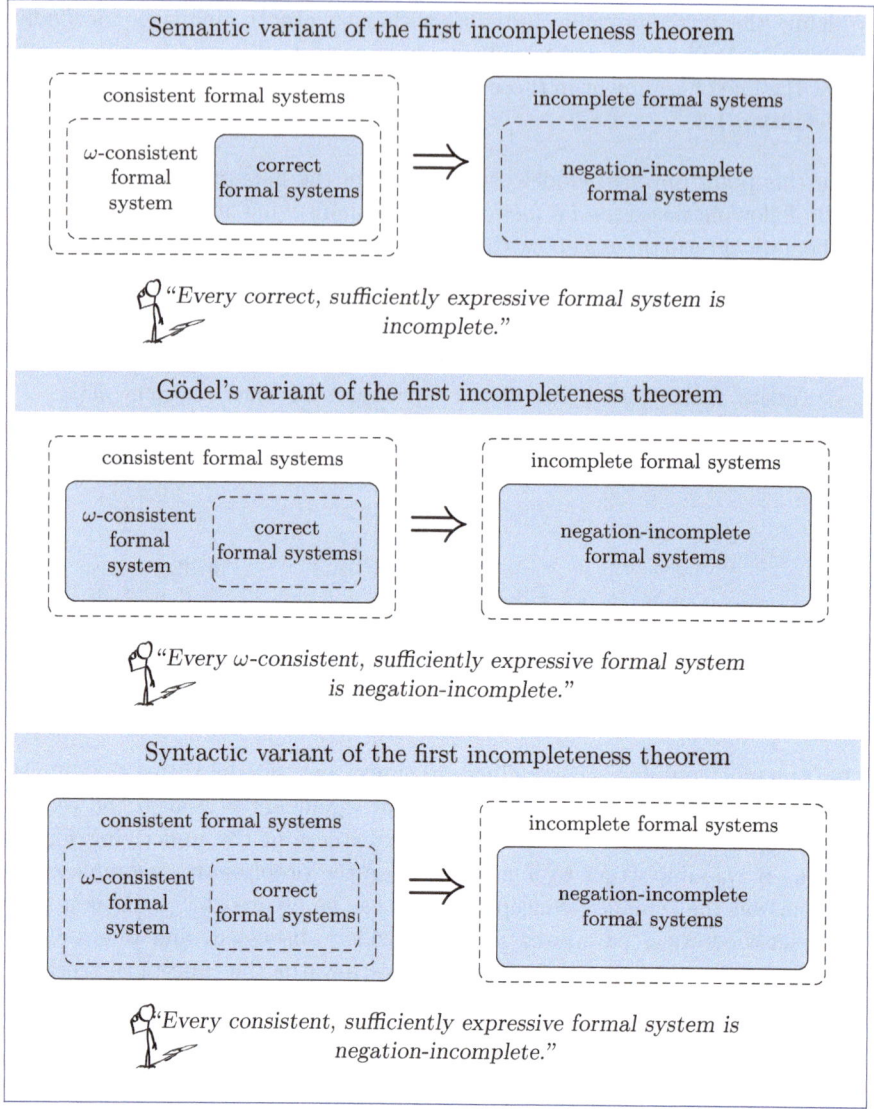

Figure 3.10: Three variants of the first incompleteness theorem. The semantic variant is the weakest, and the syntactic variant is the strongest formulation.

interpretation of formulas, thus not referencing the semantic concepts of correctness and completeness at any point. These concepts are replaced by the purely syntactic notions of consistency and negation completeness.

Gödel initially aimed to prove the syntactic variant but fell short of fully realizing his objective. He was compelled to slightly strengthen the assumption by requiring the underlying formal systems to be ω-consistent. Section 6.1 will define this term and elaborate on its meaning. Just this much in advance: Every ω-consistent formal system is consistent, but not vice versa.

3.4 Gödel, Richard and the Liar

Thus, the variant of the first incompleteness theorem proved by Gödel is slightly weaker than its syntactic counterpart but considerably stronger than the semantic variant. Since 1936, we know that the syntactic variant is also correct. That year, the American logician Barkley Rosser proved that the assumption of ω-consistency is replaceable with ordinary consistency [89, 13].

In the final section of the proof sketch, Gödel highlights that the first incompleteness theorem leads to a remarkable conclusion:

> Aus der Bemerkung, dass $[R(q)\,;\,q]$ seine eigene Unbeweisbarkeit behauptet, folgt sofort, dass $[R(q);q]$ richtig ist, denn $[R(q);q]$ ist ja unbeweisbar (weil unentscheidbar). Der im System PM unentscheidbare Satz wurde also durch metamathematische Überlegungen doch entschieden. Die genaue Analyse dieses merkwürdigen Umstandes führt zu überraschenden Resultaten, bezüglich der Widerspruchsfreiheitsbeweise formaler Systeme, die in Abschnitt 4 (Satz XI) näher behandelt werden.

> From the remark that $[R(q)\,;\,q]$ asserts its own unprovability it follows immediately that $[R(q);q]$ is true, since $[R(q);q]$ is indeed unprovable (because it is undecidable). The proposition undecidable in the system PM is thus decided by metamathematical arguments. The precise analysis of this remarkable circumstance leads to surprising results conerning consistency proofs of formal systems, which will be treated in more detail in Section 4 (Theorem XI).

Proposition XI, mentioned at the end of the paragraph, is the second incompleteness theorem. It asserts that a formal system expressive enough to formalize the proof of the first incompleteness theorem cannot prove its consistency. In the first chapter, we have already alluded to the far-reaching consequences of this theorem in philosophical terms. For most mathematicians, it manifests the impossibility of Hilbert's program: proving the consistency of classical mathematics with finite means.

4 System P

> *"Just as languages like Greek or Sanskrit are historical facts and not absolute logical necessities, it is only reasonable to assume that logics and mathematics are similarly historical, accidental forms of expression."*
>
> John von Neumann [73]

In the previous chapter, we have successfully mastered the first part of Gödel's historic paper. Having thoroughly worked through the proof sketch, we now understand Gödel's lines of reasoning in proving the first incompleteness theorem. For the exact execution of the proof, we first describe the formal system for which Gödel will prove the existence of undecidable propositions.

> 2.
> Wir gehen nun an die exakte Durchführung des oben skizzierten Beweises und geben zunächst eine genaue Beschreibung des formalen Systems *P*, für welches wir die Existenz unentscheidbarer Sätze nachweisen wollen. *P* ist im wesentlichen das System, welches

> 2.
> We pass now to the rigorous execution of the proof sketched above, and we first give a precise description of the formal system *P* for which we wish to prove the existence of undecidable propositions.

With system P, Gödel defines a formal system capable of formulating statements about the natural numbers. The title of his article already reveals the formal system he was inspired by: the *Principia Mathematica* (PM) by Russell and Whitehead. P is essentially the system obtained when the logic of PM is superposed upon the Peano axioms.

> Sätze nachweisen wollen. *P* ist im wesentlichen das System, welches man erhält, wenn man die Peanoschen Axiome mit der Logik der PM [16]) überbaut (Zahlen als Individuen, Nachfolgerrelation als undefinierten Grundbegriff).
>
> ---
> [16]) Die Hinzufügung der Peanoschen Axiome ebenso wie alle anderen am System PM angebrachten Abänderungen dienen lediglich zur Vereinfachung des Beweises und sind prinzipiell entbehrlich.

P is essentially the system which one obtains by building the logic of PM around Peano's axioms (numbers as individuals, successor relation as undefined primitive concept).[16]

---

[16] The addition of Peano's axioms, as well as all other changes made in the system PM, serve only to simplify the proof and are theoretically dispensable.

We will introduce System P in the same way as the example system E used in Section 1.2 to illustrate the basic properties of formal systems. First, we define the syntax, that is, we specify which symbol strings form a well-defined formula and which do not. Second, we introduce a semantic level by assigning a substantive meaning to the individual language elements. Third, we supply the axioms and inference rules for deriving new theorems.

4.1 Syntax

Die Grundzeichen des Systems P sind die folgenden:
I. Konstante: „$\sim$" (nicht), „$\vee$" (oder), „Π" (für alle), „0" (Null), „f" (der Nachfolger von), „(", „)" (Klammern).
II. Variable ersten Typs (für Individuen, d. h. natürliche Zahlen inklusive 0): „x_1", „y_1", „z_1", ...
Variable zweiten Typs (für Klassen von Individuen): „x_2", „y_2", „z_2", ...
Variable dritten Typs (für Klassen von Klassen von Individuen): „x_3", „y_3", „z_3", ...
usw. für jede natürliche Zahl als Typus[17].

---

[17] Es wird vorausgesetzt, daß für jeden Variablentypus abzählbar viele Zeichen zur Verfügung stehen.

The primitive symbols of the system P are the following:
I. Constants: "$\sim$" (not), "$\vee$" (or), "Π" (for all), "0" (zero), "f" (the successor of), "(", ")" (parentheses).
II. Variables of the first type (for individuals, i.e. natural numbers including 0): "x_1", "y_1", "z_1", ...
Variables of the second type (for classes of individuals): "x_2", "y_2", "z_2", ...
Variables of the third type (for classes of classes of individuals): "x_3", "y_3", "z_3", ...
– Etc., for every natural number as type.[17]

---

[17] It is assumed that, for each type, denumerably many variables are at our disposal.

4.1 Syntax

The primitive signs '∼' and 'Π' are no longer in use today. Utilizing the modern symbols '¬' (negation) and '∀' (universal quantification), the definition reads as follows:

Definition 4.1 — Primitive signs of System P

The formulas of system P are built from the following primitive signs:

- '¬', '∨', '∀'   (Logical operators and quantifiers)
- '0'   (Constant)
- 'f'   (Unary function symbol)
- '(', ')'   (Grouping symbols)
- x_i, y_i, z_i, ...   (Variables of type i)

Make sure to keep an eye on the indices of the variables! In mathematics and modern logic, it usually doesn't matter whether a variable is named x_1 or x_2. The indices have no semantic meaning and are only used to generate a sufficiently large amount of identifiers. In type theory, and thus in Gödel's system P, things are different, as the index defines the type of a variable and, therefore, its hierarchy level. Consequently, x_1 and x_2 differ not only in name but also in their substantive meaning.

In the following, we will frequently substitute variables with placeholders. If the type of a variable is irrelevant, we use placeholders such as ξ or ζ with no index. To emphasize that a variable has type i, we write ξ_i or ζ_i instead. To sum up:

$\xi, \zeta, \ldots$ represent arbitrary variables (e. g., x_1, y_2, or z_3).

$\xi_1, \zeta_1, \ldots$ represent arbitrary variables of type 1 (e. g., x_1, y_1 or z_1).

$\xi_2, \zeta_2, \ldots$ represent arbitrary variables of type 2 (e. g., x_2, y_2 or z_2).

...

Next, Gödel points out a peculiarity that distinguishes his formal system from the *Principia Mathematica*. While the latter allows the formation of expressions like $x_2(x_1, y_1)$, Gödel's system P restricts all variables of higher type to be unary predicates. For instance, it is permissible to write $x_2(x_1)$, but not $x_2(x_1, y_1)$. Gödel points out that this is not a restriction in the strict sense: Constructs of this kind are superfluous, since relations can be defined as classes of ordered pairs, and ordered pairs in turn as classes of classes.

> Anm.: Variable für zwei- und mehrstellige Funktionen (Relationen) sind als Grundzeichen überflüssig, da man Relationen als

Klassen geordneter Paare definieren kann und geordnete Paare wiederum als Klassen von Klassen, z. B. das geordnete Paar a, b durch $((a), (a, b))$, wo (x, y) bzw. (x) die Klassen bedeuten, deren einzige Elemente x, y bzw. x sind [18]).

---

[18]) Auch inhomogene Relationen können auf diese Weise definiert werden, z. B. eine Relation zwischen Individuen und Klassen als eine Klasse aus Elementen der Form: $((x_2), ((x_1), x_2))$. Alle in den PM über Relationen beweisbaren Sätze sind, wie eine einfache Überlegung lehrt, auch bei dieser Behandlungsweise beweisbar.

Remark: Variables for functions (relations) of two or more arguments are superfluous as primitive symbols, since one can define relations as classes of ordered pairs and ordered pairs, in turn, as classes of classes, e.g. define the ordered pair a, b by $((a), (a, b))$, where (x, y) denotes the class whose only elements are x and y, and (x) that whose only element is x. [18])

---

[18]) Inhomogeneous relations can also be defined in this way, e.g. a relation between individuals and classes as a class of elements of the form $((x_2), ((x_1), x_2))$. All theorems about relations provable in PM are, as is easily seen, also provable under this method of treatment.

4.1.1 Terms and Formulas

Gödel continues to broaden his terminology. Among others, he introduces distinct character combinations, referred to as expressions of the first type:

Über formal unentscheidbare Sätze der Principia Mathematica etc. 177

Unter einem Zeichen ersten Typs verstehen wir eine Zeichenkombination der Form:

$$a, \; fa, \; ffa, \; fffa \ldots \text{usw.}$$

wo a entweder 0 oder eine Variable ersten Typs ist. Im ersten Fall nennen wir ein solches Zeichen Zahlzeichen. Für $n > 1$ verstehen wir unter einem Zeichen n-ten Typs dasselbe wie Variable n-ten Typs. Zeichenkombinationen der Form $a(b)$, wo b

On formally undecidable propositions of Principia Mathematica etc. 177

By a term of the first type we mean a combination of symbols of the form:

4.1 Syntax

> $a, \; fa, \; ffa, \; fffa, \; \ldots, \text{etc.},$
>
> where a is either 0 or a variable of the first type. In the first case we call such an expression a numeral. For $n > 1$ we mean by a term of the n-th type just a variable of the n-th type.

If Gödel were publishing today, he would likely refer to expressions of the first type as *terms*.

Definition 4.2 — Terms of System P

In system P, *terms* are formed according to the following rules:

- 0 is a term.
- Every variable ξ_1 of type 1 is a term.
- If σ is a term, then f σ is also a term.

For any variable ξ_1, that is, any variable of type 1, the symbol strings

$$\xi_1, \; f\,\xi_1, \; f\,f\,\xi_1, \; f\,f\,f\,\xi_1, \ldots$$

are terms. Some examples are:

$$x_1, \; f\,x_1, \; f\,f\,x_1, \; f\,f\,f\,x_1, \ldots$$
$$y_1, \; f\,y_1, \; f\,f\,y_1, \; f\,f\,f\,y_1, \ldots$$
$$z_1, \; f\,z_1, \; f\,f\,z_1, \; f\,f\,f\,z_1, \ldots$$

Terms can also be formed without variables:

$$0, \; f\,0, \; f\,f\,0, \; f\,f\,f\,0, \ldots \tag{4.1}$$

Gödel refers to terms with no variables as *"Zahlzeichen"*. *"Zahl"* is the German word for *number*, and *"Zeichen"* means *sign* or *symbol*. The name explains itself as soon as the symbols are given a substantive meaning: Gödel identifies the symbol 0 with the natural number 0 and the symbol f with the successor operation. Hence, in terms of content, each *"Zahlzeichen"* represents a specific natural number.

Over time, the term *"Zahlzeichen"*, along with many others used by Gödel, has faded from the German vocabulary. Today, native German speakers would surely understand the individual word components but might struggle to recognize the meaning of the compound word as it was understood during Gödel's time. Today, German native speakers would likely choose the Latin word *"Nu-*

meral" instead. Most contemporary translations of Gödel's paper opt for the same word, eliminating the need for translation.

The expressions of higher types are even easier to understand, as they are identical to the variables of higher types. I. e., x_2, y_2, z_2 are *expressions of type 2*, x_3, y_3, z_3 are *expressions of type 3* and so on.

Next, Gödel introduces elementary formulas:

> Variable n-ten Typs. **Zeichenkombinationen der Form** $a(b)$, **wo** b **ein Zeichen** n-**ten und** a **ein Zeichen** $n+1$-**ten Typs ist, nennen wir Elementarformeln.** Die Klasse der Formeln definieren wir als die

> Combinations of symbols of the form $a(b)$, where b is a term of the n-th type and a is a term of the $(n+1)$st type, will be called elementary formulas.

In the remainder of this book, we will use the term *atomic formula* instead of *elementary formula*, as the former is more common today.

Definition 4.3 Atomic formulas of System P

In System P, *atomic formulas* are formed according to the following rules, where ξ_i and ζ_i denote variables of type i, and σ denotes a term:

- $\xi_2(\sigma)$
- $\xi_{n+1}(\zeta_n)$ $\hfill (n \geq 2)$

For instance, $x_2(x_1)$ and $x_3(x_2)$ are atomic formulas, but $x_3(x_1)$ is not. The established syntax rules reflect the elementary principle of type theory that a predicate of type $n+1$ expects an argument of type n.

> Elementarformeln. Die Klasse der Formeln definieren wir als die kleinste Klasse [19]), zu welcher sämtliche Elementarformeln gehören und zu welcher zugleich mit a, b stets auch $\sim(a)$, $(a) \vee (b)$, $x \Pi(a)$ gehören (wobei x eine beliebige Variable ist) [19a]). $(a) \vee (b)$ nennen wir die Disjunktion aus a und b, $\sim(a)$ die Negation und $x \Pi(a)$ eine Generalisation von a. Satzformel heißt eine Formel, in

---

[19]) Bez. dieser Definition (und analoger später vorkommender) vgl. J. Łukasiewicz und A. Tarski, Untersuchungen über den Aussagenkalkül, Comptes Rendus des séances de la Société des Sciences et des Lettres de Varsovie XXIII, 1930, Cl. III.

4.1 Syntax

> ¹⁹ᵃ⁾ $x \Pi (a)$ ist also auch dann eine Formel, wenn x in a nicht oder nicht frei vorkommt. In diesem Fall bedeutet $x \Pi (a)$ natürlich dasselbe wie a.

> We define the class of formulas as the smallest class ¹⁹⁾ to which all elementary formulas belong and to which $\sim (a)$, $(a) \vee (b)$, $x \Pi (a)$ (where x is an arbitrary variable) ¹⁹ᵃ⁾ also belong whenever a and b belong. We call $(a) \vee (b)$ the disjunction of a and b, $\sim (a)$ the negation of a, and $x \Pi (a)$ a generalization of a.
>
> ---
>
> ¹⁹⁾ With respect to this definition (and similar ones later), cf. J. Łukasiewicz and A. Tarski, "Untersuchungen über den Aussagenkalkül.", Comptes Rendus des séances de la Société des Sciences et des Lettres de Varsovie XXIII, 1930, Cl. III.
>
> ¹⁹ᵃ⁾ Thus, $x \Pi (a)$ is also a formula when x does not occur or does not occur free in a. Naturally, in this case, $x \Pi (a)$ has the same meaning as a.

In this passage, Gödel defines the rules for the construction of formulas:

Definition 4.4 — Formulas of System P

In System P, *formulas* are formed according to the following rules:

- All atomic formulas are formulas.
- Let ξ be a variable. If φ and ψ are formulas, so are:
  - $\neg(\varphi)$ (Negation)
  - $(\varphi) \vee (\psi)$ (Disjunction)
  - $\forall \xi \, (\varphi)$ (Generalization or universal quantification)

We will also employ the usual abbreviations:

$$(\varphi) \to (\psi) := (\neg(\varphi)) \vee (\psi) \quad \text{(Implication)}$$
$$(\varphi) \wedge (\psi) := \neg((\neg(\varphi)) \vee (\neg(\psi))) \quad \text{(Conjunction)}$$
$$(\varphi) \leftrightarrow (\psi) := ((\varphi) \to (\psi)) \wedge ((\psi) \to (\varphi)) \quad \text{(Equivalence)}$$
$$\exists \xi \, (\varphi) := \neg(\forall \xi \, (\neg(\varphi))) \quad \text{(Existential quantification)}$$

Based on the above definition, the following strings of symbols constitute well-defined formulas:

$$\neg(\forall x_2 \, ((x_2(f \, x_1)) \to (x_2(0)))) \tag{4.2}$$

$$(\forall x_2 \, ((x_2(f \, x_1)) \to (x_2(f \, y_1)))) \to (\forall x_2 \, ((x_2(x_1)) \to (x_2(y_1)))) \tag{4.3}$$

$$((x_2(0)) \wedge (\forall x_1 \, ((x_2(x_1)) \to (x_2(f \, x_1))))) \to (\forall x_1 \, (x_2(x_1))) \tag{4.4}$$

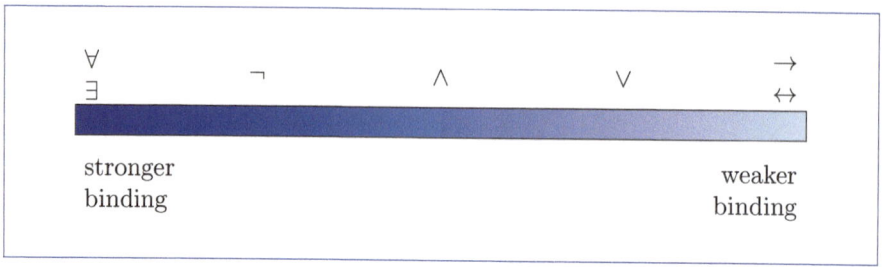

Figure 4.1: Binding rules for formulas of the formal system P

Keep in mind that the new operators are not integral parts of the formal language. They are syntactic abbreviations with the sole purpose of letting us write down formulas more concisely.

The following symbol strings do not qualify as formulas in a strict sense, as the subexpressions are not fully parenthesized:

$$\neg \forall x_2 \, (x_2(f \, x_1) \rightarrow (x_2(0))) \tag{4.5}$$

$$\forall x_2 \, (x_2(f \, x_1) \rightarrow x_2(f \, y_1)) \rightarrow \forall x_2 \, (x_2(x_1) \rightarrow (x_2(y_1))) \tag{4.6}$$

$$x_2(0) \land \forall x_1 \, (x_2(x_1) \rightarrow x_2(f \, x_1)) \rightarrow \forall x_1 \, x_2(x_1) \tag{4.7}$$

Because these formulas are more accessible for us to read than their fully bracketed counterparts (4.2) to (4.4), we will take the liberty of omitting parentheses whenever it seems appropriate. Binding rules define where we ought to fill in the omitted pairs of parentheses to obtain native formulas of system P. In particular, the following rules apply (Figure 4.1):

- '$\forall$' and '$\exists$' bind the strongest.
- '$\neg$' binds stronger than '$\land$'.
- '$\land$' binds stronger than '$\lor$'.
- '$\lor$' binds stronger than '$\rightarrow$' and '$\leftrightarrow$'.

According to these rules, the quantifier in the formula below refers only to the part left of the implication symbol:

$$\forall x_1 \, x_2(x_1) \rightarrow x_3(x_2)$$

Fully parenthesized, the formula looks like

$$(\forall x_1 \, x_2(x_1)) \rightarrow x_3(x_2)$$

and must be strictly distinguished from this variant:

$$\forall x_1 \, (x_2(x_1) \rightarrow x_3(x_2))$$

4.1 Syntax

Next, we establish binding rules for expressions where the same operator repeats multiple times. For this purpose, we declare the negation operator *right-associative* and all binary operators *left-associative*. The following examples clarify the meaning of these terms:

$$\neg\neg\neg\varphi = \neg(\neg(\neg\varphi))$$
$$\varphi \vee \psi \vee \chi = (\varphi \vee \psi) \vee \chi$$
$$\varphi \to \psi \to \chi = (\varphi \to \psi) \to \chi$$

A variable may occur as a *bound variable* or as a *free variable*. The occurrence of a variable ξ is bound if ξ is embedded in a subformula of the form $\forall \xi \, \varphi$ or $\exists \xi \, \varphi$; otherwise, it is free. The following example demonstrates that a variable may occur both free and bound within the same formula:

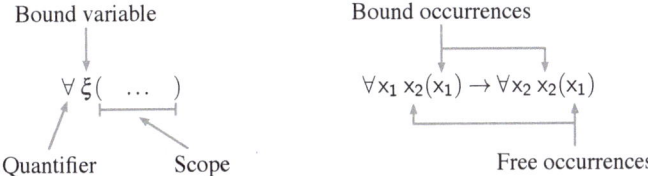

Today, most mathematicians call a formula with no free variables a *closed formula* and speak of an *open formula* otherwise. In the 1930s, these terms had not yet been coined, making Gödel's words sound quite unusual to today's readers:

> eine Generalisation von *a*. Satzformel heißt eine Formel, in der keine freie Variable vorkommt (freie Variable in der bekannten Weise definiert). Eine Formel mit genau *n*-freien Individuenvariablen (und sonst keinen freien Variablen) nennen wir *n*-stelliges Relationszeichen, für $n=1$ auch Klassenzeichen.

> A sentence is a formula in which no free variables occur (free variables being defined in the usual way). A formula with exactly *n* free individual variables (and otherwise no free variables) is called an *n*-ary predicate, for $n=1$ also a class expression.

What we now call a *closed formula* is termed *Satzformel* in Gödel's text. English-speaking authors usually translate this term to *sentential formula* or, as seen above, simply to *sentence*. A formula with exactly *n* free variables of type 1 is called *n-stelliges Relationszeichen* in the original, which translates to *n-ary relation sign*. For $n = 1$, Gödel uses the term *Klassenzeichen*, which most English-speaking authors translate to *class expressions*. Make sure to memorize the latter term. The proof sketch has already demonstrated the crucial role of class expressions in Gödel's line of reasoning.

> A formula with all variables of higher type being bound is
> - a *sentence* or *sentential formula* if it has no free variable,
> - a *relation sign* if it has at least one free variable,
> - a *class expression* if it has exactly one free variable.

4.1.2 Substitutions

> Unter Subst $a \, \binom{v}{b}$ (wo a eine Formel, v eine Variable und b ein Zeichen vom selben Typ wie v bedeutet) verstehen wir die Formel, welche aus a entsteht, wenn man darin v überall, wo es frei ist, durch b ersetzt [20]). Wir sagen, daß eine Formel a eine
>
> ---
> [20]) Falls v in a nicht als freie Variable vorkommt, soll Subst $a \, \binom{v}{b} = a$ sein. Man beachte, daß "Subst" ein Zeichen der Metamathematik ist.

> By Subst $a \, \binom{v}{b}$ (where a is a formula, v is a variable, and b is a term of the same type as v) we understand the formula which arises from a when we replace v, wherever it is free, by b [20]).
>
> ---
> [20]) In case v does not occur as a free variable in a, then Subst $a \, \binom{v}{b} = a$. One should note that "Subst" is a metamathematical symbol.

A substitution creates a formula from another by replacing all free occurrences of a variable with an expression of the same type. More precisely, for any given formula $\varphi(\xi)$ with the free variable ξ,

$$\varphi[\xi \leftarrow \sigma] \qquad (4.8)$$

denotes the formula that results from φ by replacing all free occurrences of ξ with σ. If it is evident from the context which variable is to be replaced, (4.8) is abbreviated by

$$\varphi(\sigma).$$

This notation has already been employed in the proof sketch.

The typing rules require a variable ξ to be substituted by an expression of the same type. Consequently, a variable of type 1 may be replaced by any term.

4.1 Syntax

A variable of type i with $i \geq 2$, however, may only be replaced by another variable of type i.

The subsequent substitutions adhere to the typing rules and are thus valid:

$$\varphi[x_1 \leftarrow y_1] \quad \checkmark \qquad \varphi[x_1 \leftarrow 0] \quad \checkmark \qquad \varphi[x_1 \leftarrow f\ 0] \quad \checkmark$$
$$\varphi[x_2 \leftarrow x_2] \quad \checkmark \qquad \varphi[x_2 \leftarrow y_2] \quad \checkmark \qquad \varphi[x_2 \leftarrow z_2] \quad \checkmark$$

The following three substitutions are prohibited as they create type conflicts:

$$\varphi[x_1 \leftarrow y_2] \quad \times \qquad \varphi[x_2 \leftarrow 0] \quad \times \qquad \varphi[x_2 \leftarrow f\ x_1] \quad \times$$

Let's look at some more specific examples. Since Gödel uses a notation that feels very unusual for us, all substitutions are given twice: first in Gödel's original notation and then in a modern variant:

■ Gödel's notation:

$$\text{Subst } x_2\ \Pi\ x_2(x_1)\ \tbinom{x_1}{0} = x_2\ \Pi\ x_2(0)$$
$$\text{Subst } x_2\ \Pi\ x_2(x_1)\ \tbinom{x_1}{y_1} = x_2\ \Pi\ x_2(y_1)$$

■ Modern notation:

$$(\forall x_2\ x_2(x_1))[x_1 \leftarrow 0] = \forall x_2\ x_2(0)$$
$$(\forall x_2\ x_2(x_1))[x_1 \leftarrow y_1] = \forall x_2\ x_2(y_1)$$

If the substituted variable appears both free and bound, the bound occurrences remain untouched, as the following examples demonstrate:

■ Gödel's notation:

$$\text{Subst } x_1\ \Pi\ x_2(x_1) \rightarrow x_2\ \Pi\ x_2(x_1)\ \tbinom{x_1}{y_1} = x_1\ \Pi\ x_2(x_1) \rightarrow x_2\ \Pi\ x_2(y_1)$$
$$\text{Subst } x_1\ \Pi\ x_2(x_1) \rightarrow x_2\ \Pi\ x_2(x_1)\ \tbinom{x_2}{y_2} = x_1\ \Pi\ y_2(x_1) \rightarrow x_2\ \Pi\ x_2(x_1)$$

■ Modern notation:

$$(\forall x_1\ x_2(x_1) \rightarrow \forall x_2\ x_2(x_1))[x_1 \leftarrow y_1] = (\forall x_1\ x_2(x_1) \rightarrow \forall x_2\ x_2(y_1))$$
$$(\forall x_1\ x_2(x_1) \rightarrow \forall x_2\ x_2(x_1))[x_2 \leftarrow y_2] = (\forall x_1\ y_2(x_1) \rightarrow \forall x_2\ x_2(x_1))$$

The original formula remains unaltered if the substituted variable does not occur freely anywhere.

The next example demonstrates a peculiarity:

- Gödel's notation:

$$\text{Subst } (E\ y_1)(x_2(x_1)\ \&\ y_2(y_1))\ \binom{x_1}{y_1} = (E\ y_1)(x_2(y_1)\ \&\ y_2(y_1))$$

- Modern notation:

$$(\exists y_1\ (x_2(x_1) \wedge y_2(y_1)))\ [x_1 \leftarrow y_1] = \exists y_1\ (x_2(y_1) \wedge y_2(y_1))$$

The substitution creates a so-called *collision* since it replaces a free variable with a variable that will become bound at the insertion point. In the following, mainly *collision-free substitutions* will matter.

In footnote 20, Gödel points out that *"Subst"* is a metamathematical symbol. Being not a native language element of P, it fundamentally differs from symbols such as '∨' or '¬', which are part of the object language.

Last but not least, Gödel introduces the concept of *type elevation*:

> frei ist, durch *b* ersetzt [20]). Wir sagen, daß eine Formel *a* eine Typenerhöhung einer anderen *b* ist, wenn *a* aus *b* dadurch entsteht, daß man den Typus aller in *b* vorkommenden Variablen um die gleiche Zahl erhöht.

> We say that a formula *a* is a type elevation of another formula *b* when *a* arises from *b* by raising the type of all variables occurring in *b* by the same number.

A formula ψ is obtained from a formula φ by type elevation if, for an arbitrary natural number $n \geq 1$, each variable ξ_i is replaced with the variable ξ_{i+n}. For instance, from

$$\forall x_2\ (x_2(x_1) \rightarrow x_2(y_1))$$

the following formulas can be obtained:

$$\forall x_3\ (x_3(x_2) \rightarrow x_3(y_2))$$
$$\forall x_4\ (x_4(x_3) \rightarrow x_4(y_3))$$
$$\forall x_5\ (x_5(x_4) \rightarrow x_5(y_4))$$
$$\ldots$$

4.2 Semantics

Having agreed on how the primitive symbols of system P can be combined into complex formulas, we will now assign a substantive meaning to all language

4.2 Semantics

components. Similar to the example system E discussed in Section 1.2, we define the semantics by the model relation '$\models$'. Its exact definition relies on the concept of *interpretation*, which is introduced first.

Definition 4.5 — Interpretation

An interpretation I is a mapping with the following properties:

- I maps the term 0 to the natural number 0.
- I maps every term of the form f σ to the natural number $I(\sigma) + 1$.
- I maps each variable ξ_{i+1} $(i \geq 0)$ to an element of the set $\mathcal{P}^i(\mathbb{N})$.

$\mathcal{P}^i(\mathbb{N})$ denotes the power set of order i and is defined recursively:

$$\mathcal{P}^0(\mathbb{N}) := \mathbb{N} \qquad \mathcal{P}^{i+1}(\mathbb{N}) := 2^{\mathcal{P}^i(\mathbb{N})}$$

Among others, this definition states the following:

- $I(\xi_1)$ is a natural number.

  ☞ e. g., 42

- $I(\overline{n})$ is the natural number n.

  ☞ e. g., 4 for $\overline{n} = $ f f f f 0

- $I(\xi_2)$ is a set of natural numbers.

  ☞ e. g., $\{2, 3, 5, 7, 11, 13, 17, 19, 23, \ldots\}$

- $I(\xi_3)$ is a set of sets of natural numbers.

  ☞ e. g., $\{\{0, 1\}, \{1, 2\}, \{2, 3\}, \{3, 4\}, \{4, 5\}, \ldots\}$

- $I(\xi_4)$ is a set of sets of sets of natural numbers.

  ☞ e. g., $\{\{\{0\}, \{0, 1\}\}, \{\{1\}, \{1, 2\}\}, \{\{2\}, \{2, 3\}\}, \ldots\}$

An interpretation allows us to assign a substantive meaning to each formula of P. We will write $I \models \varphi$ to express that φ becomes a substantively true statement when the variables are interpreted according to I. Formally, the model relation is defined as follows:

Definition 4.6 — Model relation

Let I be an interpretation. The model relation $I \models \varphi$ is defined inductively:

$$I \models \xi_{i+1}(\zeta_i) :\Leftrightarrow I(\zeta_i) \in I(\xi_{i+1})$$
$$I \models \neg \varphi :\Leftrightarrow I \not\models \varphi$$

$$I \models \varphi \vee \psi \;:\Leftrightarrow\; I \models \varphi \text{ or } I \models \psi$$
$$I \models \forall \xi_{i+1}\, \varphi \;:\Leftrightarrow\; \text{For all } N \in \mathcal{P}^i(\mathbb{N}),\; I_{\xi_{i+1}/N} \models \varphi$$

An interpretation I with $I \models \varphi$ is called a *model* for φ.

The notation $I_{\xi/N}$ has not been used before. It denotes the interpretation that maps ξ to N and is otherwise identical to I:

$$I_{\xi/N}(\zeta) := \begin{cases} N & \text{if } \zeta = \xi \\ I(\zeta) & \text{otherwise} \end{cases}$$

The substantive meanings of the logical operators '∧', '→', '↔', and the existential quantifier '∃' follow directly from the above definition:

$$I \models \varphi \wedge \psi \Leftrightarrow I \models \varphi \text{ and } I \models \psi$$
$$I \models \varphi \to \psi \Leftrightarrow I \models \varphi \text{ implies } I \models \psi$$
$$I \models \varphi \leftrightarrow \psi \Leftrightarrow I \models \varphi \text{ if and only if } I \models \psi$$
$$I \models \exists \xi_{i+1}\, \varphi \Leftrightarrow I_{\xi_{i+1}/N} \models \varphi \text{ holds for some } N \in \mathcal{P}^i(\mathbb{N})$$

As an example, consider the formula

$$\forall x_1\, (x_2(x_1) \vee y_2(x_1)) \tag{4.9}$$

and an interpretation I with

$$I(x_2) := \{0, 2, 4, 6, 8, 10, \ldots\} \quad (\text{☞ all even numbers})$$
$$I(y_2) := \{1, 3, 5, 7, 9, 11, \ldots\} \quad (\text{☞ all odd numbers})$$

Then, the following holds:

$$I \models x_2(x_1) \Leftrightarrow I \text{ interprets the variable } x_1 \text{ as an even number.}$$
$$I \models y_2(x_1) \Leftrightarrow I \text{ interprets the variable } x_1 \text{ as an odd number.}$$

From this, it follows for all $n \in \mathbb{N}$:

$$I_{x_1/n} \models x_2(x_1) \vee y_2(x_1)$$

That, in turn, means that formula (4.9) becomes a substantively true statement under the interpretation I:

$$I \models \forall x_1\, (x_2(x_1) \vee y_2(x_1))$$

After all, the result is not surprising, as the formula states a trivial arithmetic fact under the chosen interpretation:

"Every natural number is even or odd."

4.2 Semantics

Next, let us consider an interpretation I' with:

$$I'(x_2) := \{0, 2, 4, 6, 8, 10, \ldots\} \qquad (\text{☞ all even numbers})$$
$$I'(y_2) := \{0, 1, 4, 9, 16, 25, \ldots\} \qquad (\text{☞ all square numbers})$$

In this case, the chosen interpretation is no longer a model for the given formula:

$$I' \not\models \forall x_1 \, (x_2(x_1) \vee y_2(x_1))$$

This result again aligns with our expectations, as the formula now corresponds to the false statement:

"*Every natural number is even or a square number.*"

Now, the next two notions should be easy to understand:

Definition 4.7 — Logical Consequence, Equivalence

- ψ is a logical consequence of φ, denoted as $\varphi \models \psi$, if the following holds:

$$I \models \varphi \Rightarrow I \models \psi$$

(☞ Every model of φ is a model of ψ.)

- φ and ψ are equivalent, denoted as $\varphi \equiv \psi$, if the following holds:

$$\varphi \models \psi \text{ and } \psi \models \varphi$$

(☞ φ and ψ have the same models.)

The notion $\varphi \models \psi$ generalizes to sets of formulas. If M is such a set, then

$$M \models \psi \qquad (4.10)$$

states that if an interpretation I is a model of all formulas in M, it is also a model of ψ. If M is the empty set, then (4.10) expresses that any interpretation is a model of ψ. In this case, we call ψ a true formula and simply write

$$\models \psi$$

instead of $\emptyset \models \psi$. Again, let us examine some examples:

$$\models \exists x_2 \, \exists y_2 \, (\forall x_1 \, x_2(x_1) \vee y_2(x_1))$$
$$\models \forall x_1 \, \exists y_1 \, \neg \forall x_2 \, (x_2(x_1) \rightarrow x_2(y_1))$$
$$\models x_2(x_1) \rightarrow \exists x_1 \, x_2(x_1)$$

$$\not\models \forall x_1 \, (x_2(x_1) \vee y_2(x_1))$$
$$\not\models \exists y_1 \, \forall x_1 \, \neg \forall x_2 \, (x_2(x_1) \rightarrow x_2(y_1))$$

$$\not\models x_2(x_1) \to \forall x_1\, x_2(x_1)$$

A pivotal scenario unfolds for closed formulas. If φ is such a formula, it is either true for all interpretations ($\models \varphi$) or false for all interpretations. In the latter case, the negated formula $\neg \varphi$ is true for all interpretations ($\models \neg \varphi$). Thus, if only closed formulas are considered, there is precisely one formula that is substantially true and one that is substantially false among φ and $\neg \varphi$.

> In system P, for every closed formula φ,
> either $\models \varphi$ or $\models \neg \varphi$ holds.

Remember that such a relationship cannot be established for open formulas. For instance, the formula

$$\varphi := \forall x_1\, x_2(x_1)$$

neither satisfies $\models \varphi$ nor $\models \neg \varphi$, as both φ and $\neg \varphi$ become substantively false under certain interpretations.

4.2.1 Definition of Equality

The ability to quantify over arbitrary variables classifies Gödel's system P as a *higher-order logic*. Unlike in *first-order logic*, where quantification is confined to individual variables (variables of type 1), P allows variables of higher types to be bound. The following colloquial description of the universal quantifier elucidates the substantive meaning of quantifying over higher types:

- ∎ $\forall x_1 \ldots \ \hat{=}$ *"For all natural numbers, ..."*
- ∎ $\forall x_2 \ldots \ \hat{=}$ *"For all properties of natural numbers, ..."*
- ∎ $\forall x_3 \ldots \ \hat{=}$ *"For all properties of properties of natural numbers, ..."*

The freedom of talking about the properties of natural numbers renders system P highly expressive. For instance, it allows us to define a relation that plays a key role in mathematics: equality. In this context, to "*define*" means to specify a formula $\varphi_{\text{id}}(\xi_1, \zeta_1)$ that is true precisely when the two free variables ξ_1 and ζ_1 are interpreted as the same individual:

$$I \models \varphi_{\text{id}}(\xi_1, \zeta_1) \Leftrightarrow I(\xi_1) = I(\zeta_1)$$

For the construction of φ_{id}, we need a characterization of equality that the linguistic means of system P can capture. The solution to this problem is provided by Leibniz:

4.2 Semantics

> "The same are those that can be substituted everywhere, preserving truth."
>
> "Eadem sunt quae sibi ubique substitui possunt, salva veritate."
>
> Gottfried Wilhelm Leibniz [12]

This quote describes Leibniz's famous *principle of identity*. It states that ξ_1 and ζ_1 represent the same individual if and only if no property can distinguish them. In other words, any property applying to ξ_1 also applies to ζ_1 and vice versa. Leibniz's characterization of equality can be readily expressed within Gödel's system P by the following formula:

$$\xi_1 = \zeta_1 := \forall x_2\, (x_2(\xi_1) \leftrightarrow x_2(\zeta_1)) \tag{4.11}$$

By type elevation, the definition of equality can be generalized as follows:

$$\xi_i = \zeta_i := \forall x_{i+1}\, (x_{i+1}(\xi_i) \leftrightarrow x_{i+1}(\zeta_i)) \qquad (i \geq 1)$$

It is an essential result of mathematical logic that quantification over the individual domain, which is the set of natural numbers in the case of P, is insufficient to define equality. As a result, first-order logics, including first-order predicate logic (PL1), must incorporate the equality sign as a distinct symbol with special semantics. Consequently, PL1 and PL1 with equality must be clearly distinguished. In formal systems such as P, such differentiation is not necessary. The system is expressive enough to define equality by its own means, particularly by quantifying over a variable of higher type.

4.2.2 Definition of Natural Numbers

With the definition of equality at our fingertips, it is easy to pen down equations (4.5) to (4.7) in a concise way:

$$f\, x_1 \neq 0 \tag{4.12}$$
$$f\, x_1 = f\, y_1 \rightarrow x_1 = y_1 \tag{4.13}$$
$$x_2(0) \wedge \forall x_1\, (x_2(x_1) \rightarrow x_2(f\, x_1)) \rightarrow \forall x_1\, x_2(x_1) \tag{4.14}$$

The formulas unveil their true colors in this representation: They are three of the five Peano axioms, uniquely characterizing the natural numbers.

After familiarizing ourselves with the semantics of Gödel's system P, it is time to introduce its axioms and inference rules.

4.3 Axioms and Inference Rules

Formalizing the Peano axioms in the language of P is a prerequisite for proving theorems about natural numbers. Thus, it comes as no surprise that Gödel begins with precisely these axioms:

> Folgende Formeln (I bis V) heißen **Axiome** (sie sind mit Hilfe der in bekannter Weise definierten Abkürzungen: . , $\supset$, $\equiv$, (Ex), $=$ [21]) und mit Verwendung der üblichen Konventionen über das Weglassen von Klammern angeschrieben) [22]):
> I.
> 1. $\sim (fx_1 = 0)$
> 2. $fx_1 = fy_1 \supset x_1 = y_1$
> 3. $x_2(0) \,.\, x_1 \Pi (x_2(x_1) \supset x_2(fx_1)) \supset x_1 \Pi (x_2(x_1))$.
>
> ---
> [21]) $x_1 = y_1$ ist, wie in PM I, * 13 durch $x_2 \Pi (x_2(x_1) \supset x_2(y_1))$ definiert zu denken (ebenso für die höheren Typen).
> [22]) Um aus den angeschriebenen Schemata die Axiome zu erhalten, muß man also (in II, III, IV nach Ausführung der erlaubten Einsetzungen) noch
> 1. die Abkürzungen eliminieren,
> 2. die unterdrückten Klammern hinzufügen.
>
> Man beachte, daß die so entstehenden Ausdrücke „Formeln" in obigem Sinn sein müssen. (Vgl. auch die exakten Definitionen der metamathem. Begriffe S.182fg.)

> The following formulas (I through V) are called axioms (they are written with the help of the abbreviations: . , $\supset$, $\equiv$, (Ex), $=$ [21]), which are defined in the well-known way, and with the use of the usual conventions on the omission of parentheses [22]):
> I.
> 1. $\sim (fx_1 = 0)$
> 2. $fx_1 = fy_1 \supset x_1 = y_1$
> 3. $x_2(0) \,.\, x_1 \Pi (x_2(x_1) \supset x_2(fx_1)) \supset x_1 \Pi (x_2(x_1))$.
>
> ---
> [21]) As in PM I, *13, $x_1 = y_1$ is to be thought of as defined by $x_2\Pi(x_2(x_1) \supset x_2(y_1))$ (similarly for higher types).
> [22]) In order to obtain the axioms from the schemata as written, one must therefore (after performing the permitted substitutions in II, III, IV)
> 1. eliminate abbreviations,
> 2. add omitted parentheses.
>
> One should observe that the resulting expressions must be "formulas" in the above sense. (Cf. also the precise definitions of the metamathematical concepts on page 182 ff.)

Translating the three formulas from the original paper into modern notation yields the three formulas (4.12), (4.13) and (4.14). It is not a mistake that Gödel only mentions three of the five Peano axioms. The first two Peano axioms are implicitly formalized, as 0 is incorporated into the language of P as a constant symbol, and the successor operation f as a function symbol.

Do you recall Dedekind's isomorphism theorem from page 55? It allows us to resolve an apparent contradiction you may have already noticed. Definition 4.5

4.3 Axioms and Inference Rules

explicitly stipulates that the symbols 0 and f are interpreted as zero and the successor operation on the natural numbers, respectively. Gödel himself, however, left them undefined. Let us recall his words from page 139:

> P is essentially the system which one obtains by building the logic of PM around Peano's axioms (numbers as individuals, successor relation as undefined primitive concept). [16])

When the symbols 0 and f remain undefined, they can interpreted freely, as long as the interpretation is compatible with the axioms. The fact that they nonetheless have their intended substantive meaning follows from Dedekind's isomorphism theorem, which asserts that the Peano axioms uniquely characterize the natural numbers up to isomorphism. Thus, only one interpretation for the symbols 0 and f is possible: the interpretation of the symbol 0 as zero and of the symbol f as the successor operation.

Formulas or sets of formulas that possess a unique substantive meaning are said to be *categorical*. In the case of the Peano axioms, categoricity ensures that the symbols 0 and f are assigned their intended substantive meaning as defined in Definition 4.5. Conversely, this implies that we had no semantic freedom in this definition. Any other interpretation of 0 and f would have conflicted with the axioms of P.

Footnotes 21 and 22 are significant. In the former, Gödel defines equality as follows:
$$x_1 = y_1 := \forall x_2 \, (x_2(x_1) \rightarrow x_2(y_1)) \tag{4.15}$$
The formula is meant to hold not only for the two specific variables x_1 and y_1, but for any variable of type 1. Generalized, (4.15) reads like this:
$$\xi_1 = \zeta_1 := \forall x_2 \, (x_2(\xi_1) \rightarrow x_2(\zeta_1)) \tag{4.16}$$
Gödel uses this scheme also for higher types:
$$\xi_i = \zeta_i := \forall x_{i+1} \, (x_{i+1}(\xi_i) \rightarrow x_{i+1}(\zeta_i)) \tag{4.17}$$
This formula slightly differs from our definition, which originates directly from Leibniz's characterization of equality. On page 155, we have agreed on the following:
$$\xi_i = \zeta_i := \forall x_{i+1} \, (x_{i+1}(\xi_i) \leftrightarrow x_{i+1}(\zeta_i)) \qquad (i \geq 1)$$
Gödel's definition originates from the *Principia Mathematica*. At first sight, it seems to be weaker than Leibniz's variant. However, it is easy to check that both definitions are equivalent. If
$$x_{i+1}(\xi_i) \rightarrow x_{i+1}(\zeta_i)$$

is always true, no matter which relation is associated with x_{i+1}, then the relationship is also always true for the complementary relation. Hence, if

$$\forall x_{i+1} \, (x_{i+1}(\xi_i) \to x_{i+1}(\zeta_i)) \tag{4.18}$$

is a true formula, so is

$$\forall x_{i+1} \, (\neg x_{i+1}(\xi_i) \to \neg x_{i+1}(\zeta_i)), \tag{4.19}$$

and this formula is equivalent to

$$\forall x_{i+1} \, (x_{i+1}(\zeta_i) \to x_{i+1}(\xi_i)). \tag{4.20}$$

Together, (4.17) and (4.20) exhibit the equivalence between Gödel's definition of equality and Leibniz's principle of identity. On page 185, we will revisit the outlined derivation. In particular, we will formalize the argument within P and derive (4.20) step by step from the axioms.

In footnote 22, Gödel emphasizes that the formulas in their depicted form are no native formulas of P. To obtain those, one must therefore still

"1. eliminate abbreviations",

$\sim x_2 \, \Pi \, (\sim x_2(f \, x_1) \lor x_2(0))$
$\sim x_2 \, \Pi \, (\sim x_2(f \, x_1) \lor x_2(f \, y_1)) \lor x_2 \, \Pi \, (\sim x_2(x_1) \lor x_2(y_1))$
$\sim (\sim (\sim x_2(0) \lor \sim x_1 \, \Pi \, (\sim x_2(x_1) \lor x_2(f \, x_1)))) \lor x_1 \, \Pi \, (x_2(x_1))$

"2. add omitted parentheses":

$\sim (x_2 \, \Pi \, ((\sim (x_2(f \, x_1))) \lor (x_2(0))))$
$(\sim (x_2 \, \Pi \, ((\sim (x_2(f \, x_1))) \lor (x_2(f \, y_1))))) \lor (x_2 \, \Pi \, ((\sim (x_2(f \, x_1))) \lor (x_2(f \, y_1))))$
$(\sim (\sim ((\sim (x_2(0))) \lor (\sim (x_1 \, \Pi \, ((\sim (x_2(x_1))) \lor (x_2(f \, x_1))))))))) \lor (x_1 \, \Pi \, (x_2(x_1)))$

These formulas are the three Peano axioms expressed in the native language of the formal system P. Admittedly, they look pretty confusing in this form, and their true meanings are hardly recognizable.

Next, Gödel introduces a series of propositional axioms:

---

178   Kurt Gödel,

II.   Jede Formel, die aus den folgenden Schemata durch Einsetzung beliebiger Formeln für p, q, r entsteht.

1. $p \lor p \supset p$ 3. $p \lor q \supset q \lor p$
2. $p \supset p \lor q$ 4. $(p \supset q) \supset (r \lor p \supset r \lor q)$.

4.3 Axioms and Inference Rules 159

> 178   Kurt Gödel,
>
> II. Every formula which arises from the following schemata by substitution of arbitrary formulas for p, q, r.
>
>   1. $p \lor p \supset p$ 3. $p \lor q \supset q \lor p$
>   2. $p \supset p \lor q$ 4. $(p \supset q) \supset (r \lor p \supset r \lor q)$.

We already know these axioms from page 80, Figure 2.22. They correspond to the axioms (2), (3), (4), and (6) of the *Principia Mathematica*. Gödel also notes that these axioms are understood as schemata, with variables p, q, and r being placeholders that any other formula can replace.

Immediately after, Gödel introduces two axioms of predicate logic:

> III.   Jede Formel, die aus einem der beiden Schemata
>
>   1. $v\Pi(a) \supset \text{Subst } a \binom{v}{c}$
>   2. $v\Pi(b \lor a) \supset b \lor v\Pi(a)$
>
> dadurch entsteht, daß man für a, v, b, c folgende Einsetzungen vornimmt (und in 1. die durch „Subst" angezeigte Operation ausführt):
>
>   Für a eine beliebige Formel, für v eine beliebige Variable, für b eine Formel, in der v nicht frei vorkommt, für c ein Zeichen vom selben Typ wie v, vorausgesetzt, daß c keine Variable enthält, welche in a an einer Stelle gebunden ist, an der v frei ist [23]).
>
> ―――――
>
> [23]) c ist also entweder eine Variable oder 0 oder ein Zeichen der Form $f \ldots f u$, wo u entweder 0 oder eine Variable 1. Typs ist. Bez. des Begriffs „frei (gebunden) an einer Stelle von a" vgl. die in Fußnote [24]) zitierte Arbeit I A 5.

> III.   Every formula which results from one of the two schemata
>
>   1. $v\Pi(a) \supset \text{Subst } a \binom{v}{c}$
>   2. $v\Pi(b \lor a) \supset b \lor v\Pi(a)$
>
> by making one of the following substitutions for a, v, b, c (and carrying out in 1. the operation indicated by "Subst"):
>
>   For a an arbitrary formula; for v an arbitrary variable; for b a formula in which v does not occur free; and for c a term of the same type as v, assuming that c contains no variable which is bound at a place in a at which v is free [23]).
>
> ―――――
>
> [23]) c is therefore either a variable or 0 or a term of the form $f \ldots f u$, where u is either 0 or a variable of the first type. With respect to the concept "free (bound) at a place of a", cf. I A5 of the paper cited in footnote [24]).

In modern notation, these two axioms look like this:

$$\forall \xi_n\, \varphi \to \varphi[\xi_n \leftarrow \sigma_n] \quad \text{(if the substitution is collision-free)}$$
$$\forall \xi\, (\psi \lor \varphi) \to (\psi \lor \forall \xi\, \varphi) \quad \text{(if } \xi \text{ does not appear freely in } \psi)$$

Let's look at an example. From schema III.1, the following axiom can be derived:

$$\forall x_1\, y_2(x_1) \to y_2(y_1)$$

Keep in mind that the schema is only applicable under two conditions:

- The type of σ_n must match the type of the substituted variable. For instance, it is not permitted to choose x_1 for ξ_n and y_2 for σ_n. If it was, the formula

$$\forall x_1\, y_2(x_1) \to y_2(y_2)$$

  would be an axiom which is incompatible with the syntactic structure of P.

- The substitution must be *collision-free*, that is, the substitution must not bind a variable that occurs freely in σ_n. The following example demonstrates why this rule is crucial for the correctness of the calculus:

$$\forall x_1\, \exists y_1\, (x_1 \neq y_1) \to \exists y_1\, (y_1 \neq y_1)$$

This formula is substantively false, yet it would be an axiom if we waived the restriction to collision-free substitutions.

> IV. Jede Formel, die aus dem Schema
>
> $$1.\ (E\,u)(v\,\Pi\,(u(v) \equiv a))$$
>
> dadurch entsteht, daß man für v bzw. u beliebige Variable vom Typ n bzw. $n+1$ und für a eine Formel, die u nicht frei enthält, einsetzt. Dieses Axiom vertritt das Reduzibilitätsaxiom (Komprehensionsaxiom der Mengenlehre).

> IV. Every formula which results from the schema
>
> $$1.\ (E\,u)(v\,\Pi\,(u(v) \equiv a))$$
>
> by substituting for v (for u) an arbitrary variable of the type n (of type $n+1$), and for a any formula which does not contain u free. This axiom represents the axiom of reducability (comprehension axiom of set theory).

Here, Gödel introduces an axiom that is already familiar to us. It closely resembles to the axiom of comprehension discussed on page 103. Fortunately,

4.3 Axioms and Inference Rules

there is no need to fear antinomies, as the type system of P prevents the formation of contradictory totalities, such as the set of all sets.

Further down, the equivalence between Gödel's and Leibniz's definition of equality will be proved using Gödel's comprehension schema. It will be key to reproduce the inference from (4.18) to (4.19), outlined on page 158, within the formal system P.

> V. Jede Formel, die aus der folgenden durch Typenerhöhung entsteht (und diese Formel selbst):
>
> 1. $x_1 \Pi (x_2 (x_1) \equiv y_2 (x_1)) \supset x_2 = y_2$.
>
> Dieses Axiom besagt, daß eine Klasse durch ihre Elemente vollständig bestimmt ist.

> V. Every formula which arises from the following by type elevation (and this formula itself):
>
> 1. $x_1 \Pi (x_2 (x_1) \equiv y_2 (x_1)) \supset x_2 = y_2$.
>
> This axiom asserts that a class is completely determined by its elements.

To get a grip on this formula, suppose that the interpretation I assigns the sets $X \subseteq \mathbb{N}$ and $Y \subseteq \mathbb{N}$ to the variables x₂ and y₂, respectively. Then, the axiom expresses the following substantive relationship:

$$(x \in X \Leftrightarrow x \in Y) \Rightarrow X = Y$$

In colloquial terms, this symbolic expression reads like this: If X and Y contain the same elements ($x \in X \Leftrightarrow x \in Y$), then X and Y are identical ($X = Y$). This relationship is the *principle of extensionality* of set theory. In its general form, it states that the meaning of an expression is determined entirely by its scope, that is, by the objects it names or describes. In the context of set theory, the principle expresses that a set is solely determined by its elements.

Next, Gödel directs his attention to the logical inference apparatus:

> Eine Formel c heißt unmittelbare Folge aus a und b (bzw. aus a), wenn a die Formel $(\sim(b)) \vee (c)$ ist. (bzw. wenn c die Formel $v\Pi(a)$ ist, wo v eine beliebige Variable bedeutet). Die Klasse der beweisbaren Formeln wird definiert als die kleinste Klasse von Formeln, welche die Axiome enthält und gegen die Relation „unmittelbare Folge" abgeschlossen ist [24]).
>
> ---
> [24]) Die Einsetzungsregel wird dadurch überflüssig, daß wir alle möglichen Einsetzungen bereits in den Axiomen selbst vorgenommen haben (analog bei J. v. Neumann, Zur Hilbertschen Beweistheorie, Math. Zeitschr. 26, 1927).

> A formula c is called an immediate consequence of a and b (of a) if a is the formula $(\sim(b)) \vee (c)$ (if c is the formula $v\Pi(a)$, where v denotes an arbitrary variable). The class of provable formulas is defined as the smallest class of formulas which contains the axioms and is closed with respect to the relation "immediate consequence" [24]).
>
> ---
>
> [24]) The rule of substitution has been rendered superfluous by our having carried out all possible substitutions in the axioms themselves (as in J. v. Neumann, "Zur Hilbertschen Beweistheorie", Math. Zeitschr. 26, 1927).

Here, Gödel stipulates that the provable formulas include precisely those that are a direct consequence of the axioms or already proven formulas. According to his definition,

- ψ is an immediate consequence of the formulas φ and $\neg \varphi \vee \psi$,
- $\forall \xi\, \varphi$ is an immediate consequence of the formula φ for each variable ξ.

Writing $\neg \varphi \vee \psi$ more concisely as $\varphi \to \psi$, the two inference rules look like this:

$$\frac{\varphi \quad \varphi \to \psi}{\psi} \quad \text{(MP)} \qquad\qquad \frac{\varphi}{\forall \xi\, \varphi} \quad \text{(G)}$$

The first rule is the well-known modus ponens. The second rule is commonly referred to as the *Generalization Rule*, which can be applied to bind free variables with a universal quantifier. At this point, the axioms and inference rules are fully defined and summarized in Table 4.1.

4.4 Formal Proofs

The time has come to breathe life into Gödel's system. We begin with a straightforward example: the proof of $x_2(0) \to x_2(0)$:

$x_2(0) \to x_2(0)$

1. $\vdash\ x_2(0) \to x_2(0) \vee x_2(0)$ (II.2)
2. $\vdash\ x_2(0) \vee x_2(0) \to x_2(0)$ (II.1)
3. $\vdash\ (x_2(0) \vee x_2(0) \to x_2(0)) \to$
   $\qquad (\neg x_2(0) \vee (x_2(0) \vee x_2(0))) \to (\neg x_2(0) \vee x_2(0)))$ (II.4)
4. $\vdash\ (x_2(0) \vee x_2(0) \to x_2(0)) \to$

4.4 Formal Proofs

Table 4.1: Gödel's system P

Axiom Group I (Peano Axioms)	
$\neg(f\, x_1 = 0)$	(I.1)
$f\, x_1 = f\, y_1 \to x_1 = y_1$	(I.2)
$x_2(0) \land \forall x_1\, (x_2(x_1) \to x_2(f\, x_1)) \to \forall x_1\, x_2(x_1)$	(I.3)
Axiom Group II (Propositional Logic Axioms)	
$\varphi \lor \varphi \to \varphi$	(II.1)
$\varphi \to \varphi \lor \psi$	(II.2)
$\varphi \lor \psi \to \psi \lor \varphi$	(II.3)
$(\varphi \to \psi) \to (\chi \lor \varphi \to \chi \lor \psi)$	(II.4)
Axiom Group III (Predicate Logic Axioms)	
$\forall \xi_n\, \varphi \to \varphi[\xi_n \leftarrow \sigma_n]$ (if the substitution is collision-free)	(III.1)
$\forall \xi\, (\psi \lor \varphi) \to (\psi \lor \forall \xi\, \varphi)$ (if ξ does not occur freely in ψ)	(III.2)
Axiom Group IV (Comprehension)	
$\exists \xi_{n+1}\, \forall \zeta_n\, (\xi_{n+1}(\zeta_n) \leftrightarrow \varphi)$ (if ξ_{n+1} does not occur freely in φ)	(IV.1)
Axiom Group V (Extensionality)	
$\forall x_1\, (x_2(x_1) \leftrightarrow y_2(x_1)) \to x_2 = y_2$	(V.1)
Inference Rules	
$\dfrac{\varphi,\ \varphi \to \psi}{\psi}$ (MP) $\qquad \dfrac{\varphi}{\forall \xi\, \varphi}$ (G)	

$$
\begin{aligned}
&((x_2(0) \to x_2(0) \lor x_2(0)) \to (x_2(0) \to x_2(0))) &&\text{(Def, 3)}\\
&5.\ \vdash\ (x_2(0) \to x_2(0) \lor x_2(0)) \to (x_2(0) \to x_2(0)) &&\text{(MP, 2,4)}\\
&6.\ \vdash\ x_2(0) \to x_2(0) &&\text{(MP, 1,5)}
\end{aligned}
$$

The proof chain commences with three axioms, derived from the schemata II.2, II.1, and II.4 by applying the following substitutions:

$$[\varphi \leftarrow x_2(0), \psi \leftarrow x_2(0)] \qquad \text{(applied to II.2)}$$
$$[\varphi \leftarrow x_2(0)] \qquad \text{(applied to II.1)}$$
$$[\varphi \leftarrow x_2(0) \vee x_2(0), \psi \leftarrow x_2(0), \chi \leftarrow \neg x_2(0)] \qquad \text{(applied to II.4)}$$

The transition from 3. to 4. is not a genuine derivation step. Despite the different appearances of both formulas, they are factually the same, as only the subexpressions

$$\neg x_2(0) \vee (x_2(0) \vee x_2(0)) \text{ and } \neg x_2(0) \vee x_2(0)$$

have been replaced with their respective equivalent representations

$$x_2(0) \to x_2(0) \vee x_2(0) \text{ and } x_2(0) \to x_2(0).$$

The formulas in the last two lines were derived by repeatedly applying the modus ponens inference rule to previous members of the proof chain.

The next proof follows the same scheme but leads to a slightly different theorem:

$x_2(f\ 0) \to x_2(f\ 0)$

1. $\vdash\ x_2(f\ 0) \to x_2(f\ 0) \vee x_2(f\ 0)$ (II.2)
2. $\vdash\ x_2(f\ 0) \vee x_2(f\ 0) \to x_2(f\ 0)$ (II.1)
3. $\vdash\ (x_2(f\ 0) \vee x_2(f\ 0) \to x_2(f\ 0)) \to$
   $\quad (\neg x_2(f\ 0) \vee (x_2(f\ 0) \vee x_2(f\ 0)) \to (\neg x_2(f\ 0) \vee x_2(f\ 0)))$ (II.4)
4. $\vdash\ (x_2(f\ 0) \vee x_2(f\ 0) \to x_2(f\ 0)) \to$
   $\quad ((x_2(f\ 0) \to x_2(f\ 0) \vee x_2(f\ 0)) \to (x_2(f\ 0) \to x_2(f\ 0)))$ (Def, 3)
5. $\vdash\ (x_2(f\ 0) \to x_2(f\ 0) \vee x_2(f\ 0)) \to (x_2(f\ 0) \to x_2(f\ 0))$ (MP, 2,4)
6. $\vdash\ x_2(f\ 0) \to x_2(f\ 0)$ (MP, 1,5)

In fact, both proofs are structurally identical. Using the same scheme, the theorem $\varphi \to \varphi$ can be derived for any formula φ. Therefore, we will replace subformulas with placeholders whenever possible. The outcome is a *proof template* that turns into an actual proof by replacing the placeholders accordingly.

For the above example, the proof template looks like this:

$\varphi \to \varphi$ (H.0)

4.4 Formal Proofs

1. $\vdash \varphi \to \varphi \vee \varphi$ (II.2)
2. $\vdash \varphi \vee \varphi \to \varphi$ (II.1)
3. $\vdash (\varphi \vee \varphi \to \varphi) \to ((\neg\varphi \vee (\varphi \vee \varphi)) \to (\neg\varphi \vee \varphi))$ (II.4)
4. $\vdash (\varphi \vee \varphi \to \varphi) \to ((\varphi \to \varphi \vee \varphi) \to (\varphi \to \varphi))$ (Def, 3)
5. $\vdash (\varphi \to \varphi \vee \varphi) \to (\varphi \to \varphi)$ (MP, 2,4)
6. $\vdash \varphi \to \varphi$ (MP, 1,5)

To simplify the derivation of theorems, we will supplement the inference rules of system P with the barbara syllogism, which we will refer to as the *modus barbara* (MB) in the proofs below. This rule implements classical *chain inference*, expressible in our notation as such:

$$\frac{\varphi \to \psi \qquad \psi \to \chi}{\varphi \to \chi} \quad \text{(MB)}$$

It is important to note that the new inference rule does not require any alteration to the formal system P, as it is replicable through existing means. Every proof employing the modus barbara can be translated into another proof that no longer relies on this rule. The following proof fragment shows how the formula $\varphi \to \chi$ can be derived natively from $\varphi \to \psi$ and $\psi \to \chi$ within P.

Justification of (MB)

1. $\varphi \to \psi$
2. $\psi \to \chi$
3. $\vdash (\psi \to \chi) \to (\neg\varphi \vee \psi \to \neg\varphi \vee \chi)$ (II.4)
4. $\vdash (\psi \to \chi) \to ((\varphi \to \psi) \to (\varphi \to \chi))$ (Def, 3)
5. $\vdash (\varphi \to \psi) \to (\varphi \to \chi)$ (MP, 2,4)
6. $\vdash \varphi \to \chi$ (MP, 1,5)

The new inference rule allows for a more concise presentation of formal proofs. Using the new rule, the proof template for our example theorem shrinks to just three lines:

$\varphi \to \varphi$ (H.0)

1. $\vdash \varphi \to \varphi \vee \varphi$ (II.2)
2. $\vdash \varphi \vee \varphi \to \varphi$ (II.1)
3. $\vdash \varphi \to \varphi$ (MB, 1,2)

Three additional inference rules can be justified in a similar manner:

$$\frac{\varphi \to \psi}{\chi \vee \varphi \to \chi \vee \psi} \quad \text{(DL)} \qquad \frac{\varphi \to \psi}{\varphi \vee \chi \to \psi \vee \chi} \quad \text{(DR)}$$

$$\frac{\varphi \to \psi}{(\chi \to \varphi) \to (\chi \to \psi)} \quad \text{(IL)}$$

Justification of (DL)

1. $\varphi \to \psi$
2. $\vdash$ $(\varphi \to \psi) \to (\chi \vee \varphi \to \chi \vee \psi)$ (II.4)
3. $\vdash$ $\chi \vee \varphi \to \chi \vee \psi$ (MP, 1,2)

Justification of (DR)

1. $\varphi \to \psi$
2. $\vdash$ $\chi \vee \varphi \to \chi \vee \psi$ (DL, 1)
3. $\vdash$ $\varphi \vee \chi \to \chi \vee \varphi$ (II.3)
4. $\vdash$ $\varphi \vee \chi \to \chi \vee \psi$ (MB, 3,2)
5. $\vdash$ $\chi \vee \psi \to \psi \vee \chi$ (II.3)
6. $\vdash$ $\varphi \vee \chi \to \psi \vee \chi$ (MB, 4,5)

Justification of (IL)

1. $\varphi \to \psi$
2. $\vdash$ $(\neg\chi \vee \varphi) \to (\neg\chi \vee \psi)$ (DL, 1)
3. $\vdash$ $(\chi \to \varphi) \to (\chi \to \psi)$ (Def, 2)

4.4.1 Propositional Logic Theorems

We will proceed to derive additional theorems in P, commencing with the proof of the propositional logic theorems listed in Table 4.2.

The theorems (H.x) originate from the renowned textbook *Principles of Theoretical Logic* by David Hilbert and Wilhelm Ackermann, first published in 1928. The book has seen six editions and, for many years, stood as a premier textbook in mathematical logic. In the initial three editions, the two authors utilized the same axiom system as Gödel, allowing for a one-to-one replication

4.4 Formal Proofs

Table 4.2: Theorems of system P

	Theorem Group 1: Propositional Logic	
(H.1)	$(\varphi \to \psi) \to ((\chi \to \varphi) \to (\chi \to \psi))$	
(H.2)	$\neg \varphi \lor \varphi$	
(H.3)	$\varphi \lor \neg \varphi$	
(H.4)	$\varphi \to \neg \neg \varphi$	identical with Frege's axiom (F.6)
(H.5)	$\neg \neg \varphi \to \varphi$	identical with Frege's axiom (F.5)
(H.6)	$(\varphi \to \psi) \to (\neg \psi \to \neg \varphi)$	identical with Frege's axiom (F.4)
(H.7)	$\neg(\varphi \land \psi) \to \neg \varphi \lor \neg \psi$	
(H.8)	$\neg \varphi \lor \neg \psi \to \neg(\varphi \land \psi)$	
(H.9)	$\neg(\varphi \lor \psi) \to \neg \varphi \land \neg \psi$	
(H.10)	$\neg \varphi \land \neg \psi \to \neg(\varphi \lor \psi)$	
(H.11)	$\varphi \land \psi \to \psi \land \varphi$	
(H.12)	$\varphi \land \psi \to \varphi$	
(H.13)	$\varphi \land \psi \to \psi$	
(H.14)	$\varphi \lor (\psi \lor \chi) \to \psi \lor (\varphi \lor \chi)$	
(F.3)	$(\varphi \to (\psi \to \chi)) \to (\psi \to (\varphi \to \chi))$	
(H.15)	$\varphi \lor (\psi \lor \chi) \to (\varphi \lor \psi) \lor \chi$	
(H.16)	$(\varphi \lor \psi) \lor \chi \to \varphi \lor (\psi \lor \chi)$	
(H.17)	$(\varphi \land \psi) \land \chi \to \varphi \land (\psi \land \chi)$	
(H.18)	$\varphi \to (\psi \to \varphi \land \psi)$	
(A.1)	$(\varphi \to (\psi \to \chi)) \to (\varphi \land \psi \to \chi)$	
(A.2)	$(\varphi \land \psi \to \chi) \to (\varphi \to (\psi \to \chi))$	
(A.3)	$\varphi \lor (\varphi \lor \psi) \to \varphi \lor \psi$	
(A.4)	$(\varphi \to (\varphi \to \psi)) \to (\varphi \to \psi)$	
(H.19)	$\varphi \lor (\psi \land \chi) \to (\varphi \lor \psi) \land (\varphi \lor \chi)$	
(H.20)	$(\varphi \lor \psi) \land (\varphi \lor \chi) \to \varphi \lor (\psi \land \chi)$	
(A.5)	$\varphi \land (\psi \lor \chi) \to (\varphi \land \psi) \lor (\varphi \land \chi)$	
(A.6)	$(\varphi \land \psi) \lor (\varphi \land \chi) \to \varphi \land (\psi \lor \chi)$	
(F.1)	$\varphi \to (\psi \to \varphi)$	
(F.2)	$(\varphi \to (\psi \to \chi)) \to ((\varphi \to \psi) \to (\varphi \to \chi))$	
(A.7)	$\neg \varphi \to (\varphi \to \psi)$	

of the proofs in the formal system P. In subsequent editions, however, the axiom system was replaced by another. As a result, the original proof sequences are no longer present in those editions.

Despite its age, Hilbert and Ackermann's book is still a worthy read, retaining much of the clarity and rigor that distinguished it from the beginning. However, there are a few aspects to consider when reading the original text. For instance, Hilbert and Ackermann used the expression $\varphi\psi$ as shorthand for the disjunction $(\varphi \vee \psi)$ rather than the conjunction $(\varphi \wedge \psi)$, as it is common today.

In addition to the examples from Hilbert and Ackermann, Table 4.2 also includes theorems prefixed with the letters F or A. The theorems labeled (F.x) are part of Frege's axiom system, discussed on page 37. The theorems labeled (A.x) do not originate from any historical source.

$(\varphi \to \psi) \to ((\chi \to \varphi) \to (\chi \to \psi))$ (H.1)

1. $\vdash (\varphi \to \psi) \to (\neg \chi \vee \varphi \to \neg \chi \vee \psi)$ (II.4)
2. $\vdash (\varphi \to \psi) \to ((\chi \to \varphi) \to (\chi \to \psi))$ (Def, 1)

$\neg \varphi \vee \varphi$ (H.2)

1. $\vdash \varphi \to \varphi \vee \varphi$ (II.2)
2. $\vdash \varphi \vee \varphi \to \varphi$ (II.1)
3. $\vdash \varphi \to \varphi$ (MB, 1,2)
4. $\vdash \neg \varphi \vee \varphi$ (Def, 3)

$\varphi \vee \neg \varphi$ (H.3)

1. $\vdash \neg \varphi \vee \varphi$ (H.2)
2. $\vdash \neg \varphi \vee \varphi \to \varphi \vee \neg \varphi$ (II.3)
3. $\vdash \varphi \vee \neg \varphi$ (MP, 1,2)

$\varphi \to \neg \neg \varphi$ (H.4)

1. $\vdash \neg \varphi \vee \neg \neg \varphi$ (H.3)
2. $\vdash \varphi \to \neg \neg \varphi$ (Def, 1)

$\neg \neg \varphi \to \varphi$ (H.5)

4.4 Formal Proofs

1. $\vdash \neg\varphi \to \neg\neg\neg\varphi$ (H.4)
2. $\vdash \varphi \lor \neg\varphi \to \varphi \lor \neg\neg\neg\varphi$ (DL, 1)
3. $\vdash \varphi \lor \neg\varphi$ (H.3)
4. $\vdash \varphi \lor \neg\neg\neg\varphi$ (MP, 3,2)
5. $\vdash \varphi \lor \neg\neg\neg\varphi \to \neg\neg\neg\varphi \lor \varphi$ (II.3)
6. $\vdash \neg\neg\neg\varphi \lor \varphi$ (MP, 4,5)
7. $\vdash \neg\neg\varphi \to \varphi$ (Def, 6)

$(\varphi \to \psi) \to (\neg\psi \to \neg\varphi)$ (H.6)

1. $\vdash \psi \to \neg\neg\psi$ (H.4)
2. $\vdash \neg\varphi \lor \psi \to \neg\varphi \lor \neg\neg\psi$ (DL, 1)
3. $\vdash \neg\varphi \lor \neg\neg\psi \to \neg\neg\psi \lor \neg\varphi$ (II.3)
4. $\vdash \neg\varphi \lor \psi \to \neg\neg\psi \lor \neg\varphi$ (MB, 2,3)
5. $\vdash (\varphi \to \psi) \to (\neg\psi \to \neg\varphi)$ (Def, 4)

The formulas (H.4), (H.5), and (H.6) are already known to us from Section 2.1.2. They are part of the axiom system of the *Begriffsschrift* and identical to the formulas (F.6), (F.5), and (F.4) on page 37. Formula (H.6) justifies several new inference rules that will come in very handy for us:

$$\frac{\varphi \to \psi}{\neg\psi \to \neg\varphi} \text{ (INV)} \qquad \frac{\neg\varphi \to \neg\psi}{\psi \to \varphi} \text{ (INV)}$$

$$\frac{\varphi \to \psi}{\chi \land \varphi \to \chi \land \psi} \text{ (KL)} \qquad \frac{\varphi \to \psi}{\varphi \land \chi \to \psi \land \chi} \text{ (KR)}$$

Due to their substantive proximity, we use the same abbreviation (INV) for the first two inference rules.

Justification of (INV) (First variant)

1. $\varphi \to \psi$
2. $\vdash (\varphi \to \psi) \to (\neg\psi \to \neg\varphi)$ (H.6)
3. $\vdash \neg\psi \to \neg\varphi$ (MP, 1,2)

Justification of (INV) (Second variant)

1. $\neg\varphi \to \neg\psi$

2. ⊢ $(\neg\varphi \to \neg\psi) \to (\neg\neg\psi \to \neg\neg\varphi)$ \hfill (H.6)
3. ⊢ $\neg\neg\psi \to \neg\neg\varphi$ \hfill (MP, 1,2)
4. ⊢ $\psi \to \neg\neg\psi$ \hfill (H.4)
5. ⊢ $\psi \to \neg\neg\varphi$ \hfill (MB, 4,3)
6. ⊢ $\neg\neg\varphi \to \varphi$ \hfill (H.5)
7. ⊢ $\psi \to \varphi$ \hfill (MB, 5,6)

Justification of (KL)

1. $\varphi \to \psi$
2. ⊢ $\neg\psi \to \neg\varphi$ \hfill (INV, 1)
3. ⊢ $\neg\chi \vee \neg\psi \to \neg\chi \vee \neg\varphi$ \hfill (DL, 2)
4. ⊢ $(\neg\chi \vee \neg\psi \to \neg\chi \vee \neg\varphi) \to (\neg(\neg\chi \vee \neg\varphi) \to \neg(\neg\chi \vee \neg\psi))$ \hfill (H.6)
5. ⊢ $\neg(\neg\chi \vee \neg\varphi) \to \neg(\neg\chi \vee \neg\psi)$ \hfill (MP, 3,4)
6. ⊢ $\chi \wedge \varphi \to \chi \wedge \psi$ \hfill (Def, 5)

Justification of (KR)

1. $\varphi \to \psi$
2. ⊢ $\neg\psi \to \neg\varphi$ \hfill (INV, 1)
3. ⊢ $\neg\psi \vee \neg\chi \to \neg\varphi \vee \neg\chi$ \hfill (DR, 2)
4. ⊢ $(\neg\psi \vee \neg\chi \to \neg\varphi \vee \neg\chi) \to (\neg(\neg\varphi \vee \neg\chi) \to \neg(\neg\psi \vee \neg\chi))$ \hfill (H.6)
5. ⊢ $\neg(\neg\varphi \vee \neg\chi) \to \neg(\neg\psi \vee \neg\chi)$ \hfill (MP, 3,4)
6. ⊢ $\varphi \wedge \chi \to \psi \wedge \chi$ \hfill (Def, 5)

$\neg(\varphi \wedge \psi) \to \neg\varphi \vee \neg\psi$ \hfill (H.7)

1. ⊢ $\neg\neg(\neg\varphi \vee \neg\psi) \to \neg\varphi \vee \neg\psi$ \hfill (H.5)
2. ⊢ $\neg(\varphi \wedge \psi) \to \neg\varphi \vee \neg\psi$ \hfill (Def, 1)

$\neg\varphi \vee \neg\psi \to \neg(\varphi \wedge \psi)$ \hfill (H.8)

1. ⊢ $\neg\varphi \vee \neg\psi \to \neg\neg(\neg\varphi \vee \neg\psi)$ \hfill (H.4)
2. ⊢ $\neg\varphi \vee \neg\psi \to \neg(\varphi \wedge \psi)$ \hfill (Def, 1)

$\neg(\varphi \vee \psi) \to \neg\varphi \wedge \neg\psi$ \hfill (H.9)

1. ⊢ $\neg\neg\varphi \to \varphi$ \hfill (H.5)

4.4 Formal Proofs

2. $\vdash \neg\neg\varphi \vee \neg\neg\psi \to \varphi \vee \neg\neg\psi$ (DR, 1)
3. $\vdash \neg\neg\psi \to \psi$ (H.5)
4. $\vdash \varphi \vee \neg\neg\psi \to \varphi \vee \psi$ (DL, 3)
5. $\vdash \neg\neg\varphi \vee \neg\neg\psi \to \varphi \vee \psi$ (MB, 2,4)
6. $\vdash \neg(\varphi \vee \psi) \to \neg(\neg\neg\varphi \vee \neg\neg\psi)$ (INV, 5)
7. $\vdash \neg(\varphi \vee \psi) \to \neg\varphi \wedge \neg\psi$ (Def, 6)

$\neg\varphi \wedge \neg\psi \to \neg(\varphi \vee \psi)$ (H.10)

1. $\vdash \varphi \to \neg\neg\varphi$ (H.4)
2. $\vdash \varphi \vee \psi \to \neg\neg\varphi \vee \psi$ (DR, 1)
3. $\vdash \psi \to \neg\neg\psi$ (H.4)
4. $\vdash \neg\neg\varphi \vee \psi \to \neg\neg\varphi \vee \neg\neg\psi$ (DL, 3)
5. $\vdash \varphi \vee \psi \to \neg\neg\varphi \vee \neg\neg\psi$ (MB, 2,4)
6. $\vdash \neg(\neg\neg\varphi \vee \neg\neg\psi) \to \neg(\varphi \vee \psi)$ (INV, 5)
7. $\vdash \neg\varphi \wedge \neg\psi \to \neg(\varphi \vee \psi)$ (Def, 6)

$\varphi \wedge \psi \to \psi \wedge \varphi$ (H.11)

1. $\vdash \neg\psi \vee \neg\varphi \to \neg\varphi \vee \neg\psi$ (II.3)
2. $\vdash \neg(\neg\varphi \vee \neg\psi) \to \neg(\neg\psi \vee \neg\varphi)$ (INV, 1)
3. $\vdash \varphi \wedge \psi \to \psi \wedge \varphi$ (Def, 2)

$\varphi \wedge \psi \to \varphi$ (H.12)

1. $\vdash \neg\varphi \to \neg\varphi \vee \neg\psi$ (II.2)
2. $\vdash \neg(\neg\varphi \vee \neg\psi) \to \neg\neg\varphi$ (INV, 1)
3. $\vdash \varphi \wedge \psi \to \neg\neg\varphi$ (Def, 2)
4. $\vdash \neg\neg\varphi \to \varphi$ (H.5)
5. $\vdash \varphi \wedge \psi \to \varphi$ (MB, 3,4)

$\varphi \wedge \psi \to \psi$ (H.13)

1. $\vdash \psi \wedge \varphi \to \psi$ (H.12)
2. $\vdash \varphi \wedge \psi \to \psi \wedge \varphi$ (H.11)
3. $\vdash \varphi \wedge \psi \to \psi$ (MB, 2,1)

The forthcoming theorem matches the fifth propositional logic axiom of the *Principia Mathematica*. The formal proof originates from the 1926 article *Axiomatic Investigation of the Propositional Calculus of the Principia Mathematica* [5], in which Paul Bernays summarized the results of his 1918 habilitation thesis [4]. Through this derivation, Bernays demonstrated that the fifth axiom of the *Principia* is redundant and can be safely deleted from the list of axioms.

$$\varphi \vee (\psi \vee \chi) \to \psi \vee (\varphi \vee \chi) \tag{H.14}$$

1. $\vdash \chi \to \chi \vee \varphi$ (II.2)
2. $\vdash \chi \vee \varphi \to \varphi \vee \chi$ (II.3)
3. $\vdash \chi \to \varphi \vee \chi$ (MB, 1,2)
4. $\vdash (\chi \to \varphi \vee \chi) \to (\psi \vee \chi \to \psi \vee (\varphi \vee \chi))$ (II.4)
5. $\vdash \psi \vee \chi \to \psi \vee (\varphi \vee \chi)$ (MP, 3,4)
6. $\vdash (\psi \vee \chi \to \psi \vee (\varphi \vee \chi)) \to (\varphi \vee (\psi \vee \chi) \to \varphi \vee (\psi \vee (\varphi \vee \chi)))$ (II.4)
7. $\vdash \varphi \vee (\psi \vee \chi) \to \varphi \vee (\psi \vee (\varphi \vee \chi))$ (MP, 5,6)
8. $\vdash \varphi \vee (\psi \vee (\varphi \vee \chi)) \to (\psi \vee (\varphi \vee \chi)) \vee \varphi$ (II.3)
9. $\vdash \varphi \vee (\psi \vee \chi) \to (\psi \vee (\varphi \vee \chi)) \vee \varphi$ (MB, 7,8)
10. $\vdash \varphi \to \varphi \vee \chi$ (II.2)
11. $\vdash \varphi \vee \chi \to (\varphi \vee \chi) \vee \psi$ (II.2)
12. $\vdash (\varphi \vee \chi) \vee \psi \to \psi \vee (\varphi \vee \chi)$ (II.3)
13. $\vdash \varphi \vee \chi \to \psi \vee (\varphi \vee \chi)$ (MB, 11,12)
14. $\vdash \varphi \to \psi \vee (\varphi \vee \chi)$ (MB, 10,13)
15. $\vdash (\varphi \to \psi \vee (\varphi \vee \chi)) \to$
    $((\psi \vee (\varphi \vee \chi)) \vee \varphi \to (\psi \vee (\varphi \vee \chi)) \vee (\psi \vee (\varphi \vee \chi)))$ (II.4)
16. $\vdash (\psi \vee (\varphi \vee \chi)) \vee \varphi \to (\psi \vee (\varphi \vee \chi)) \vee (\psi \vee (\varphi \vee \chi))$ (MP, 14,15)
17. $\vdash (\psi \vee (\varphi \vee \chi)) \vee (\psi \vee (\varphi \vee \chi)) \to \psi \vee (\varphi \vee \chi)$ (II.1)
18. $\vdash (\psi \vee (\varphi \vee \chi)) \vee \varphi \to \psi \vee (\varphi \vee \chi)$ (MB, 16,17)
19. $\vdash \varphi \vee (\psi \vee \chi) \to \psi \vee (\varphi \vee \chi)$ (MB, 9,18)

From this theorem, the third axiom of the *Begriffsschrift* is quickly derivable. It differs from (H.14) only in notation.

$$(\varphi \to (\psi \to \chi)) \to (\psi \to (\varphi \to \chi)) \tag{F.3}$$

1. $\vdash \neg\varphi \vee (\neg\psi \vee \chi) \to \neg\psi \vee (\neg\varphi \vee \chi)$ (H.14)
2. $\vdash \neg\varphi \vee (\psi \to \chi) \to \neg\psi \vee (\varphi \to \chi)$ (Def, 1)
3. $\vdash (\varphi \to (\psi \to \chi)) \to (\psi \to (\varphi \to \chi))$ (Def, 2)

4.4 Formal Proofs

This theorem justifies another handy inference rule, referred to as the *Exchange Rule* and abbreviated as ER.

$$\frac{\varphi \to (\psi \to \chi)}{\psi \to (\varphi \to \chi)} \tag{ER}$$

Justification of (ER)

1. $\varphi \to (\psi \to \chi)$
2. $\vdash (\varphi \to (\psi \to \chi)) \to (\psi \to (\varphi \to \chi))$ (F.3)
3. $\vdash \psi \to (\varphi \to \chi)$ (MP, 1,2)

$\varphi \vee (\psi \vee \chi) \to (\varphi \vee \psi) \vee \chi$ (H.15)

1. $\vdash \psi \vee \chi \to \chi \vee \psi$ (II.3)
2. $\vdash \varphi \vee (\psi \vee \chi) \to \varphi \vee (\chi \vee \psi)$ (DL, 1)
3. $\vdash \varphi \vee (\chi \vee \psi) \to \chi \vee (\varphi \vee \psi)$ (H.14)
4. $\vdash \varphi \vee (\psi \vee \chi) \to \chi \vee (\varphi \vee \psi)$ (MB, 2,3)
5. $\vdash \chi \vee (\varphi \vee \psi) \to (\varphi \vee \psi) \vee \chi$ (II.3)
6. $\vdash \varphi \vee (\psi \vee \chi) \to (\varphi \vee \psi) \vee \chi$ (MB, 4,5)

$(\varphi \vee \psi) \vee \chi \to \varphi \vee (\psi \vee \chi)$ (H.16)

1. $\vdash (\varphi \vee \psi) \vee \chi \to \chi \vee (\varphi \vee \psi)$ (II.3)
2. $\vdash \chi \vee (\varphi \vee \psi) \to \varphi \vee (\chi \vee \psi)$ (H.14)
3. $\vdash (\varphi \vee \psi) \vee \chi \to \varphi \vee (\chi \vee \psi)$ (MB, 1,2)
4. $\vdash \chi \vee \psi \to \psi \vee \chi$ (II.3)
5. $\vdash \varphi \vee (\chi \vee \psi) \to \varphi \vee (\psi \vee \chi)$ (DL, 4)
6. $\vdash (\varphi \vee \psi) \vee \chi \to \varphi \vee (\psi \vee \chi)$ (MB, 3,5)

$(\varphi \wedge \psi) \wedge \chi \to \varphi \wedge (\psi \wedge \chi)$ (H.17)

1. $\vdash \neg\neg(\neg\psi \vee \neg\chi) \to \neg\psi \vee \neg\chi$ (H.5)
2. $\vdash \neg\varphi \vee \neg\neg(\neg\psi \vee \neg\chi) \to \neg\varphi \vee (\neg\psi \vee \neg\chi)$ (DL, 1)
3. $\vdash \neg\varphi \vee (\neg\psi \vee \neg\chi) \to (\neg\varphi \vee \neg\psi) \vee \neg\chi$ (H.15)
4. $\vdash \neg\varphi \vee \neg\neg(\neg\psi \vee \neg\chi) \to (\neg\varphi \vee \neg\psi) \vee \neg\chi$ (MB, 2,3)
5. $\vdash \neg\varphi \vee \neg\psi \to \neg\neg(\neg\varphi \vee \neg\psi)$ (H.4)
6. $\vdash \neg\chi \vee (\neg\varphi \vee \neg\psi) \to \neg\chi \vee \neg\neg(\neg\varphi \vee \neg\psi)$ (DL, 5)

7. ⊢ (¬φ ∨ ¬ψ) ∨ ¬χ → ¬χ ∨ (¬φ ∨ ¬ψ)   (II.3)
8. ⊢ (¬φ ∨ ¬ψ) ∨ ¬χ → ¬χ ∨ ¬¬(¬φ ∨ ¬ψ)   (MB, 7,6)
9. ⊢ ¬χ ∨ ¬¬(¬φ ∨ ¬ψ) → ¬¬(¬φ ∨ ¬ψ) ∨ ¬χ   (II.3)
10. ⊢ (¬φ ∨ ¬ψ) ∨ ¬χ → ¬¬(¬φ ∨ ¬ψ) ∨ ¬χ   (MB, 8,9)
11. ⊢ (¬φ ∨ ¬¬(¬ψ ∨ ¬χ)) → ¬¬(¬φ ∨ ¬ψ) ∨ ¬χ   (MB, 4,10)
12. ⊢ ¬(¬¬(¬φ ∨ ¬ψ) ∨ ¬χ) → ¬(¬φ ∨ ¬¬(¬ψ ∨ ¬χ))   (INV, 11)
13. ⊢ ¬(¬(φ ∧ ψ) ∨ ¬χ) → ¬(¬φ ∨ ¬(ψ ∧ χ))   (Def, 12)
14. ⊢ (φ ∧ ψ) ∧ χ → φ ∧ (ψ ∧ χ)   (Def, 13)

φ → (ψ → φ ∧ ψ)   (H.18)

1. ⊢ (¬φ ∨ ¬ψ) ∨ ¬(¬φ ∨ ¬ψ)   (H.3)
2. ⊢ (¬φ ∨ ¬ψ) ∨ ¬(¬φ ∨ ¬ψ) → ¬φ ∨ (¬ψ ∨ ¬(¬φ ∨ ¬ψ))   (H.16)
3. ⊢ ¬φ ∨ (¬ψ ∨ ¬(¬φ ∨ ¬ψ))   (MP, 1,2)
4. ⊢ ¬φ ∨ (¬ψ ∨ (φ ∧ ψ))   (Def, 3)
5. ⊢ φ → (ψ → φ ∧ ψ)   (Def, 4)

(φ → (ψ → χ)) → (φ ∧ ψ → χ)   (A.1)

1. ⊢ (¬φ ∨ ¬ψ) → ¬¬(¬φ ∨ ¬ψ)   (H.4)
2. ⊢ (¬φ ∨ ¬ψ) ∨ χ → ¬¬(¬φ ∨ ¬ψ) ∨ χ   (DR, 1)
3. ⊢ ¬φ ∨ (¬ψ ∨ χ) → (¬φ ∨ ¬ψ) ∨ χ   (H.15)
4. ⊢ ¬φ ∨ (¬ψ ∨ χ) → ¬¬(¬φ ∨ ¬ψ) ∨ χ   (MB, 3,2)
5. ⊢ (φ → (ψ → χ)) → (¬(¬φ ∨ ¬ψ) → χ)   (Def, 4)
6. ⊢ (φ → (ψ → χ)) → (φ ∧ ψ → χ)   (Def, 5)

(φ ∧ ψ → χ) → (φ → (ψ → χ))   (A.2)

1. ⊢ ¬¬(¬φ ∨ ¬ψ) → (¬φ ∨ ¬ψ)   (H.5)
2. ⊢ ¬¬(¬φ ∨ ¬ψ) ∨ χ → (¬φ ∨ ¬ψ) ∨ χ   (DR, 1)
3. ⊢ (¬φ ∨ ¬ψ) ∨ χ → ¬φ ∨ (¬ψ ∨ χ)   (H.16)
4. ⊢ ¬¬(¬φ ∨ ¬ψ) ∨ χ → ¬φ ∨ (¬ψ ∨ χ)   (MB, 2,3)
5. ⊢ (¬(¬φ ∨ ¬ψ) → χ) → (φ → (ψ → χ))   (Def, 4)
6. ⊢ (φ ∧ ψ → χ) → (φ → (ψ → χ))   (Def, 5)

φ ∨ (φ ∨ ψ) → φ ∨ ψ   (A.3)

1. ⊢ φ ∨ (φ ∨ ψ) → (φ ∨ φ) ∨ ψ   (H.15)

4.4 Formal Proofs

2. $\vdash (\varphi \vee \varphi) \to \varphi$ (II.1)
3. $\vdash (\varphi \vee \varphi) \vee \psi \to \varphi \vee \psi$ (DR, 2)
4. $\vdash \varphi \vee (\varphi \vee \psi) \to \varphi \vee \psi$ (MB, 1,3)

$$(\varphi \to (\varphi \to \psi)) \to (\varphi \to \psi) \tag{A.4}$$

1. $\vdash \neg\varphi \vee (\neg\varphi \vee \psi) \to \neg\varphi \vee \psi$ (A.3)
2. $\vdash (\varphi \to (\varphi \to \psi)) \to (\varphi \to \psi)$ (Def, 1)

$$\varphi \vee (\psi \wedge \chi) \to (\varphi \vee \psi) \wedge (\varphi \vee \chi) \tag{H.19}$$

1. $\vdash \psi \wedge \chi \to \psi$ (H.12)
2. $\vdash \varphi \vee (\psi \wedge \chi) \to \varphi \vee \psi$ (DL, 1)
3. $\vdash \psi \wedge \chi \to \chi$ (H.13)
4. $\vdash \varphi \vee (\psi \wedge \chi) \to \varphi \vee \chi$ (DL, 3)
5. $\vdash \varphi \vee \psi \to (\varphi \vee \chi \to (\varphi \vee \psi) \wedge (\varphi \vee \chi))$ (H.18)
6. $\vdash \varphi \vee (\psi \wedge \chi) \to (\varphi \vee \chi \to (\varphi \vee \psi) \wedge (\varphi \vee \chi))$ (MB, 2,5)
7. $\vdash \varphi \vee \chi \to (\varphi \vee (\psi \wedge \chi) \to (\varphi \vee \psi) \wedge (\varphi \vee \chi))$ (ER, 6)
8. $\vdash \varphi \vee (\psi \wedge \chi) \to (\varphi \vee (\psi \wedge \chi) \to (\varphi \vee \psi) \wedge (\varphi \vee \chi))$ (MB, 4,7)
9. $\vdash (\varphi \vee (\psi \wedge \chi) \to (\varphi \vee (\psi \wedge \chi) \to (\varphi \vee \psi) \wedge (\varphi \vee \chi))) \to$
    $(\varphi \vee (\psi \wedge \chi) \to (\varphi \vee \psi) \wedge (\varphi \vee \chi))$ (A.4)
10. $\vdash \varphi \vee (\psi \wedge \chi) \to (\varphi \vee \psi) \wedge (\varphi \vee \chi)$ (MP, 8,9)

$$(\varphi \vee \psi) \wedge (\varphi \vee \chi) \to \varphi \vee (\psi \wedge \chi) \tag{H.20}$$

1. $\vdash \psi \to (\chi \to \psi \wedge \chi)$ (H.18)
2. $\vdash (\chi \to \psi \wedge \chi) \to (\varphi \vee \chi \to \varphi \vee (\psi \wedge \chi))$ (II.4)
3. $\vdash \psi \to (\varphi \vee \chi \to \varphi \vee (\psi \wedge \chi))$ (MB, 1,2)
4. $\vdash \varphi \vee \chi \to (\psi \to \varphi \vee (\psi \wedge \chi))$ (ER, 3)
5. $\vdash (\psi \to \varphi \vee (\psi \wedge \chi)) \to (\varphi \vee \psi \to \varphi \vee (\varphi \vee (\psi \wedge \chi)))$ (II.4)
6. $\vdash \varphi \vee \chi \to (\varphi \vee \psi \to \varphi \vee (\varphi \vee (\psi \wedge \chi)))$ (MB, 4,5)
7. $\vdash \varphi \vee (\varphi \vee (\psi \wedge \chi)) \to \varphi \vee (\psi \wedge \chi)$ (A.3)
8. $\vdash (\varphi \vee \psi \to \varphi \vee (\varphi \vee (\psi \wedge \chi))) \to (\varphi \vee \psi \to \varphi \vee (\psi \wedge \chi))$ (IL, 7)
9. $\vdash \varphi \vee \chi \to (\varphi \vee \psi \to \varphi \vee (\psi \wedge \chi))$ (MB, 6,8)
10. $\vdash \varphi \vee \psi \to (\varphi \vee \chi \to \varphi \vee (\psi \wedge \chi))$ (ER, 9)
11. $\vdash (\varphi \vee \psi \to (\varphi \vee \chi \to \varphi \vee (\psi \wedge \chi))) \to$
    $((\varphi \vee \psi) \wedge (\varphi \vee \chi) \to \varphi \vee (\psi \wedge \chi))$ (A.1)
12. $\vdash (\varphi \vee \psi) \wedge (\varphi \vee \chi) \to \varphi \vee (\psi \wedge \chi)$ (MP, 10,11)

$\varphi \wedge (\psi \vee \chi) \to (\varphi \wedge \psi) \vee (\varphi \wedge \chi)$ (A.5)

1. $\vdash (\neg\varphi \vee \neg\psi) \wedge (\neg\varphi \vee \neg\chi) \to \neg\varphi \vee (\neg\psi \wedge \neg\chi)$ (H.20)
2. $\vdash \neg(\neg\varphi \vee (\neg\psi \wedge \neg\chi)) \to \neg((\neg\varphi \vee \neg\psi) \wedge (\neg\varphi \vee \neg\chi))$ (INV, 1)
3. $\vdash \neg((\neg\varphi \vee \neg\psi) \wedge (\neg\varphi \vee \neg\chi)) \to \neg(\neg\varphi \vee \neg\psi) \vee \neg(\neg\varphi \vee \neg\chi)$ (H.7)
4. $\vdash \neg(\neg\varphi \vee (\neg\psi \wedge \neg\chi)) \to \neg(\neg\varphi \vee \neg\psi) \vee \neg(\neg\varphi \vee \neg\chi)$ (MB, 2,3)
5. $\vdash \neg(\neg\varphi \vee (\neg\psi \wedge \neg\chi)) \to (\varphi \wedge \psi) \vee (\varphi \wedge \chi)$ (Def, 4)
6. $\vdash \neg\psi \wedge \neg\chi \to \neg(\psi \vee \chi)$ (H.10)
7. $\vdash \neg\varphi \vee (\neg\psi \wedge \neg\chi) \to \neg\varphi \vee \neg(\psi \vee \chi)$ (DL, 6)
8. $\vdash \neg(\neg\varphi \vee \neg(\psi \vee \chi)) \to \neg(\neg\varphi \vee (\neg\psi \wedge \neg\chi))$ (INV, 7)
9. $\vdash \varphi \wedge (\psi \vee \chi) \to \neg(\neg\varphi \vee (\neg\psi \wedge \neg\chi))$ (Def, 8)
10. $\vdash \varphi \wedge (\psi \vee \chi) \to (\varphi \wedge \psi) \vee (\varphi \wedge \chi)$ (MB, 9,5)

$(\varphi \wedge \psi) \vee (\varphi \wedge \chi) \to \varphi \wedge (\psi \vee \chi)$ (A.6)

1. $\vdash \neg\varphi \vee (\neg\psi \wedge \neg\chi) \to (\neg\varphi \vee \neg\psi) \wedge (\neg\varphi \vee \neg\chi)$ (H.19)
2. $\vdash \neg((\neg\varphi \vee \neg\psi) \wedge (\neg\varphi \vee \neg\chi)) \to \neg(\neg\varphi \vee (\neg\psi \wedge \neg\chi))$ (INV, 1)
3. $\vdash \neg(\neg\varphi \vee \neg\psi) \vee \neg(\neg\varphi \vee \neg\chi) \to \neg((\neg\varphi \vee \neg\psi) \wedge (\neg\varphi \vee \neg\chi))$ (H.8)
4. $\vdash \neg(\neg\varphi \vee \neg\psi) \vee \neg(\neg\varphi \vee \neg\chi) \to \neg(\neg\varphi \vee (\neg\psi \wedge \neg\chi))$ (MB, 3,2)
5. $\vdash (\varphi \wedge \psi) \vee (\varphi \wedge \chi) \to \neg(\neg\varphi \vee (\neg\psi \wedge \neg\chi))$ (Def, 4)
6. $\vdash \neg(\psi \vee \chi) \to \neg\psi \wedge \neg\chi$ (H.9)
7. $\vdash \neg\varphi \vee \neg(\psi \vee \chi) \to \neg\varphi \vee (\neg\psi \wedge \neg\chi)$ (DL, 6)
8. $\vdash \neg(\neg\varphi \vee (\neg\psi \wedge \neg\chi)) \to \neg(\neg\varphi \vee \neg(\psi \vee \chi))$ (INV, 7)
9. $\vdash \neg(\neg\varphi \vee (\neg\psi \wedge \neg\chi)) \to \varphi \wedge (\psi \vee \chi)$ (Def, 8)
10. $\vdash (\varphi \wedge \psi) \vee (\varphi \wedge \chi) \to \varphi \wedge (\psi \vee \chi)$ (MB, 5,9)

The following two theorems correspond to the remaining axioms of the *Begriffsschrift*, namely Frege's axioms (F.1) and (F.2).

$\varphi \to (\psi \to \varphi)$ (F.1)

1. $\vdash \varphi \to \varphi \vee \neg\psi$ (II.2)
2. $\vdash \varphi \vee \neg\psi \to \neg\psi \vee \varphi$ (II.3)
3. $\vdash \varphi \to \neg\psi \vee \varphi$ (MB, 1,2)
4. $\vdash \varphi \to (\psi \to \varphi)$ (Def, 3)

4.4 Formal Proofs

$$(\varphi \to (\psi \to \chi)) \to ((\varphi \to \psi) \to (\varphi \to \chi)) \tag{F.2}$$

1. $\vdash \neg\varphi \lor \varphi$ (H.2)
2. $\vdash \neg\varphi \lor \varphi \to (\neg\varphi \lor \neg\psi \to (\neg\varphi \lor \varphi) \land (\neg\varphi \lor \neg\psi))$ (H.18)
3. $\vdash \neg\varphi \lor \neg\psi \to (\neg\varphi \lor \varphi) \land (\neg\varphi \lor \neg\psi)$ (MP, 1,2)
4. $\vdash (\neg\varphi \lor \varphi) \land (\neg\varphi \lor \neg\psi) \to \neg\varphi \lor (\varphi \land \neg\psi)$ (H.20)
5. $\vdash \neg\varphi \lor \neg\psi \to \neg\varphi \lor (\varphi \land \neg\psi)$ (MB, 3,4)
6. $\vdash \neg\varphi \lor (\varphi \land \neg\psi) \to (\varphi \land \neg\psi) \lor \neg\varphi$ (II.3)
7. $\vdash \neg\varphi \lor \neg\psi \to (\varphi \land \neg\psi) \lor \neg\varphi$ (MB, 5,6)
8. $\vdash \psi \to \neg\neg\psi$ (H.4)
9. $\vdash \neg\varphi \lor \psi \to \neg\varphi \lor \neg\neg\psi$ (DL, 8)
10. $\vdash \neg(\neg\varphi \lor \neg\neg\psi) \to \neg(\neg\varphi \lor \psi)$ (INV, 9)
11. $\vdash \varphi \land \neg\psi \to \neg(\varphi \to \psi)$ (Def, 10)
12. $\vdash (\varphi \land \neg\psi) \lor \neg\varphi \to \neg(\varphi \to \psi) \lor \neg\varphi$ (DR, 11)
13. $\vdash \neg\varphi \lor \neg\psi \to \neg(\varphi \to \psi) \lor \neg\varphi$ (MB, 7,12)
14. $\vdash \neg\varphi \lor (\neg\psi \lor \chi) \to (\neg\varphi \lor \neg\psi) \lor \chi$ (H.15)
15. $\vdash (\varphi \to (\psi \to \chi)) \to (\neg\varphi \lor \neg\psi) \lor \chi$ (Def, 14)
16. $\vdash (\neg\varphi \lor \neg\psi) \lor \chi \to (\neg(\varphi \to \psi) \lor \neg\varphi) \lor \chi$ (DR, 13)
17. $\vdash (\neg(\varphi \to \psi) \lor \neg\varphi) \lor \chi \to \neg(\varphi \to \psi) \lor (\neg\varphi \lor \chi)$ (H.16)
18. $\vdash (\neg\varphi \lor \neg\psi) \lor \chi \to \neg(\varphi \to \psi) \lor (\neg\varphi \lor \chi)$ (MB, 16,17)
19. $\vdash (\neg\varphi \lor \neg\psi) \lor \chi \to ((\varphi \to \psi) \to (\varphi \to \chi))$ (Def, 18)
20. $\vdash (\varphi \to (\psi \to \chi)) \to ((\varphi \to \psi) \to (\varphi \to \chi))$ (MB, 15,19)

At this juncture, we have achieved a significant intermediate result: All propositional logic axioms of the *Begriffsschrift* are theorems of P. Consequently, any propositional formula derivable within the logical calculus of the *Begriffsschrift* is also provable in Gödel's formal system P. The reverse also holds, implying that both propositional logic inference apparatuses are equivalent.

The next theorem fulfills our promise from page 117 where we have claimed that in each formal system that includes the ordinary propositional logic apparatus, any other formula is derivable from a contradictory pair of formulas. The following theorem elucidates the rationale behind this phenomenon:

$$\neg\varphi \to (\varphi \to \psi) \tag{A.7}$$

1. $\vdash \neg\neg\psi \to \psi$ (H.5)
2. $\vdash \neg\neg\psi \lor \neg\varphi \to \psi \lor \neg\varphi$ (DR, 1)
3. $\vdash \psi \lor \neg\varphi \to \neg\varphi \lor \psi$ (II.3)

4. ⊢ ¬¬ψ ∨ ¬φ → ¬φ ∨ ψ (MB, 2,3)
5. ⊢ (¬ψ → ¬φ) → (φ → ψ) (Def, 4)
6. ⊢ ¬φ → (¬ψ → ¬φ) (F.1)
7. ⊢ ¬φ → (φ → ψ) (MB, 6,5)

Now, if for some formula φ, both φ and $\neg\varphi$ were theorems, any other formula ψ would be derivable by applying the modus ponens twice on (A.7):

 From φ and $\neg\varphi$ any other formula ψ can be derived.

1. φ
2. $\neg\varphi$
3. ⊢ ¬φ → (φ → ψ) (A.7)
4. ⊢ φ → ψ (MP, 2,3)
5. ⊢ ψ (MP, 1,4)

4.4.2 Hypothesis-Based Proving

In this section, we will introduce a descriptive tool for presenting proof templates in a more compact and concise manner. The reduction builds upon the idea of permitting not only axioms and previously proven theorems within a proof chain but also allowing other formulas intended as *hypotheses*. The following example illustrates a proof involving a hypothesis:

The hypothesis $\forall x_1\, x_2(x_1)$ implies $\forall y_1\, x_2(y_1)$

1. $\forall x_1\, x_2(x_1)$ (Hyp)
2. ⊢ $\forall x_1\, x_2(x_1) \to x_2(y_1)$ (III.1)
3. ⊢ $x_2(y_1)$ (MP, 1,2)
4. ⊢ $\forall y_1\, x_2(y_1)$ (G, 3)

The formula $\forall y_1\, x_2(y_1)$ is clearly not a theorem of P since the proof relies on the assumption that $\forall x_1\, x_2(x_1)$ is a theorem. Strictly speaking, the following has been proven:

"If $\forall x_1\, x_2(x_1)$ *is a theorem, then* $\forall y_1\, x_2(y_1)$ *is also a theorem.*"

The implication operator describes such if-then relationships within the object language. Therefore, it is tempting to assume that the above derivation

4.4 Formal Proofs

sequence is convertable into a proof for the theorem

$$\forall x_1\, x_2(x_1) \to \forall y_1\, x_2(y_1).$$

With some effort, this is indeed possible:

$\forall x_1\, x_2(x_1) \to \forall y_1\, x_2(y_1)$

1.	$\vdash\ \forall x_1\, x_2(x_1) \to \forall x_1\, x_2(x_1)$	(H.0)
2.	$\vdash\ \forall x_1\, x_2(x_1) \to x_2(y_1)$	(III.1)
3.	$\vdash\ (\forall x_1\, x_2(x_1) \to x_2(y_1)) \to (\forall x_1\, x_2(x_1) \to (\forall x_1\, x_2(x_1) \to x_2(y_1)))$	(F.1)
4.	$\vdash\ \forall x_1\, x_2(x_1) \to (\forall x_1\, x_2(x_1) \to x_2(y_1))$	(MP, 2,3)
5.	$\vdash\ (\forall x_1\, x_2(x_1) \to (\forall x_1\, x_2(x_1) \to x_2(y_1))) \to$ $((\forall x_1\, x_2(x_1) \to \forall x_1\, x_2(x_1)) \to (\forall x_1\, x_2(x_1) \to x_2(y_1)))$	(F.2)
6.	$\vdash\ (\forall x_1\, x_2(x_1) \to \forall x_1\, x_2(x_1)) \to (\forall x_1\, x_2(x_1) \to x_2(y_1))$	(MP, 4,5)
7.	$\vdash\ \forall x_1\, x_2(x_1) \to x_2(y_1)$	(MP, 1,6)
8.	$\vdash\ \forall y_1\, (\forall x_1\, x_2(x_1) \to x_2(y_1))$	(G, 7)
9.	$\vdash\ \forall y_1\, (\neg \forall x_1\, x_2(x_1) \vee x_2(y_1))$	(Def, 8)
10.	$\vdash\ \forall y_1\, (\neg \forall x_1\, x_2(x_1) \vee x_2(y_1)) \to (\neg \forall x_1\, x_2(x_1) \vee \forall y_1\, x_2(y_1))$	(III.2)
11.	$\vdash\ \neg \forall x_1\, x_2(x_1) \vee \forall y_1\, x_2(y_1)$	(MP, 9,10)
12.	$\vdash\ \forall x_1\, x_2(x_1) \to \forall y_1\, x_2(y_1)$	(Def, 11)

The derivation is based on the idea of translating the original proof chain

$$\psi_1, \psi_2, \psi_3, \psi_4$$

into a chain of the following form:

$$\forall x_1\, x_2(x_1) \to \psi_1, \ldots, \forall x_1\, x_2(x_1) \to \psi_2, \ldots, \forall x_1\, x_2(x_1) \to \psi_3, \ldots, \forall x_1\, x_2(x_1) \to \psi_4$$

Proceeding from one line to the next requires several intermediate steps, each marked in gray within the proof. From the above example, we can readily observe how to construct these intermediate steps:

- The hypothesis φ is transformed into $\varphi \to \varphi$ via (H.0).
- Axioms are translated into the appropriate form via (F.1).
- The application of the modus ponens is simulated using (F.2).
- The application of the generalization rule is simulated using (III.2).

For the sake of clarity, we will adopt the following notation for proofs that depend on hypotheses in the described form:

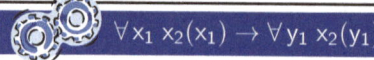
$\forall x_1\, x_2(x_1) \to \forall y_1\, x_2(y_1)$

1.	$\forall x_1\, x_2(x_1)$	(Hyp)
2.	$\vdash \forall x_1\, x_2(x_1) \to x_2(y_1)$	(III.1)
3.	$\vdash x_2(y_1)$	(MP, 1,2)
4.	$\vdash \forall y_1\, x_2(y_1)$	(G, 3)
5.	$\vdash \forall x_1\, x_2(x_1) \to \forall y_1\, x_2(y_1)$	(DT)

The acronym DT refers to the *Deduction Theorem*, a fundamental theorem in mathematical logic. It establishes a general relationship between the semantic inference relation and the syntactic implication operator, formally justifying the representation of a proof chain in the form utilized by our example. Further details about this theorem are provided in [70].

Earlier, we have stated that any formula qualifies as a potential hypothesis. We will now show that this is only partially correct. To quickly identify the issue, let's return to the derivation sequence from page 179. In line 10, we have generated the instance

$$\forall y_1\, (\neg \forall x_1\, x_2(x_1) \vee x_2(y_1)) \to (\neg \forall x_1\, x_2(x_1) \vee \forall y_1\, x_2(y_1))$$

from axiom schema (III.2) to shift the left universal quantifier to the right. We could only form this instance because the variable y_1 did not appear freely in the hypothesis. This observation reveals a significant limitation of hypothesis-based proofs: We must avoid binding variables via the generalization rule if they appear freely in the hypotheses. Only then is it possible to form the instances of (III.2) needed to transform the proof into a valid derivation sequence.

The following example illustrates the significance of this rule. When ignored, it is easy to prove substantively false statements:

A putative proof of $x_2(x_1) \to \forall x_1\, x_2(x_1)$

1.	$x_2(x_1)$	(Hyp)
2.	$\vdash \forall x_1\, x_2(x_1)$	(G, 1)
3.	$\vdash x_2(x_1) \to \forall x_1\, x_2(x_1)$	(DT)

At first glance, the proof appears correct, yet it produces the formula

$$x_2(x_1) \to \forall x_1\, x_2(x_1)$$

which we have already identified as substantively false on page 153. The error is quickly identified. The free variable x_1 of the hypothesis $x_2(x_1)$ gets bound in the second line via the generalization rule.

4.4 Formal Proofs

Next, we will explore a property of the formal system P that plays a crucial role in the main part of Gödel's proof. We already know that Gödel proves the incompleteness of P; in particular, he proves that for at least one formula φ, neither φ nor $\neg\varphi$ can be derived from the axioms. Thus, the following question arises: Can the formal system P become contradictory if one of the two formulas, for instance φ, is added to the axioms? The answer is no and readily justifiable with the knowledge acquired above.

The argument runs as follows: If the system extended by φ were contradictory, then the formula $\neg\varphi$ would be derivable, too. However, because $\neg\varphi$ was not derivable before, the proof must utilize the new axiom, implying that φ appears somewhere in the proof chain. If the formula φ is put to the front, then the proof must have the following general shape:

$$\varphi, \psi_1, \ldots, \psi_n, \neg\varphi \tag{4.21}$$

Similar to what we did above, we can translate this proof into a proof of the original (unextended) system by prefixing each member of the proof chain with φ as a hypothesis:

$$\varphi \rightarrow \varphi, \ldots, \varphi \rightarrow \psi_1, \ldots, \varphi \rightarrow \psi_n, \ldots, \varphi \rightarrow \neg\varphi \tag{4.22}$$

The result is a proof for $\varphi \rightarrow \neg\varphi$, which can be extended as follows:

 $\varphi \rightarrow \neg\varphi$ implies $\neg\varphi$

1. $\vdash \varphi \rightarrow \neg\varphi$
2. $\vdash \neg\varphi \vee \neg\varphi \rightarrow \neg\varphi$ (II.1)
3. $\vdash (\varphi \rightarrow \neg\varphi) \rightarrow \neg\varphi$ (Def, 2)
4. $\vdash \neg\varphi$ (MP, 1,3)

Consequently, $\neg\varphi$ would be a theorem of P, contrary to the assumption. Overall, this means that it is safe to add an undecidable formula to the axioms without endangering the consistency of the formal system:

Theorem 4.8

If φ is undecidable in P, then P remains consistent when supplementing the axioms with φ.

Note that Theorem 4.8 applies to any formula, but our argument is only correct if φ contains no free variables. If φ is an open formula, the proof is a bit more involved but follows the same line of reasoning. In particular, it must then be argued over the *universal closure* $\forall \xi \forall \zeta \ldots \varphi$ to transition from (4.21) to (4.22).

Table 4.3: Theorems of system P (continued)

	Theorem Group 2 (Predicate Logic)	
(P.1)	$\forall \xi \, (\varphi \to \psi) \to (\forall \xi \, \varphi \to \forall \xi \, \psi)$	
(P.2)	$\forall \xi \, (\varphi \to \psi) \to (\exists \xi \, \varphi \to \exists \xi \, \psi)$	
(P.3)	$\varphi[\xi \leftarrow \sigma] \to \exists \xi \, \varphi$	(if the substitution is collision-free)
(P.4)	$\varphi \to \exists \xi \, \varphi$	
(P.5)	$\forall \xi \, (\psi \lor \varphi) \to \forall \xi \, \psi \lor \varphi$	(if ξ does not occur freely in φ)
(P.6)	$\forall \xi \, (\psi \to \varphi) \to (\psi \to \forall \xi \, \varphi)$	(if ξ does not occur freely in ψ)
(P.7)	$\forall \xi \, (\psi \to \varphi) \to (\exists \xi \, \psi \to \varphi)$	(if ξ does not occur freely in ψ)
(P.8)	$\forall \xi \, \varphi \to \forall \zeta \, \varphi[\xi \leftarrow \zeta]$	(if the substitution is collision-free)
(P.9)	$\exists \xi \, \varphi \to \exists \zeta \, \varphi[\xi \leftarrow \zeta]$	(if the substitution is collision-free)
(P.10)	$x_2(0) \to (\forall x_1 \, (x_2(x_1) \to x_2(f \, x_1)) \to \forall x_1 \, x_2(x_1))$	

4.4.3 Predicate Logic Theorems

Next, we will prove a series of theorems from predicate logic, as summarized in Table 4.3.

$\forall \xi \, (\varphi \to \psi) \to (\forall \xi \, \varphi \to \forall \xi \, \psi)$ (P.1)

1.		$\forall \xi \, (\varphi \to \psi)$	(Hyp)
2.		$\forall \xi \, \varphi$	(Hyp)
3.	$\vdash$	$\forall \xi \, (\varphi \to \psi) \to (\varphi \to \psi)$	(III.1)
4.	$\vdash$	$\varphi \to \psi$	(MP, 1,3)
5.	$\vdash$	$\forall \xi \, \varphi \to \varphi$	(III.1)
6.	$\vdash$	φ	(MP, 2,5)
7.	$\vdash$	ψ	(MP, 6,4)
8.	$\vdash$	$\forall \xi \, \psi$	(G, 7)
9.	$\vdash$	$\forall \xi \, \varphi \to \forall \xi \, \psi$	(DT)
10.	$\vdash$	$\forall \xi \, (\varphi \to \psi) \to (\forall \xi \, \varphi \to \forall \xi \, \psi)$	(DT)

$\forall \xi \, (\varphi \to \psi) \to (\exists \xi \, \varphi \to \exists \xi \, \psi)$ (P.2)

1.		$\forall \xi \, (\varphi \to \psi)$	(Hyp)
2.	$\vdash$	$\forall \xi \, (\varphi \to \psi) \to (\varphi \to \psi)$	(III.1)
3.	$\vdash$	$\varphi \to \psi$	(MP, 1,2)

4.4 Formal Proofs

4. $\vdash \neg\psi \to \neg\varphi$	(INV, 3)
5. $\vdash \forall \xi\, (\neg\psi \to \neg\varphi)$	(G, 4)
6. $\vdash \forall \xi\, (\neg\psi \to \neg\varphi) \to (\forall \xi\, \neg\psi \to \forall \xi\, \neg\varphi)$	(P.1)
7. $\vdash \forall \xi\, \neg\psi \to \forall \xi\, \neg\varphi$	(MP, 5,6)
8. $\vdash \neg\forall \xi\, \neg\varphi \to \neg\forall \xi\, \neg\psi$	(INV, 7)
9. $\vdash \exists \xi\, \varphi \to \exists \xi\, \psi$	(Def, 8)
10. $\vdash \forall \xi\, (\varphi \to \psi) \to (\exists \xi\, \varphi \to \exists \xi\, \psi)$	(DT)

(P.1) and (P.2) justify two additional inference rules:

$$\dfrac{\forall \xi\,(\varphi \to \psi)}{\forall \xi\, \varphi \to \forall \xi\, \psi} \quad \text{(A)} \qquad\qquad \dfrac{\forall \xi\,(\varphi \to \psi)}{\exists \xi\, \varphi \to \exists \xi\, \psi} \quad \text{(E)}$$

$\varphi[\xi \leftarrow \sigma] \to \exists \xi\, \varphi$ (if the substitution is collision-free) (P.3)

1. $\vdash \forall \xi\, \neg\varphi \to \neg\varphi[\xi \leftarrow \sigma]$	(III.1)
2. $\vdash \neg\neg\varphi[\xi \leftarrow \sigma] \to \neg\forall \xi\, \neg\varphi$	(INV, 1)
3. $\vdash \neg\neg\varphi[\xi \leftarrow \sigma] \to \exists \xi\, \varphi$	(Def, 2)
4. $\vdash \varphi[\xi \leftarrow \sigma] \to \neg\neg\varphi[\xi \leftarrow \sigma]$	(H.4)
5. $\vdash \varphi[\xi \leftarrow \sigma] \to \exists \xi\, \varphi$	(MB, 4,3)

$\varphi \to \exists \xi\, \varphi$ (P.4)

1. $\vdash \varphi[\xi \leftarrow \xi] \to \exists \xi\, \varphi$	(P.3)
2. $\vdash \varphi \to \exists \xi\, \varphi$	($\varphi[\xi \leftarrow \xi] = \varphi$)

$\forall \xi\, (\psi \vee \varphi) \to \forall \xi\, \psi \vee \varphi$ (if ξ does not occur freely in φ) (P.5)

1. $\forall \xi\, (\psi \vee \varphi)$	(Hyp)
2. $\vdash \forall \xi\, (\psi \vee \varphi) \to \psi \vee \varphi$	(III.1)
3. $\vdash \psi \vee \varphi$	(MP, 1,2)
4. $\vdash \psi \vee \varphi \to \varphi \vee \psi$	(II.3)
5. $\vdash \varphi \vee \psi$	(MP, 3,4)
6. $\vdash \forall \xi\, (\varphi \vee \psi)$	(G, 5)
7. $\vdash \forall \xi\, (\varphi \vee \psi) \to \varphi \vee \forall \xi\, \psi$	(III.2)
8. $\vdash \varphi \vee \forall \xi\, \psi$	(MP, 6,7)
9. $\vdash \varphi \vee \forall \xi\, \psi \to \forall \xi\, \psi \vee \varphi$	(II.3)
10. $\vdash \forall \xi\, \psi \vee \varphi$	(MP, 8,9)

11. ⊢ $\forall \xi \, (\psi \vee \varphi) \to \forall \xi \, \psi \vee \varphi$ (DT)

$\forall \xi \, (\psi \to \varphi) \to (\psi \to \forall \xi \, \varphi)$ (if ξ does not occur freely in φ) (P.6)

1. ⊢ $\forall \xi \, (\neg \psi \vee \varphi) \to \neg \psi \vee \forall \xi \, \varphi$ (III.2)
2. ⊢ $\forall \xi \, (\psi \to \varphi) \to (\psi \to \forall \xi \, \varphi)$ (Def, 1)

$\forall \xi \, (\psi \to \varphi) \to (\exists \xi \, \psi \to \varphi)$ (if ξ does not occur freely in φ) (P.7)

1. $\forall \xi \, (\psi \to \varphi)$ (Hyp)
2. ⊢ $\forall \xi \, (\neg \psi \vee \varphi) \to \forall \xi \, \neg \psi \vee \varphi$ (P.5)
3. ⊢ $\forall \xi \, (\psi \to \varphi) \to \forall \xi \, \neg \psi \vee \varphi$ (Def, 2)
4. ⊢ $\forall \xi \, \neg \psi \vee \varphi$ (MP, 1,3)
5. ⊢ $\forall \xi \, \neg \psi \to \neg \neg \forall \xi \, \neg \psi$ (H.4)
6. ⊢ $\forall \xi \, \neg \psi \vee \varphi \to \neg \neg \forall \xi \, \neg \psi \vee \varphi$ (DR, 5)
7. ⊢ $\neg \neg \forall \xi \, \neg \psi \vee \varphi$ (MP, 4,6)
8. ⊢ $\neg \forall \xi \, \neg \psi \to \varphi$ (Def, 7)
9. ⊢ $\exists \xi \, \psi \to \varphi$ (Def, 8)
10. ⊢ $\forall \xi \, (\psi \to \varphi) \to (\exists \xi \, \psi \to \varphi)$ (DT)

The last three proven theorems are variants of axiom (III.2), permitting us to rearrange quantifiers under certain conditions. To represent proof chains more concisely, we will utilize the last two in the form of inference rules:

$$\frac{\forall \xi \, (\psi \to \varphi)}{\psi \to \forall \xi \, \varphi} \quad \text{(if } \xi \text{ does not occur freely in } \varphi\text{)} \qquad \text{(BA)}$$

$$\frac{\forall \xi \, (\psi \to \varphi)}{\exists \xi \, \psi \to \varphi} \quad \text{(if } \xi \text{ does not occur freely in } \varphi\text{)} \qquad \text{(BE)}$$

The following two theorems articulate the principle of *bounded renaming*, stating that we may replace a quantified variable with another as long as no collisions occur.

$\forall \xi \, \varphi \to \forall \zeta \, \varphi[\xi \leftarrow \zeta]$ (if the substitution is collision-free) (P.8)

1. $\forall \xi \, \varphi$ (Hyp)
2. ⊢ $\forall \xi \, \varphi \to \varphi[\xi \leftarrow \zeta]$ (III.1)
3. ⊢ $\varphi[\xi \leftarrow \zeta]$ (MP, 1,2)
4. ⊢ $\forall \zeta \, \varphi[\xi \leftarrow \zeta]$ (G, 3)
5. ⊢ $\forall \xi \, \varphi \to \forall \zeta \, \varphi[\xi \leftarrow \zeta]$ (DT)

4.4 Formal Proofs

> $\exists \xi\, \varphi \to \exists \zeta\, \varphi[\xi \leftarrow \zeta]$ (if the substitution is collision-free) (P.9)

1. ⊢ $\forall \zeta\, \neg\varphi[\xi \leftarrow \zeta] \to \forall \xi\, \neg\varphi$ (P.8)
2. ⊢ $\neg\forall \xi\, \neg\varphi \to \neg\forall \zeta\, \neg\varphi[\xi \leftarrow \zeta]$ (INV)
3. ⊢ $\exists \xi\, \varphi \to \exists \zeta\, \varphi[\xi \leftarrow \zeta]$ (Def, 2)

The final theorem provides an alternative formulation of the principle of induction, eliminating the need for the conjunction operator '∧':

> $x_2(0) \to (\forall x_1\, (x_2(x_1) \to x_2(f\, x_1)) \to \forall x_1\, x_2(x_1))$ (P.10)

1. ⊢ $x_2(0) \land \forall x_1\, (x_2(x_1) \to x_2(f\, x_1)) \to \forall x_1\, x_2(x_1)$ (I.3)
2. ⊢ $(x_2(0) \land \forall x_1\, (x_2(x_1) \to x_2(f\, x_1)) \to \forall x_1\, x_2(x_1)) \to$
   $(x_2(0) \to (\forall x_1\, (x_2(x_1) \to x_2(f\, x_1)) \to \forall x_1\, x_2(x_1)))$ (A.2)
3. ⊢ $x_2(0) \to (\forall x_1\, (x_2(x_1) \to x_2(f\, x_1)) \to \forall x_1\, x_2(x_1))$ (MP, 1,2)

4.4.4 Theorems About Equality

In this section, we will prove the theorems listed in Table 4.4, all dealing with equality. Always keep in mind that the equality operator is not a native language element of P but merely the abbreviation for the following expression:

$$\xi_i = \zeta_i \;:=\; \forall x_{i+1}\, (x_{i+1}(\xi_i) \to x_{i+1}(\zeta_i)) \tag{4.23}$$

We start by proving this definition equivalent to the formula obtained from Leibniz's Principle of Identity on page 155:

$$\xi_i = \zeta_i \;:=\; \forall x_{i+1}\, (x_{i+1}(\xi_i) \leftrightarrow x_{i+1}(\zeta_i))$$

To establish equivalence, it has to be shown that the implication operator in (4.23) is reversible. The following theorem reveals that this is indeed the case.

> $\xi_i = \zeta_i \to \forall x_{i+1}\, (x_{i+1}(\zeta_i) \to x_{i+1}(\xi_i))$ (G.1)

1. $\xi_1 = \zeta_1$ (Hyp)
2. ⊢ $\forall x_2\, (x_2(\xi_1) \to x_2(\zeta_1))$ (Def, 1)
3. ⊢ $\forall x_2\, (x_2(\xi_1) \to x_2(\zeta_1)) \to (y_2(\xi_1) \to y_2(\zeta_1))$ (III.1)
4. ⊢ $y_2(\xi_1) \to y_2(\zeta_1)$ (MP, 2,3)
5. $\forall \xi_1\, (y_2(\xi_1) \leftrightarrow \neg x_2(\xi_1))$ (Hyp)

Table 4.4: Theorems of system P (continued)

	Theorem Group 3: Equality
(G.1)	$\xi_i = \zeta_i \to \forall x_{i+1} \, (x_{i+1}(\zeta_i) \to x_{i+1}(\xi_i))$
(G.2)	$\sigma_i = \sigma_i$
(PM 13.16)	$\sigma_i = \tau_i \to \tau_i = \sigma_i$
(PM 13.17)	$\sigma_i = \tau_i \to (\tau_i = \rho_i \to \sigma_i = \rho_i)$
(G.3)	$\sigma_i = \tau_i \to (\sigma_i = \rho_i \to \tau_i = \rho_i)$
(G.4)	$\sigma_i = \tau_i \to (\rho_i = \tau_i \to \sigma_i = \rho_i)$
(G.5)	$\sigma_i = \tau_i \to (\tau_i \neq \rho_i \to \sigma_i \neq \rho_i)$
(G.6)	$\sigma_i = \tau_i \to (\sigma_i \neq \rho_i \to \tau_i \neq \rho_i)$
(G.7)	$\sigma_i = \tau_i \to (\rho_i \neq \tau_i \to \sigma_i \neq \rho_i)$
(G.8)	$\sigma_1 = \tau_1 \to f\,\sigma_1 = f\,\tau_1$

6. $\vdash \forall \xi_1 \, ((y_2(\xi_1) \to \neg x_2(\xi_1)) \land (\neg x_2(\xi_1) \to y_2(\xi_1)))$ \hfill (Def, 5)
7. $\vdash \forall \xi_1 \, ((y_2(\xi_1) \to \neg x_2(\xi_1)) \land (\neg x_2(\xi_1) \to y_2(\xi_1))) \to$
   $\quad ((y_2(\xi_1) \to \neg x_2(\xi_1)) \land (\neg x_2(\xi_1) \to y_2(\xi_1)))$ \hfill (III.1)
8. $\vdash (y_2(\xi_1) \to \neg x_2(\xi_1)) \land (\neg x_2(\xi_1) \to y_2(\xi_1))$ \hfill (MP, 6,7)
9. $\vdash (y_2(\xi_1) \to \neg x_2(\xi_1)) \land (\neg x_2(\xi_1) \to y_2(\xi_1)) \to (\neg x_2(\xi_1) \to y_2(\xi_1))$
   \hfill (H.13)
10. $\vdash \neg x_2(\xi_1) \to y_2(\xi_1)$ \hfill (MP, 8,9)
11. $\vdash \neg x_2(\xi_1) \to y_2(\zeta_1)$ \hfill (MB, 10,4)
12. $\vdash \forall \xi_1 \, ((y_2(\xi_1) \to \neg x_2(\xi_1)) \land (\neg x_2(\xi_1) \to y_2(\xi_1))) \to$
    $\quad ((y_2(\zeta_1) \to \neg x_2(\zeta_1)) \land (\neg x_2(\zeta_1) \to y_2(\zeta_1)))$ \hfill (III.1)
13. $\vdash (y_2(\zeta_1) \to \neg x_2(\zeta_1)) \land (\neg x_2(\zeta_1) \to y_2(\zeta_1))$ \hfill (MP, 6,12)
14. $\vdash (y_2(\zeta_1) \to \neg x_2(\zeta_1)) \land (\neg x_2(\zeta_1) \to y_2(\zeta_1)) \to (y_2(\zeta_1) \to \neg x_2(\zeta_1))$
    \hfill (H.12)
15. $\vdash y_2(\zeta_1) \to \neg x_2(\zeta_1)$ \hfill (MP, 13,14)
16. $\vdash \neg x_2(\xi_1) \to \neg x_2(\zeta_1)$ \hfill (MB, 11,15)
17. $\vdash x_2(\zeta_1) \to x_2(\xi_1)$ \hfill (INV, 16)
18. $\vdash \forall x_2 \, (x_2(\zeta_1) \to x_2(\xi_1))$ \hfill (G, 17)
19. $\vdash \forall \xi_1 \, (y_2(\xi_1) \leftrightarrow \neg x_2(\xi_1)) \to \forall x_2 \, (x_2(\zeta_1) \to x_2(\xi_1))$ \hfill (DT)
20. $\vdash \forall y_2 \, (\forall \xi_1 \, (y_2(\xi_1) \leftrightarrow \neg x_2(\xi_1)) \to \forall x_2 \, (x_2(\zeta_1) \to x_2(\xi_1)))$ \hfill (G, 19)
21. $\vdash \forall y_2 \, (\neg \forall \xi_1 \, (y_2(\xi_1) \leftrightarrow \neg x_2(\xi_1)) \lor \forall x_2 \, (x_2(\zeta_1) \to x_2(\xi_1)))$ \hfill (Def, 20)
22. $\vdash \forall y_2 \, (\neg \forall \xi_1 \, (y_2(\xi_1) \leftrightarrow \neg x_2(\xi_1)) \lor \forall x_2 \, (x_2(\zeta_1) \to x_2(\xi_1))) \to$
    $\quad \forall y_2 \, \neg \forall \xi_1 \, (y_2(\xi_1) \leftrightarrow \neg x_2(\xi_1)) \lor \forall x_2 \, (x_2(\zeta_1) \to x_2(\xi_1))$ \hfill (P.5)

4.4 Formal Proofs

23. ⊢	$\forall y_2 \neg \forall \xi_1 \, (y_2(\xi_1) \leftrightarrow \neg x_2(\xi_1)) \vee \forall x_2 \, (x_2(\zeta_1) \to x_2(\xi_1))$	(MP, 21,22)
24. ⊢	$\neg \forall y_2 \neg \forall \xi_1 \, (y_2(\xi_1) \leftrightarrow \neg x_2(\xi_1)) \to \forall x_2 \, (x_2(\zeta_1) \to x_2(\xi_1))$	(Def, 23)
25. ⊢	$\exists y_2 \forall \xi_1 \, (y_2(\xi_1) \leftrightarrow \neg x_2(\xi_1)) \to \forall x_2 \, (x_2(\zeta_1) \to x_2(\xi_1))$	(Def, 24)
26. ⊢	$\exists y_2 \forall \xi_1 \, (y_2(\xi_1) \leftrightarrow \neg x_2(\xi_1))$	(IV.1)
27. ⊢	$\forall x_2 \, (x_2(\zeta_1) \to x_2(\xi_1))$	(MP, 26,25)
28. ⊢	$\xi_1 = \zeta_1 \to \forall x_2 \, (x_2(\zeta_1) \to x_2(\xi_1))$	(DT)

The proof proceeds analogously for the higher types.

$\sigma_i = \sigma_i$ (G.2)

1. ⊢	$x_2(\sigma_1) \to x_2(\sigma_1)$	(H.0)
2. ⊢	$\forall x_2 \, (x_2(\sigma_1) \to x_2(\sigma_1))$	(G, 1)
3. ⊢	$\sigma_1 = \sigma_1$	(Def, 2)

The proof proceeds analogously for the higher types.

The following two theorems are part of the first volume of the *Principia Mathematica*. They postulate the symmetry and transitivity of the equality relation.

$\sigma_i = \tau_i \to \tau_i = \sigma_i$ (PM 13.16)

1.	$\sigma_1 = \tau_1$	(Hyp)
2. ⊢	$\sigma_1 = \tau_1 \to \forall x_2 \, (x_2(\tau_1) \to x_2(\sigma_1))$	(G.1)
3. ⊢	$\forall x_2 \, (x_2(\tau_1) \to x_2(\sigma_1))$	(MP, 1,2)
4. ⊢	$\tau_1 = \sigma_1$	(Def, 3)
5. ⊢	$\sigma_1 = \tau_1 \to \tau_1 = \sigma_1$	(DT)

The proof proceeds analogously for the higher types.

$\sigma_i = \tau_i \to (\tau_i = \rho_i \to \sigma_i = \rho_i)$ (PM 13.17)

1.	$\sigma_1 = \tau_1$	(Hyp)
2.	$\tau_1 = \rho_1$	(Hyp)
3. ⊢	$\forall x_2 \, (x_2(\sigma_1) \to x_2(\tau_1))$	(Def, 1)
4. ⊢	$\forall x_2 \, (x_2(\sigma_1) \to x_2(\tau_1)) \to (x_2(\sigma_1) \to x_2(\tau_1))$	(III.1)
5. ⊢	$x_2(\sigma_1) \to x_2(\tau_1)$	(MP, 3,4)
6. ⊢	$\forall x_2 \, (x_2(\tau_1) \to x_2(\rho_1))$	(Def, 2)
7. ⊢	$\forall x_2 \, (x_2(\tau_1) \to x_2(\rho_1)) \to (x_2(\tau_1) \to x_2(\rho_1))$	(III.1)

8. $\vdash x_2(\tau_1) \to x_2(\rho_1)$		(MP, 6,7)
9. $\vdash x_2(\sigma_1) \to x_2(\rho_1)$		(MB, 5,8)
10. $\vdash \forall x_2 \, (x_2(\sigma_1) \to x_2(\rho_1))$		(G, 9)
11. $\vdash \sigma_1 = \rho_1$		(Def, 10)
12. $\vdash \tau_1 = \rho_1 \to \sigma_1 = \rho_1$		(DT)
13. $\vdash \sigma_1 = \tau_1 \to (\tau_1 = \rho_1 \to \sigma_1 = \rho_1)$		(DT)

The proof proceeds analogously for the higher types.

⚙️ $\sigma_i = \tau_i \to (\sigma_i = \rho_i \to \tau_i = \rho_i)$ (G.3)

1. $\vdash \tau_i = \sigma_i \to (\sigma_i = \rho_i \to \tau_i = \rho_i)$ (PM 13.17)
2. $\vdash \sigma_i = \tau_i \to \tau_i = \sigma_i$ (PM 13.16)
3. $\vdash \sigma_i = \tau_i \to (\sigma_i = \rho_i \to \tau_i = \rho_i)$ (MB, 2,1)

⚙️ $\sigma_i = \tau_i \to (\rho_i = \tau_i \to \sigma_i = \rho_i)$ (G.4)

1. $\vdash \sigma_i = \tau_i \to (\tau_i = \rho_i \to \sigma_i = \rho_i)$ (PM 13.17)
2. $\vdash \tau_i = \rho_i \to (\sigma_i = \tau_i \to \sigma_i = \rho_i)$ (ER, 1)
3. $\vdash \rho_i = \tau_i \to \tau_i = \rho_i$ (PM 13.16)
4. $\vdash \rho_i = \tau_i \to (\sigma_i = \tau_i \to \sigma_i = \rho_i)$ (MB, 3,2)
5. $\vdash \sigma_i = \tau_i \to (\rho_i = \tau_i \to \sigma_i = \rho_i)$ (ER, 4)

The three most recently proven theorems justify the following inference rules, all labeled with the same abbreviation (GL) due to their structural similarity:

$$\frac{\sigma_i = \tau_i \quad \tau_i = \rho_i}{\sigma_i = \rho_i} \qquad \frac{\sigma_i = \tau_i \quad \sigma_i = \rho_i}{\tau_i = \rho_i} \qquad \frac{\sigma_i = \tau_i \quad \rho_i = \tau_i}{\sigma_i = \rho_i} \qquad \text{(GL)}$$

The subsequent three theorems appear nearly identical, differing only in using the crossed-out equal sign instead of the standard equality operator. Formally, $\varphi \neq \psi$ is the abbreviation for $\neg(\varphi = \psi)$.

⚙️ $\sigma_i = \tau_i \to (\tau_i \neq \rho_i \to \sigma_i \neq \rho_i)$ (G.5)

1. $\sigma_i = \tau_i$ (Hyp)
2. $\sigma_i = \rho_i$ (Hyp)
3. $\vdash \tau_i = \rho_i$ (GL, 1,2)
4. $\vdash \sigma_i = \rho_i \to \tau_i = \rho_i$ (DT)

4.4 Formal Proofs

5. $\vdash \tau_i \neq \rho_i \to \sigma_i \neq \rho_i$ (INV, 4)
6. $\vdash \sigma_i = \tau_i \to (\tau_i \neq \rho_i \to \sigma_i \neq \rho_i)$ (DT)

$$\sigma_i = \tau_i \to (\sigma_i \neq \rho_i \to \tau_i \neq \rho_i) \qquad (G.6)$$

1. $\sigma_i = \tau_i$ (Hyp)
2. $\tau_i = \rho_i$ (Hyp)
3. $\vdash \sigma_i = \rho_i$ (GL, 1,2)
4. $\vdash \tau_i = \rho_i \to \sigma_i = \rho_i$ (DT)
5. $\vdash \sigma_i \neq \rho_i \to \tau_i \neq \rho_i$ (INV, 4)
6. $\vdash \sigma_i = \tau_i \to (\sigma_i \neq \rho_i \to \tau_i \neq \rho_i)$ (DT)

$$\sigma_i = \tau_i \to (\rho_i \neq \tau_i \to \sigma_i \neq \rho_i) \qquad (G.7)$$

1. $\sigma_i = \tau_i$ (Hyp)
2. $\sigma_i = \rho_i$ (Hyp)
3. $\vdash \rho_i = \tau_i$ (GL, 1,2)
4. $\vdash \sigma_i = \rho_i \to \rho_i = \tau_i$ (DT)
5. $\vdash \rho_i \neq \tau_i \to \sigma_i \neq \rho_i$ (INV, 4)
6. $\vdash \sigma_i = \tau_i \to (\rho_i \neq \tau_i \to \sigma_i \neq \rho_i)$ (DT)

The proven theorems lead straight to the following inference rules:

$$\frac{\sigma_i = \tau_i \quad \tau_i \neq \rho_i}{\sigma_i \neq \rho_i} \qquad \frac{\sigma_i = \tau_i \quad \sigma_i \neq \rho_i}{\tau_i \neq \rho_i} \qquad \frac{\sigma_i = \tau_i \quad \rho_i \neq \tau_i}{\sigma_i \neq \rho_i} \qquad \text{(UG)}$$

The next theorem is the reversal of the second Peano axiom (I.2) with the following substantive meaning: If two numbers are equal, so are their successors.

$$\sigma_1 = \tau_1 \to (f\ \sigma_1 = f\ \tau_1) \qquad (G.8)$$

1. $\xi_1 = \zeta_1$ (Hyp)
2. $\vdash \forall x_2\ (x_2(\xi_1) \to x_2(\zeta_1))$ (Def, 1)
3. $\vdash \forall x_2\ (x_2(\xi_1) \to x_2(\zeta_1)) \to (y_2(\xi_1) \to y_2(\zeta_1))$ (III.1)
4. $\vdash y_2(\xi_1) \to y_2(\zeta_1)$ (MP, 2,3)
5. $\forall \xi_1\ (y_2(\xi_1) \leftrightarrow x_2(f\ \xi_1))$ (Hyp)
6. $\vdash \forall \xi_1\ ((y_2(\xi_1) \to x_2(f\ \xi_1)) \wedge (x_2(f\ \xi_1) \to y_2(\xi_1)))$ (Def, 5)
7. $\vdash \forall \xi_1\ ((y_2(\xi_1) \to x_2(f\ \xi_1)) \wedge (x_2(f\ \xi_1) \to y_2(\xi_1))) \to$

$$((y_2(\xi_1) \to x_2(f\ \xi_1)) \wedge (x_2(f\ \xi_1) \to y_2(\xi_1)))) \quad \text{(III.1)}$$

8. $\vdash\ (y_2(\xi_1) \to x_2(f\ \xi_1)) \wedge (x_2(f\ \xi_1) \to y_2(\xi_1))$ \hfill (MP, 6,7)
9. $\vdash\ (y_2(\xi_1) \to x_2(f\ \xi_1)) \wedge (x_2(f\ \xi_1) \to y_2(\xi_1)) \to (x_2(f\ \xi_1) \to y_2(\xi_1))$ \hfill (H.13)
10. $\vdash\ x_2(f\ \xi_1) \to y_2(\xi_1)$ \hfill (MP, 8,9)
11. $\vdash\ x_2(f\ \xi_1) \to y_2(\varsigma_1)$ \hfill (MB, 10,4)
12. $\vdash\ \forall \xi_1\ ((y_2(\xi_1) \to x_2(f\ \xi_1)) \wedge (x_2(f\ \xi_1) \to y_2(\xi_1))) \to$
    $((y_2(\varsigma_1) \to x_2(f\ \varsigma_1)) \wedge (x_2(f\ \varsigma_1) \to y_2(\varsigma_1)))$ \hfill (III.1)
13. $\vdash\ (y_2(\varsigma_1) \to x_2(f\ \varsigma_1)) \wedge (x_2(f\ \varsigma_1) \to y_2(\varsigma_1))$ \hfill (MP, 6,12)
14. $\vdash\ (y_2(\varsigma_1) \to x_2(f\ \varsigma_1)) \wedge (x_2(f\ \varsigma_1) \to y_2(\varsigma_1)) \to (y_2(\varsigma_1) \to x_2(f\ \varsigma_1))$ \hfill (H.12)
15. $\vdash\ y_2(\varsigma_1) \to x_2(f\ \varsigma_1)$ \hfill (MP, 13,14)
16. $\vdash\ x_2(f\ \xi_1) \to x_2(f\ \varsigma_1)$ \hfill (MB, 11,15)
17. $\vdash\ \forall x_2\ (x_2(f\ \xi_1) \to x_2(f\ \varsigma_1))$ \hfill (G, 16)
18. $\vdash\ \forall \xi_1\ (y_2(\xi_1) \leftrightarrow x_2(f\ \xi_1)) \to \forall x_2\ (x_2(f\ \xi_1) \to x_2(f\ \varsigma_1))$ \hfill (DT)
19. $\vdash\ \forall y_2\ (\forall \xi_1\ (y_2(\xi_1) \leftrightarrow x_2(f\ \xi_1)) \to \forall x_2\ (x_2(f\ \xi_1) \to x_2(f\ \varsigma_1)))$ \hfill (G, 18)
20. $\vdash\ \forall y_2\ (\neg \forall \xi_1\ (y_2(\xi_1) \leftrightarrow x_2(f\ \xi_1)) \vee \forall x_2\ (x_2(f\ \xi_1) \to x_2(f\ \varsigma_1)))$ \hfill (Def, 19)
21. $\vdash\ \forall y_2\ (\neg \forall \xi_1\ (y_2(\xi_1) \leftrightarrow x_2(f\ \xi_1)) \vee \forall x_2\ (x_2(f\ \xi_1) \to x_2(f\ \varsigma_1))) \to$
    $\forall y_2\ \neg \forall \xi_1\ (y_2(\xi_1) \leftrightarrow x_2(f\ \xi_1)) \vee \forall x_2\ (x_2(f\ \xi_1) \to x_2(f\ \varsigma_1))$ \hfill (P.5)
22. $\vdash\ \forall y_2\ \neg \forall \xi_1\ (y_2(\xi_1) \leftrightarrow x_2(f\ \xi_1)) \vee \forall x_2\ (x_2(f\ \xi_1) \to x_2(f\ \varsigma_1))$ \hfill (MP, 20,21)
23. $\vdash\ \neg \forall y_2\ \neg \forall \xi_1\ (y_2(\xi_1) \leftrightarrow x_2(f\ \xi_1)) \to \forall x_2\ (x_2(f\ \xi_1) \to x_2(f\ \varsigma_1))$ \hfill (Def, 22)
24. $\vdash\ \exists y_2\ \forall \xi_1\ (y_2(\xi_1) \leftrightarrow x_2(f\ \xi_1)) \to \forall x_2\ (x_2(f\ \xi_1) \to x_2(f\ \varsigma_1))$ \hfill (Def, 23)
25. $\vdash\ \exists y_2\ \forall \xi_1\ (y_2(\xi_1) \leftrightarrow x_2(f\ \xi_1))$ \hfill (IV.1)
26. $\vdash\ \forall x_2\ (x_2(f\ \xi_1) \to x_2(f\ \varsigma_1))$ \hfill (MP, 25,24)
27. $\vdash\ \xi_1 = \varsigma_1 \to \forall x_2\ (x_2(f\ \xi_1) \to x_2(f\ \varsigma_1))$ \hfill (DT)
28. $\vdash\ \xi_1 = \varsigma_1 \to f\ \xi_1 = f\ \varsigma_1$ \hfill (Def, 27)

4.4.5 Numerical Theorems

Before moving on with the proof of numerical theorems, let us focus on a previously unaddressed footnote on page 140. The focus of our interest is Gödel's footnote 16:

---

[16]) The addition of Peano's axioms, as well as all other changes made in the system PM, serve only to simplify the proof and are theoretically dispensable.

In this footnote, Gödel highlights a property of system P that we have already exploited in the context of the deduction theorem. We have demonstrated

4.4 Formal Proofs

how to transform a derivation sequence that utilizes a hypothesis ψ to prove a formula φ into a standard derivation sequence for the formula

$$\psi \rightarrow \varphi.$$

Similarly, we can translate any derivation sequence that proves a formula φ by using the Peano axioms (I.1) to (I.3) into a derivation sequence for the formula

$$(I.1) \wedge (I.2) \wedge (I.3) \rightarrow \varphi.$$

Gödel refers to this property when he says that the addition of Peano's axioms, as well as all other changes made in the system PM, serve only to simplify the proof and are theoretically dispensable.

Similarly, non-native operations, such as addition or multiplication, can be defined within P. To see how, let us assume that P provides function symbols alongside predicates. Then, we can consider an expression such as

$$x \times \overline{2} = x + x$$

as a placeholder for the formula

$$\varphi_{\text{ADD.1}} \wedge \varphi_{\text{ADD.2}} \wedge \varphi_{\text{MUL.1}} \wedge \varphi_{\text{MUL.2}} \rightarrow x \times \overline{2} = x + x \qquad (4.24)$$

where $\varphi_{\text{ADD.1}}$ to $\varphi_{\text{MUL.2}}$ are defined as follows:

$$\sigma + 0 = \sigma \qquad \text{(ADD.1)}$$
$$\sigma + \text{f}\,\tau = \text{f}\,(\sigma + \tau) \qquad \text{(ADD.2)}$$

$$\sigma \times 0 = 0 \qquad \text{(MUL.1)}$$
$$\sigma \times \text{f}\,\tau = (\sigma \times \tau) + \sigma \qquad \text{(MUL.2)}$$

In practice, however, we would soon run into problems. For one thing, we would have to consistently deal with complex formulas, and each of those formulas needed to be laboriously dissected before an interesting proof step could be carried out. For another thing, P does not offer freely definable function symbols. Hence, we cannot write down formula (4.24) in the given form. We would be compelled to encode the functions as relations, further complicating our task.

We will take a path reminiscent of Peano's *Arithmetices Principia* to work around the technical difficulties and make addition and multiplication integral parts of our formal language. In particular, we will treat the symbols '+' and '×' as native language elements and utilize the formulas (ADD.1), (ADD.2), (MUL.1), and (MUL.2) as additional axioms. The resulting formal system will be referred to as system P' to distinguish it from Gödel's system P formally.

Table 4.5: Theorems of system P'

	Theorem Group 4: Numerics
(M.3.2e)	$\sigma_1 = \tau_1 \to \sigma_1 + \rho_1 = \tau_1 + \rho_1$
(M.3.2f)	$\sigma_1 = 0 + \sigma_1$
(M.3.2g)	$f\,\sigma_1 + \tau_1 = f\,(\sigma_1 + \tau_1)$
(M.3.2h)	$\sigma_1 + \tau_1 = \tau_1 + \sigma_1$
(N.1)	$\sigma_1 = \tau_1 \to \sigma_1 + \sigma_1 = \tau_1 + \tau_1$
(M.3.5b)	$\sigma_1 \times \overline{1} = \sigma_1$
(M.3.5c)	$\sigma_1 \times \overline{2} = \sigma_1 + \sigma_1$
(N.2)	$f\,f\,(\sigma_1 \times \overline{2}) = f\,\sigma_1 + f\,\sigma_1$
(N.3)	$f\,f\,(\sigma_1 \times \overline{2}) = (f\,\sigma_1) \times \overline{2}$
(M.3.5h)	$\sigma_1 \ne 0 \to \exists z_1\,(\sigma_1 = f\,z_1)$
(N.4)	$\overline{1} \ne \sigma_1 \times \overline{2}$
(N.5)	$\sigma_1 = 0 \to \sigma_1 \times \overline{2} = 0$
(N.6)	$f\,f\,\sigma = \tau_1 \times \overline{2} \to \tau_1 \ne 0$
(N.7)	$\exists z_1\,\sigma_1 = z_1 \times \overline{2} \to \exists z_1\,f\,f\,\sigma_1 = z_1 \times \overline{2}$
(N.8)	$\exists z_1\,f\,f\,\sigma_1 = z_1 \times \overline{2} \to \exists z_1\,\sigma_1 = z_1 \times \overline{2}$

The theorems we aim to prove in P' are summarized in Table 4.5. All theorems marked with (M.x) are taken from [70], and the derivation sequences are adaptations of the proofs presented in this book.

$$\sigma_1 = \tau_1 \to \sigma_1 + \rho_1 = \tau_1 + \rho_1 \qquad \text{(M.3.2e)}$$

Let $\psi(\xi_1) := (\sigma_1 = \tau_1 \to \sigma_1 + \xi_1 = \tau_1 + \xi_1)$

1.		$\sigma_1 = \tau_1$	(Hyp)
2.	$\vdash$	$\sigma_1 + 0 = \sigma_1$	(ADD.1)
3.	$\vdash$	$\tau_1 + 0 = \tau_1$	(ADD.1)
4.	$\vdash$	$\sigma_1 + 0 = \tau_1$	(GL, 2,1)
5.	$\vdash$	$\sigma_1 + 0 = \tau_1 + 0$	(GL, 4,3)
6.	$\vdash$	$\sigma_1 = \tau_1 \to \sigma_1 + 0 = \tau_1 + 0$	(DT)
7.	$\vdash$	$\psi(0)$ *At this point, the induction start is proven*	(Def, 6)
8.		$\sigma_1 = \tau_1 \to \sigma_1 + \xi_1 = \tau_1 + \xi_1$	(Hyp)
9.		$\sigma_1 = \tau_1$	(Hyp)

4.4 Formal Proofs

10. ⊢	$\sigma_1 + \xi_1 = \tau_1 + \xi_1$	(MP, 9,8)
11. ⊢	$\sigma_1 + f\,\xi_1 = f\,(\sigma_1 + \xi_1)$	(ADD.2)
12. ⊢	$\tau_1 + f\,\xi_1 = f\,(\tau_1 + \xi_1)$	(ADD.2)
13. ⊢	$\sigma_1 + \xi_1 = \tau_1 + \xi_1 \to f\,(\sigma_1 + \xi_1) = f\,(\tau_1 + \xi_1)$	(G.8)
14. ⊢	$f\,(\sigma_1 + \xi_1) = f\,(\tau_1 + \xi_1)$	(MP, 10,13)
15. ⊢	$\sigma_1 + f\,\xi_1 = f\,(\sigma_1 + \xi_1) \to$	
	$\quad (f\,(\sigma_1 + \xi_1) = f\,(\tau_1 + \xi_1) \to \sigma_1 + f\,\xi_1 = f\,(\tau_1 + \xi_1))$	(PM 13.17)
16. ⊢	$f\,(\sigma_1 + \xi_1) = f\,(\tau_1 + \xi_1) \to \sigma_1 + f\,\xi_1 = f\,(\tau_1 + \xi_1)$	(MP, 11,15)
17. ⊢	$\sigma_1 + f\,\xi_1 = f\,(\tau_1 + \xi_1)$	(MP, 14,16)
18. ⊢	$\sigma_1 + f\,\xi_1 = \tau_1 + f\,\xi_1$	(GL, 17,12)
19. ⊢	$\sigma_1 = \tau_1 \to \sigma_1 + f\,\xi_1 = \tau_1 + f\,\xi_1$	(DT)
20. ⊢	$\psi(f\,\xi_1)$	(Def, 19)
21. ⊢	$(\sigma_1 = \tau_1 \to \sigma_1 + \xi_1 = \tau_1 + \xi_1) \to \psi(f\,\xi_1)$	(DT)
22. ⊢	$\psi(\xi_1) \to \psi(f\,\xi_1)$	(Def, 21)
23. ⊢	$\forall \xi_1\,(\psi(\xi_1) \to \psi(f\,\xi_1))$	(G, 22)

☛ *At this point, the induction step is proven*

24. ⊢	$\psi(0) \to (\forall \xi_1\,(\psi(\xi_1) \to \psi(f\,\xi_1)) \to \forall \xi_1\,\psi(\xi_1))$	(P.10)
25. ⊢	$\forall \xi_1\,(\psi(\xi_1) \to \psi(f\,\xi_1)) \to \forall \xi_1\,\psi(\xi_1)$	(MP, 7,24)
26. ⊢	$\forall \xi_1\,\psi(\xi_1)$	(MP, 23,25)
27. ⊢	$\forall \xi_1\,(\sigma_1 = \tau_1 \to \sigma_1 + \xi_1 = \tau_1 + \xi_1)$	(Def, 26)
28. ⊢	$\forall \xi_1\,(\sigma_1 = \tau_1 \to \sigma_1 + \xi_1 = \tau_1 + \xi_1) \to$	
	$\quad (\sigma_1 = \tau_1 \to \sigma_1 + \rho_1 = \tau_1 + \rho_1)$	(III.1)
29. ⊢	$\sigma_1 = \tau_1 \to \sigma_1 + \rho_1 = \tau_1 + \rho_1$	(MP, 27,28)

$\sigma_1 = 0 + \sigma_1$ (M.3.2f)

Let $\psi(\xi_1) := (\xi_1 = 0 + \xi_1)$

1. ⊢	$0 + 0 = 0$	(ADD.1)
2. ⊢	$0 + 0 = 0 \to 0 = 0 + 0$	(PM 13.16)
3. ⊢	$0 = 0 + 0$	(MP, 1,2)
4. ⊢	$\psi(0)$	(Def, 3)

☛ *At this point, the induction start is proven*

5.	$\xi_1 = 0 + \xi_1$	(Hyp)
6. ⊢	$\psi(\xi_1)$	(Def, 5)
7. ⊢	$0 + f\,\xi_1 = f\,(0 + \xi_1)$	(ADD.2)
8. ⊢	$\xi_1 = 0 + \xi_1 \to f\,\xi_1 = f\,(0 + \xi_1)$	(G.8)
9. ⊢	$f\,\xi_1 = f\,(0 + \xi_1)$	(MP, 5,8)
10. ⊢	$f\,\xi_1 = f\,(0 + \xi_1) \to (0 + f\,\xi_1 = f\,(0 + \xi_1) \to f\,\xi_1 = 0 + f\,\xi_1)$	(G.4)

11. ⊢ $0 + f\,\xi_1 = f\,(0 + \xi_1) \to f\,\xi_1 = 0 + f\,\xi_1$ (MP, 9,10)
12. ⊢ $f\,\xi_1 = 0 + f\,\xi_1$ (GL, 9,7)
13. ⊢ $\xi_1 = 0 + \xi_1 \to f\,\xi_1 = 0 + f\,\xi_1$ (DT)
14. ⊢ $\forall \xi_1\,(\xi_1 = 0 + \xi_1 \to f\,\xi_1 = 0 + f\,\xi_1)$ (G, 13)
15. ⊢ $\forall \xi_1\,(\psi(\xi_1) \to \psi(f\,\xi_1))$ 👉 *At this point, the induction step is proven* (Def, 14)
16. ⊢ $\psi(0) \to (\forall \xi_1\,(\psi(\xi_1) \to \psi(f\,\xi_1)) \to \forall \xi_1\,\psi(\xi_1))$ (P.10)
17. ⊢ $\forall \xi_1\,(\psi(\xi_1) \to \psi(f\,\xi_1)) \to \forall \xi_1\,\psi(\xi_1)$ (MP, 4,16)
18. ⊢ $\forall \xi_1\,\psi(\xi_1)$ (MP, 15,17)
19. ⊢ $\forall \xi_1\,(\xi_1 = 0 + \xi_1)$ (Def, 18)
20. ⊢ $\forall \xi_1\,(\xi_1 = 0 + \xi_1) \to (\sigma_1 = 0 + \sigma_1)$ (III.1)
21. ⊢ $\sigma_1 = 0 + \sigma_1$ (MP 19,20)

$f\,\sigma_1 + \tau_1 = f\,(\sigma_1 + \tau_1)$ (M.3.2g)

Let $\psi(\xi_1) := (f\,\sigma_1 + \xi_1 = f\,(\sigma_1 + \xi_1))$

1. ⊢ $f\,\sigma_1 + 0 = f\,\sigma_1$ (ADD.1)
2. ⊢ $\sigma_1 + 0 = \sigma_1$ (ADD.1)
3. ⊢ $\sigma_1 + 0 = \sigma_1 \to f\,(\sigma_1 + 0) = f\,\sigma_1$ (G.8)
4. ⊢ $f\,(\sigma_1 + 0) = f\,\sigma_1$ (MP, 2,3)
5. ⊢ $f\,\sigma_1 + 0 = f\,(\sigma_1 + 0)$ (GL, 1,4)
6. ⊢ $\psi(0)$ 👉 *At this point, the induction start is proven* (Def, 5)
7. $f\,\sigma_1 + \xi_1 = f\,(\sigma_1 + \xi_1)$ (Hyp)
8. ⊢ $f\,\sigma_1 + \xi_1 = f\,(\sigma_1 + \xi_1) \to f\,(f\,\sigma_1 + \xi_1) = f\,f\,(\sigma_1 + \xi_1)$ (G.8)
9. ⊢ $f\,(f\,\sigma_1 + \xi_1) = f\,f\,(\sigma_1 + \xi_1)$ (MP, 7,8)
10. ⊢ $f\,\sigma_1 + f\,\xi_1 = f\,(f\,\sigma_1 + \xi_1)$ (ADD.2)
11. ⊢ $f\,\sigma_1 + f\,\xi_1 = f\,f\,(\sigma_1 + \xi_1)$ (GL, 10,9)
12. ⊢ $\sigma_1 + f\,\xi_1 = f\,(\sigma_1 + \xi_1)$ (ADD.2)
13. ⊢ $\sigma_1 + f\,\xi_1 = f\,(\sigma_1 + \xi_1) \to f\,(\sigma_1 + f\,\xi_1) = f\,f\,(\sigma_1 + \xi_1)$ (G.8)
14. ⊢ $f\,(\sigma_1 + f\,\xi_1) = f\,f\,(\sigma_1 + \xi_1)$ (MP, 12,13)
15. ⊢ $f\,\sigma_1 + f\,\xi_1 = f\,(\sigma_1 + f\,\xi_1)$ (GL, 11,14)
16. ⊢ $f\,\sigma_1 + \xi_1 = f\,(\sigma_1 + \xi_1) \to f\,\sigma_1 + f\,\xi_1 = f\,(\sigma_1 + f\,\xi_1)$ (DT)
17. ⊢ $\psi(\xi_1) \to \psi(f\,\xi_1)$ (Def, 16)
18. ⊢ $\forall \xi_1\,(\psi(\xi_1) \to \psi(f\,\xi_1))$ 👉 *At this point, the induction step is proven* (G, 17)
19. ⊢ $\psi(0) \to (\forall \xi_1\,(\psi(\xi_1) \to \psi(f\,\xi_1)) \to \forall \xi_1\,\psi(\xi_1))$ (P.10)
20. ⊢ $\forall \xi_1\,(\psi(\xi_1) \to \psi(f\,\xi_1)) \to \forall \xi_1\,\psi(\xi_1)$ (MP, 6,19)

4.4 Formal Proofs

21. $\vdash \forall \xi_1 \, \psi(\xi_1)$ (MP, 18,20)
22. $\vdash \forall \xi_1 \, \psi(\xi_1) \to \psi(\tau_1)$ (III.1)
23. $\vdash \psi(\tau_1)$ (MP 21, 22)
24. $\vdash \mathsf{f}\,\sigma_1 + \tau_1 = \mathsf{f}\,(\sigma_1 + \tau_1)$ (Def, 23)

$\sigma_1 + \tau_1 = \tau_1 + \sigma_1$ (M.3.2h)

Let $\psi(\xi_1) := (\sigma_1 + \xi_1 = \xi_1 + \sigma_1)$

1. $\vdash \sigma_1 + 0 = \sigma_1$ (ADD.1)
2. $\vdash \sigma_1 = 0 + \sigma_1$ (M3.2f)
3. $\vdash \sigma_1 + 0 = 0 + \sigma_1$ (GL, 1,2)
4. $\vdash \psi(0)$ (Def, 3)
5. $\quad \sigma_1 + \xi_1 = \xi_1 + \sigma_1$ (Hyp)
6. $\quad \vdash \sigma_1 + \mathsf{f}\,\xi_1 = \mathsf{f}\,(\sigma_1 + \xi_1)$ (ADD.2)
7. $\quad \vdash \mathsf{f}\,\xi_1 + \sigma_1 = \mathsf{f}\,(\xi_1 + \sigma_1)$ (M3.2g)
8. $\quad \vdash \sigma_1 + \xi_1 = \xi_1 + \sigma_1 \to \mathsf{f}\,(\sigma_1 + \xi_1) = \mathsf{f}\,(\xi_1 + \sigma_1)$ (G.8)
9. $\quad \vdash \mathsf{f}\,(\sigma_1 + \xi_1) = \mathsf{f}\,(\xi_1 + \sigma_1)$ (MP, 5,8)
10. $\vdash \sigma_1 + \mathsf{f}\,\xi_1 = \mathsf{f}\,(\xi_1 + \sigma_1)$ (GL, 6,9)
11. $\vdash \sigma_1 + \mathsf{f}\,\xi_1 = \mathsf{f}\,\xi_1 + \sigma_1$ (GL, 10,7)
12. $\vdash \sigma_1 + \xi_1 = \xi_1 + \sigma_1 \to \sigma_1 + \mathsf{f}\,\xi_1 = \mathsf{f}\,\xi_1 + \sigma_1$ (DT)
13. $\vdash \psi(\xi_1) \to \psi(\mathsf{f}\,\xi_1)$ (Def, 12)
14. $\vdash \forall \xi_1 (\psi(\xi_1) \to \psi(\mathsf{f}\,\xi_1))$ (G, 13)
15. $\vdash \psi(0) \to (\forall \xi_1 (\psi(\xi_1) \to \psi(\mathsf{f}\,\xi_1)) \to \forall \xi_1 \, \psi(\xi_1))$ (P.10)
16. $\vdash \forall \xi_1 (\psi(\xi_1) \to \psi(\mathsf{f}\,\xi_1)) \to \forall \xi_1 \, \psi(\xi_1)$ (MP, 4,15)
17. $\vdash \forall \xi_1 \, \psi(\xi_1)$ (MP, 14,16)
18. $\vdash \forall \xi_1 \, \sigma_1 + \xi_1 = \xi_1 + \sigma_1$ (Def, 17)
19. $\vdash \forall \xi_1 \, \sigma_1 + \xi_1 = \xi_1 + \sigma_1 \to \sigma_1 + \tau_1 = \tau_1 + \sigma_1$ (III.1)
20. $\vdash \sigma_1 + \tau_1 = \tau_1 + \sigma_1$ (MP, 18,19)

$\sigma_1 = \tau_1 \to \sigma_1 + \sigma_1 = \tau_1 + \tau_1$ (N.1)

1. $\quad \sigma_1 = \tau_1$ (Hyp)
2. $\quad \vdash \sigma_1 = \tau_1 \to \sigma_1 + \sigma_1 = \tau_1 + \sigma_1$ (M.3.2e)
3. $\quad \vdash \sigma_1 + \sigma_1 = \tau_1 + \sigma_1$ (MP, 1,2)
4. $\quad \vdash \tau_1 + \sigma_1 = \sigma_1 + \tau_1$ (M.3.2h)
5. $\quad \vdash \sigma_1 + \sigma_1 = \sigma_1 + \tau_1$ (GL, 3,4)
6. $\quad \vdash \sigma_1 = \tau_1 \to \sigma_1 + \tau_1 = \tau_1 + \tau_1$ (M.3.2e)

7. $\vdash \sigma_1 + \tau_1 = \tau_1 + \tau_1$ (MP, 1,6)
8. $\vdash \sigma_1 + \sigma_1 = \tau_1 + \tau_1$ (GL, 5,7)
9. $\vdash \sigma_1 = \tau_1 \to \sigma_1 + \sigma_1 = \tau_1 + \tau_1$ (DT)

$\sigma_1 \times \bar{1} = \sigma_1$ (M.3.5b)

1. $\vdash \sigma_1 \times \mathsf{f}\, 0 = \sigma_1 \times 0 + \sigma_1$ (MUL.2)
2. $\vdash \sigma_1 \times 0 = 0$ (MUL.1)
3. $\vdash \sigma_1 \times 0 = 0 \to \sigma_1 \times 0 + \sigma_1 = 0 + \sigma_1$ (M.3.2e)
4. $\vdash \sigma_1 \times 0 + \sigma_1 = 0 + \sigma_1$ (MP, 2,3)
5. $\vdash \sigma_1 \times \mathsf{f}\, 0 = 0 + \sigma_1$ (GL, 1,4)
6. $\vdash \sigma_1 = 0 + \sigma_1$ (M.3.2f)
7. $\vdash \sigma_1 = 0 + \sigma_1 \to 0 + \sigma_1 = \sigma_1$ (PM 13.16)
8. $\vdash 0 + \sigma_1 = \sigma_1$ (MP, 6,7)
9. $\vdash \sigma_1 \times \mathsf{f}\, 0 = \sigma_1$ (GL, 5,8)
10. $\vdash \sigma_1 \times \bar{1} = \sigma_1$ (Def, 9)

$\sigma_1 \times \bar{2} = \sigma_1 + \sigma_1$ (M.3.5c)

1. $\vdash \sigma_1 \times \mathsf{f}\, \bar{1} = \sigma_1 \times \bar{1} + \sigma_1$ (MUL.2)
2. $\vdash \sigma_1 \times \bar{1} = \sigma_1$ (M.3.5b)
3. $\vdash \sigma_1 \times \bar{1} = \sigma_1 \to \sigma_1 \times \bar{1} + \sigma_1 = \sigma_1 + \sigma_1$ (M.3.2e)
4. $\vdash \sigma_1 \times \bar{1} + \sigma_1 = \sigma_1 + \sigma_1$ (MP, 2,3)
5. $\vdash \sigma_1 \times \mathsf{f}\, \bar{1} = \sigma_1 + \sigma_1$ (GL, 1,4)
6. $\vdash \sigma_1 \times \bar{2} = \sigma_1 + \sigma_1$ (Def, 5)

$\mathsf{f}\,\mathsf{f}\,(\sigma_1 \times \bar{2}) = \mathsf{f}\, \sigma_1 + \mathsf{f}\, \sigma_1$ (N.2)

1. $\vdash \sigma_1 \times \bar{2} = \sigma_1 + \sigma_1$ (M.3.5c)
2. $\vdash \sigma_1 \times \bar{2} = \sigma_1 + \sigma_1 \to \mathsf{f}\,(\sigma_1 \times \bar{2}) = \mathsf{f}\,(\sigma_1 + \sigma_1)$ (G.8)
3. $\vdash \mathsf{f}\,(\sigma_1 \times \bar{2}) = \mathsf{f}\,(\sigma_1 + \sigma_1)$ (MP, 1,2)
4. $\vdash \sigma_1 + \mathsf{f}\, \sigma_1 = \mathsf{f}\,(\sigma_1 + \sigma_1)$ (ADD.2)
5. $\vdash \mathsf{f}\,(\sigma_1 \times \bar{2}) = \sigma_1 + \mathsf{f}\, \sigma_1$ (GL, 3,4)
6. $\vdash \mathsf{f}\,(\sigma_1 \times \bar{2}) = \sigma_1 + \mathsf{f}\, \sigma_1 \to \mathsf{f}\,\mathsf{f}\,(\sigma_1 \times \bar{2}) = \mathsf{f}\,(\sigma_1 + \mathsf{f}\, \sigma_1)$ (G.8)
7. $\vdash \mathsf{f}\,\mathsf{f}\,(\sigma_1 \times \bar{2}) = \mathsf{f}\,(\sigma_1 + \mathsf{f}\, \sigma_1)$ (MP, 5,6)
8. $\vdash \mathsf{f}\, \sigma_1 + \mathsf{f}\, \sigma_1 = \mathsf{f}\,(\sigma_1 + \mathsf{f}\, \sigma_1)$ (M.3.2g)
9. $\vdash \mathsf{f}\,\mathsf{f}\,(\sigma_1 \times \bar{2}) = \mathsf{f}\, \sigma_1 + \mathsf{f}\, \sigma_1$ (GL, 7,8)

4.4 Formal Proofs

⚙️ $f\,f\,(\sigma_1 \times \overline{2}) = (f\,\sigma_1) \times \overline{2}$ (N.3)

1. ⊢ $f\,f\,(\sigma_1 \times \overline{2}) = (f\,\sigma_1) + (f\,\sigma_1)$ (N.2)
2. ⊢ $(f\,\sigma_1) \times \overline{2} = (f\,\sigma_1) + (f\,\sigma_1)$ (M.3.5c)
3. ⊢ $f\,f\,(\sigma_1 \times \overline{2}) = (f\,\sigma_1) \times \overline{2}$ (GL, 1,2)

⚙️ $\sigma_1 \neq 0 \rightarrow \exists z_1\,(\sigma_1 = f\,z_1)$ (M.3.5h)

Let $\psi(\xi_1) := (\xi_1 \neq 0 \rightarrow \exists z_1\,(\xi_1 = f\,z_1))$

1. ⊢ $0 = 0$ (G.2)
2. ⊢ $0 = 0 \rightarrow (\forall z_1\,0 \neq f\,z_1 \rightarrow 0 = 0)$ (F.1)
3. ⊢ $\forall z_1\,0 \neq f\,z_1 \rightarrow 0 = 0$ (MP, 1,2)
4. ⊢ $0 \neq 0 \rightarrow \neg\forall z_1\,0 \neq f\,z_1$ (INV, 3)
5. ⊢ $0 \neq 0 \rightarrow \exists z_1\,0 = f\,z_1$ (Def, 4)
6. ⊢ $\psi(0)$ 👉 *At this point, the induction start is proven* (Def, 5)
7. ⊢ $f\,\xi_1 = f\,\xi_1$ (G, 2)
8. ⊢ $(f\,\xi_1 = f\,\zeta_1)[\zeta_1 \leftarrow \xi_1] \rightarrow \exists z_1\,f\,\xi_1 = f\,z_1$ (P.3)
9. ⊢ $\exists z_1\,f\,\xi_1 = f\,z_1$ (MP, 7,8)
10. ⊢ $\exists z_1\,f\,\xi_1 = f\,z_1 \rightarrow (f\,\xi_1 \neq 0 \rightarrow \exists z_1\,f\,\xi_1 = f\,z_1)$ (F.1)
11. ⊢ $f\,\xi_1 \neq 0 \rightarrow \exists z_1\,(f\,\xi_1 = f\,z_1)$ (MP, 9,10)
12. ⊢ $\psi(f\,\xi_1)$ (Def, 11)
13. ⊢ $\psi(f\,\xi_1) \rightarrow (\psi(\xi_1) \rightarrow \psi(f\,\xi_1))$ (F.1)
14. ⊢ $\psi(\xi_1) \rightarrow \psi(f\,\xi_1)$ (MP, 12,13)
15. ⊢ $\forall \xi_1\,(\psi(\xi_1) \rightarrow \psi(f\,\xi_1))$ 👉 *At this point, the induction step is proven* (G, 14)
16. ⊢ $\psi(0) \rightarrow (\forall \xi_1\,(\psi(\xi_1) \rightarrow \psi(f\,\xi_1)) \rightarrow \forall \xi_1\,\psi(\xi_1))$ (P.10)
17. ⊢ $\forall \xi_1\,(\psi(\xi_1) \rightarrow \psi(f\,\xi_1)) \rightarrow \forall \xi_1\,\psi(\xi_1)$ (MP, 6,16)
18. ⊢ $\forall \xi_1\,\psi(\xi_1)$ (MP, 15,17)
19. ⊢ $\forall \xi\,(\xi_1 \neq 0 \rightarrow \exists z_1\,(\xi_1 = f\,z_1))$ (Def, 18)
20. ⊢ $\forall \xi\,(\xi_1 \neq 0 \rightarrow \exists z_1\,(\xi_1 = f\,z_1)) \rightarrow (\sigma_1 \neq 0 \rightarrow \exists z_1\,(\sigma_1 = f\,z_1))$ (III.1)
21. ⊢ $\sigma_1 \neq 0 \rightarrow \exists z_1\,(\sigma_1 = f\,z_1)$ (MP, 19,20)

⚙️ $\overline{1} \neq \sigma_1 \times \overline{2}$ (N.4)

Let $\psi(\xi_1) := \xi_1 \times \overline{2} \neq \overline{1}$

1. ⊢ $0 \times \overline{2} = 0 + 0$ (M.3.5c)

2. ⊢ $0 = 0 + 0$ (M.3.2f)
3. ⊢ $0 \times \overline{2} = 0$ (GL, 1,2)
4. ⊢ $\overline{1} \neq 0$ (I.1)
5. ⊢ $0 \times \overline{2} \neq \overline{1}$ (UG, 3,4)
6. ⊢ $\psi(0)$ *At this point, the induction start is proven* (Def, 5)
7. $f\, \xi_1 \times \overline{2} = f\, 0$ (Hyp)
8. ⊢ $f\, f\, (\xi_1 \times \overline{2}) = f\, \xi_1 \times \overline{2}$ (N.3)
9. ⊢ $f\, f\, (\xi_1 \times \overline{2}) = f\, 0$ (GL, 8,7)
10. ⊢ $f\, f\, (\xi_1 \times \overline{2}) = f\, 0 \to f\, (\xi_1 \times \overline{2}) = 0$ (I.2)
11. ⊢ $f\, (\xi_1 \times \overline{2}) = 0$ (MP, 9,10)
12. ⊢ $f\, \xi_1 \times \overline{2} = f\, 0 \to f\, (\xi_1 \times \overline{2}) = 0$ (DT)
13. ⊢ $f\, (\xi_1 \times \overline{2}) \neq 0 \to f\, \xi_1 \times \overline{2} \neq f\, 0$ (INV, 12)
14. ⊢ $f\, (\xi_1 \times \overline{2}) \neq 0$ (I.1)
15. ⊢ $f\, \xi_1 \times \overline{2} \neq f\, 0$ (MP, 14,13)
16. ⊢ $\psi(f\, \xi_1)$ (Def, 15)
17. ⊢ $\psi(f\, \xi_1) \to (\psi(\xi_1) \to \psi(f\, \xi_1))$ (F.1)
18. ⊢ $\psi(\xi_1) \to \psi(f\, \xi_1)$ (MP, 16,17)
19. ⊢ $\forall \xi_1\, (\psi(\xi_1) \to \psi(f\, \xi_1))$ *At this point, the induction step is proven* (G, 18)
20. ⊢ $\psi(0) \to (\forall \xi_1\, (\psi(\xi_1) \to \psi(f\, \xi_1)) \to \forall \xi_1\, \psi(\xi_1))$ (P.10)
21. ⊢ $\forall \xi_1\, (\psi(\xi_1) \to \psi(f\, \xi_1)) \to \forall \xi_1\, \psi(\xi_1)$ (MP, 6,20)
22. ⊢ $\forall \xi_1\, \psi(\xi_1)$ (MP, 19,21)
23. ⊢ $\forall \xi_1\, \psi(\xi_1) \to \psi(\sigma_1)$ (III.1)
24. ⊢ $\psi(\sigma_1)$ (MP 22,23)
25. ⊢ $\sigma_1 \times \overline{2} \neq \overline{1}$ (Def, 24)
26. ⊢ $\overline{1} = \sigma_1 \times \overline{2} \to \sigma_1 \times \overline{2} = \overline{1}$ (PM 13.16)
27. ⊢ $\sigma_1 \times \overline{2} \neq \overline{1} \to \overline{1} \neq \sigma_1 \times \overline{2}$ (INV, 26)
28. ⊢ $\overline{1} \neq \sigma_1 \times \overline{2}$ (MP, 25,27)

$\sigma_1 = 0 \to \sigma_1 \times \overline{2} = 0$ (N.5)

1. $\sigma_1 = 0$ (Hyp)
2. ⊢ $\sigma_1 = 0 \to \sigma_1 + \sigma_1 = 0 + \sigma_1$ (M.3.2e)
3. ⊢ $\sigma_1 + \sigma_1 = 0 + \sigma_1$ (MP, 1,2)
4. ⊢ $\sigma_1 = 0 + \sigma_1$ (M.3.2f)
5. ⊢ $0 = 0 + \sigma_1$ (GL, 1,4)
6. ⊢ $0 = \sigma_1 + \sigma_1$ (GL, 5,3)

4.4 Formal Proofs

7. ⊢ $\sigma_1 \times \overline{2} = \sigma_1 + \sigma_1$ (M3.5c)
8. ⊢ $\sigma_1 \times \overline{2} = 0$ (GL, 7,6)
9. ⊢ $\sigma_1 = 0 \to \sigma_1 \times \overline{2} = 0$ (DT)

---

f f $\sigma_1 = \tau_1 \times \overline{2} \to \tau_1 \neq 0$ (N.6)

1.   f f $\sigma_1 = \tau_1 \times \overline{2}$ (Hyp)
2. ⊢ f f $\sigma_1 \neq 0$ (I.1)
3. ⊢ $\tau_1 \times \overline{2} \neq 0$ (UG, 1,2)
4. ⊢ $\tau_1 = 0 \to \tau_1 \times \overline{2} = 0$ (N.5)
5. ⊢ $\tau_1 \times \overline{2} \neq 0 \to \tau_1 \neq 0$ (INV, 4)
6. ⊢ $\tau_1 \neq 0$ (MP, 3,5)
7. ⊢ f f $\sigma_1 = \tau_1 \times \overline{2} \to \tau_1 \neq 0$ (DT)

$\exists z_1\; \sigma_1 = z_1 \times \overline{2} \to \exists z_1\; \text{f f}\; \sigma_1 = z_1 \times \overline{2}$ (N.7)

1. $\sigma_1 = z_1 \times \overline{2}$ (Hyp)
2. ⊢ f f $(z_1 \times \overline{2}) = f\; z_1 \times \overline{2}$ (N.3)
3. ⊢ $\sigma_1 = z_1 \times \overline{2} \to f\; \sigma_1 = f\; (z_1 \times \overline{2})$ (G.8)
4. ⊢ $f\; \sigma_1 = f\; (z_1 \times \overline{2})$ (MP, 1,3)
5. ⊢ $f\; \sigma_1 = f\; (z_1 \times \overline{2}) \to f\; f\; \sigma_1 = f\; f\; (z_1 \times \overline{2})$ (G.8)
6. ⊢ f f $\sigma_1 = $ f f $(z_1 \times \overline{2})$ (MP, 4,5)
7. ⊢ f f $\sigma_1 = f\; z_1 \times \overline{2}$ (GL, 6,2)
8. ⊢ $\sigma_1 = z_1 \times \overline{2} \to f\; f\; \sigma_1 = f\; z_1 \times \overline{2}$ (DT)
9. ⊢ f f $\sigma_1 = f\; z_1 \times \overline{2} \to \exists z_1\; \text{f f}\; \sigma_1 = z_1 \times \overline{2}$ (P.3)
10. ⊢ $\sigma_1 = z_1 \times \overline{2} \to \exists z_1\; \text{f f}\; \sigma_1 = z_1 \times \overline{2}$ (MB, 8,9)
11. ⊢ $\forall z_1\; (\sigma_1 = z_1 \times \overline{2} \to \exists z_1\; \text{f f}\; \sigma_1 = z_1 \times \overline{2})$ (G, 10)
12. ⊢ $\exists z_1\; \sigma_1 = z_1 \times \overline{2} \to \exists z_1\; \text{f f}\; \sigma_1 = z_1 \times \overline{2}$ (BE, 11)

$\exists z_1\; \text{f f}\; \sigma_1 = z_1 \times \overline{2} \to \exists z_1\; \sigma_1 = z_1 \times \overline{2}$ (N.8)

1.   f f $\sigma_1 = z_1 \times \overline{2}$ (Hyp)
2. ⊢ f f $\sigma_1 \neq 0$ (I.1)
3. ⊢ $z_1 \times \overline{2} \neq 0$ (UG, 1,2)
4. ⊢ $z_1 = 0 \to z_1 \times \overline{2} = 0$ (N.5)
5. ⊢ $z_1 \times \overline{2} \neq 0 \to z_1 \neq 0$ (INV, 4)
6. ⊢ $z_1 \neq 0$ (MP, 3,5)
7. ⊢ $z_1 \neq 0 \to \exists y_1\; z_1 = f\; y_1$ (M.3.5h)

8. ⊢ $\exists y_1\, z_1 = f\, y_1$ (MP, 6,7)
9. $\quad z_1 = f\, y_1$ (Hyp)
10. ⊢ $z_1 = f\, y_1 \to z_1 + z_1 = f\, y_1 + f\, y_1$ (N.1)
11. ⊢ $z_1 + z_1 = f\, y_1 + f\, y_1$ (MP, 9,10)
12. ⊢ $z_1 \times \overline{2} = z_1 + z_1$ (M.3.5c)
13. ⊢ $z_1 \times \overline{2} = f\, y_1 + f\, y_1$ (GL, 12,11)
14. ⊢ $f\, f\, (y_1 \times \overline{2}) = f\, y_1 + f\, y_1$ (N.2)
15. ⊢ $z_1 \times \overline{2} = f\, f\, (y_1 \times \overline{2})$ (GL, 13,14)
16. ⊢ $f\, f\, \sigma_1 = f\, f\, (y_1 \times \overline{2})$ (GL, 1,15)
17. ⊢ $z_1 = f\, y_1 \to f\, f\, \sigma_1 = f\, f\, (y_1 \times \overline{2})$ (DT)
18. ⊢ $\forall y_1\, (z_1 = f\, y_1 \to f\, f\, \sigma_1 = f\, f\, (y_1 \times \overline{2}))$ (G, 17)
19. ⊢ $\exists y_1\, z_1 = f\, y_1 \to \exists y_1\, f\, f\, \sigma_1 = f\, f\, (y_1 \times \overline{2})$ (E, 18)
20. ⊢ $\exists y_1\, f\, f\, \sigma_1 = f\, f\, (y_1 \times \overline{2})$ (MP, 8,19)
21. ⊢ $\exists y_1\, f\, f\, \sigma_1 = f\, f\, (y_1 \times \overline{2}) \to \exists z_1\, f\, f\, \sigma_1 = f\, f\, (z_1 \times \overline{2})$ (P.9)
22. ⊢ $\exists z_1\, f\, f\, \sigma_1 = f\, f\, (z_1 \times \overline{2})$ (MP, 20,21)
23. ⊢ $f\, f\, \sigma_1 = z_1 \times \overline{2} \to \exists z_1\, f\, f\, \sigma_1 = f\, f\, (z_1 \times \overline{2})$ (DT)
24. ⊢ $f\, f\, \sigma_1 = f\, f\, (z_1 \times \overline{2}) \to f\, \sigma_1 = f\, (z_1 \times \overline{2})$ (I.2)
25. ⊢ $f\, \sigma_1 = f\, (z_1 \times \overline{2}) \to \sigma_1 = z_1 \times \overline{2}$ (I.2)
26. ⊢ $f\, f\, \sigma_1 = f\, f\, (z_1 \times \overline{2}) \to \sigma_1 = z_1 \times \overline{2}$ (MB, 24,25)
27. ⊢ $\forall z_1\, (f\, f\, \sigma_1 = f\, f\, (z_1 \times \overline{2}) \to \sigma_1 = z_1 \times \overline{2})$ (G, 26)
28. ⊢ $\exists z_1\, f\, f\, \sigma_1 = f\, f\, (z_1 \times \overline{2}) \to \exists z_1\, \sigma_1 = z_1 \times \overline{2}$ (E, 27)
29. ⊢ $f\, f\, \sigma_1 = z_1 \times \overline{2} \to \exists z_1\, \sigma_1 = z_1 \times \overline{2}$ (MB, 23,28)
30. ⊢ $\forall z_1\, (f\, f\, \sigma_1 = z_1 \times \overline{2} \to \exists z_1\, \sigma_1 = z_1 \times \overline{2})$ (G, 29)
31. ⊢ $\exists z_1\, f\, f\, \sigma_1 = z_1 \times \overline{2} \to \exists z_1\, \sigma_1 = z_1 \times \overline{2}$ (BE, 30)

At this point, our journey into the depths of the formal system P comes to an end. We have developed a profound understanding of how mathematical statements can be formalized and mechanically proved within this system.

4.5 The Arithmetization of Syntax

Gödel continues by discussing a principle we have already covered in the proof sketch: the *arithmetization of syntax*. This term refers to the concept of making the syntactic relations between the objects of a formal system visible on the arithmetic level. The formulas and proofs of a formal system are translated into natural numbers, called *Gödel numbers* today. Gödel calculates them according to the following scheme:

4.5 The Arithmetization of Syntax

> Wir ordnen nun den Grundzeichen des Systems P in folgender Weise eineindeutig natürliche Zahlen zu:
>
> Über formal unentscheidbare Sätze der Principia Mathematica etc. 179
>
> | "0" ... 1 | "∨" ... 7 | "(" ... 11 |
> |---|---|---|
> | "f" ... 3 | "Π" ... 9 | ")" ... 13 |
> | "∼" ... 5 | | |
>
> ferner den Variablen n-ten Typs die Zahlen der Form p^n (wo p eine Primzahl > 13 ist). Dadurch entspricht jeder endlichen Reihe von Grundzeichen (also auch jeder Formel) in eineindeutiger Weise eine endliche Reihe natürlicher Zahlen. Die endlichen Reihen natür-

> We now set up a one-to-one correspondence of natural numbers to the primitive symbols of the system P in the following manner:
>
> On formally undecidable propositions of Principia Mathematica etc. 179
>
> | "0" ... 1 | "∨" ... 7 | "(" ... 11 |
> |---|---|---|
> | "f" ... 3 | "Π" ... 9 | ")" ... 13 |
> | "∼" ... 5 | | |
>
> and furthermore, to the variables of n-th type we assign the numbers of the form p^n (where p is a prime number > 13). Thus, to every finite sequence of primitive symbols (hence also to every formula), there corresponds in a one-to-one fashion a finite sequence of positive integers.

In this paragraph, Gödel associates each elementary symbol of P with a unique natural number, allowing him to interpret each formula as a finite sequence of those numbers. As examples, let us consider the following formulas:

$$\varphi_1 := (x_1 \to x_1 \lor x_1) \to (x_1 \to x_1)$$
$$\varphi_2 := x_1 \to x_1 \lor x_1$$
$$\varphi_3 := x_1 \to x_1$$

φ_1, φ_2, and φ_3 are typed variants of the formulas we have already employed in the proof sketch for precisely this purpose. Before the transformation can be performed, we need to translate φ_1, φ_2, and φ_3 back into native formulas of P. Eliminating the implication operators and adding the missing parentheses leads to the following intermediate result:

$$\varphi_1 = (\neg((\neg(x_1)) \lor ((x_1) \lor (x_1)))) \lor ((\neg(x_1)) \lor (x_1))$$
$$\varphi_2 = (\neg(x_1)) \lor ((x_1) \lor (x_1))$$

$$\varphi_3 = (\neg(x_1)) \vee (x_1)$$

Now, suppose every elementary symbol is associated with a unique natural number. In that case, every finite series of basic symbols corresponds one-to-one to a finite series of natural numbers. For our example formulas, the number series look like this:

- Formula φ_1

11 11 5 17 13 11 17 7 17 13 13 11 5 17 13 11 13
↕ ↕ ↕ ↕ ↕ ↕ ↕ ↕ ↕ ↕ ↕ ↕ ↕ ↕ ↕ ↕ ↕
(¬ ((¬ (x₁)) ∨ ((x₁) ∨ (x₁)))) ∨ ((¬ (x₁)) ∨ (x₁))
↕ ↕ ↕ ↕ ↕ ↕ ↕ ↕ ↕ ↕ ↕ ↕ ↕ ↕ ↕ ↕ ↕
5 11 11 13 7 11 13 11 13 13 7 11 11 13 7 17 13

- Formula φ_2

11 11 13 7 11 13 11 13
↕ ↕ ↕ ↕ ↕ ↕ ↕ ↕
(¬ (x₁)) ∨ ((x₁) ∨ (x₁))
↕ ↕ ↕ ↕ ↕ ↕ ↕ ↕ ↕
5 17 13 11 17 7 17 13

- Formula φ_3

11 11 13 7 17
↕ ↕ ↕ ↕ ↕
(¬ (x₁)) ∨ (x₁)
↕ ↕ ↕ ↕ ↕
5 17 13 11 13

Next, Gödel merges the number sequences into a joint number.

> eine endliche Reihe natürlicher Zahlen. Die endlichen Reihen natürlicher Zahlen bilden wir nun (wieder eineindeutig) auf natürliche Zahlen ab, indem wir der Reihe $n_1, n_2, \ldots n_k$ die Zahl $2^{n_1} \cdot 3^{n_2} \ldots p_k^{n_k}$ entsprechen lassen, wo p_k die k-te Primzahl (der Größe nach) bedeutet. Dadurch ist nicht nur jedem Grundzeichen, sondern auch jeder endlichen Reihe von solchen in eineindeutiger Weise eine natürliche Zahl zugeordnet. Die dem Grundzeichen (bzw. der Grund-

> We map (again in a one-to-one fashion) the finite sequences of positive integers into the natural numbers by letting the numbers $2^{n_1} \cdot 3^{n_2} \ldots p_k^{n_k}$ correspond to the sequence $n_1, n_2, \ldots, n_k$, where p_k denotes the k-th prime number (according to magnitude). Hence, a natural number is correlated in one-to-one fashion not only to every primitive symbol but also to every finite sequence of such symbols.

4.5 The Arithmetization of Syntax

Gödel maps the sequence $n_1, n_2, \ldots, n_k$ to the joint number

$$p_1^{n_1} \cdot p_2^{n_2} \cdot \ldots \cdot p_k^{n_k},$$

where p_k denotes the k-th prime number in order of increasing magnitude. In this way, not only each elementary symbol but also every finite series of such symbols maps to a natural number one-to-one. Following this procedure, our example formulas transform into the following Gödel numbers:

■ Formula φ_1

$$\ulcorner \varphi_1 \urcorner = 2^{11} \cdot 3^5 \cdot 5^{11} \cdot 7^{11} \cdot 11^5 \cdot 13^{11} \cdot 17^{17} \cdot 19^{13} \cdot 23^{13} \cdot 29^7 \cdot$$
$$31^{11} \cdot 37^{11} \cdot 41^{17} \cdot 43^{13} \cdot 47^7 \cdot 53^{11} \cdot 59^{17} \cdot 61^{13} \cdot 67^{13} \cdot$$
$$71^{13} \cdot 73^{13} \cdot 79^7 \cdot 83^{11} \cdot 89^{11} \cdot 97^5 \cdot 101^{11} \cdot 103^{17} \cdot 107^{13} \cdot$$
$$109^{13} \cdot 113^7 \cdot 127^{11} \cdot 131^{17} \cdot 137^{13} \cdot 139^{13}$$

= 98309379256648010998694098902047834543010949148965052893476416558776
0149790629953990100301888201811202185529376396287792454556842260327
1461314855823696701470181326383622907879874345384405093365121480254
2606695752816130144290770329562152908504880727763395243142926831508
930977258048714101801556158883533852421307745350090002994417043551
1563232961123225428224495596561177042500120056839402576653277961218
0284242787644799194382885094015196581215058017583778114540931791960
5991992883644234963645633909259827532348412476038404279207059441596
1699263621075695105590114380569321086125034573349907548131167324193
897246706665735005161274900000000000

■ Formula φ_2

$$\ulcorner \varphi_2 \urcorner = 2^{11} \cdot 3^5 \cdot 5^{11} \cdot 7^{17} \cdot 11^{13} \cdot 13^{13} \cdot 17^7 \cdot 19^{11} \cdot 23^{11} \cdot 29^{17}$$
$$\cdot 31^{13} \cdot 37^7 \cdot 41^{11} \cdot 43^{17} \cdot 47^{13} \cdot 53^{13}$$

= 20816340182285939507081673729054187946247451498229633758212038396303
446170442178439761447248068335821422022983848089520217611732762285
566587724756574094163404165368055069635537980037484799584785429655
38288262952126681596509700000000000

■ Formula φ_3

$$\ulcorner \varphi_3 \urcorner = 2^{11} \cdot 3^5 \cdot 5^{11} \cdot 7^{17} \cdot 11^{13} \cdot 13^{13} \cdot 17^7 \cdot 19^{11} \cdot 23^{17} \cdot 29^{13}$$

= 40890336639361224855361068360540692836688628695014618709774435684319
150646260028601787377978471947272625342947000000000000

These early examples make one very clear: Gödel takes no account of the size of the numbers. Even for short formulas, the Gödel numbers become so huge that writing them down as decimal numbers becomes a formidable task.

The numbers grow even more dramatically when proofs are encoded. Remeber that a proof is a finite sequence

$$\varphi_1, \varphi_2, \varphi_3, \ldots, \varphi_k$$

of formulas. To translate these into a natural number, we first determine the Gödel numbers of the individual formulas. After that, we use the number of the i-th formula as the exponent of the i-th prime number p_i and form a joint product:

$$\ulcorner \varphi_1, \varphi_2, \varphi_3, \ldots, \varphi_k \urcorner := p_1^{\ulcorner \varphi_1 \urcorner} \cdot p_2^{\ulcorner \varphi_2 \urcorner} \cdot p_3^{\ulcorner \varphi_3 \urcorner} \cdot \ldots \cdot p_k^{\ulcorner \varphi_k \urcorner}$$

Encoding our example formulas this way yields the following number:

$2^{2^{11} 3^5 5^{11} 7^{11} 11^5 13^{11} 17^{17} 19^{13} 23^{13} 29^7 31^{13} 37^{11} 41^{17} 43^{13} 47^7 53^{11} 59^{17} 61^{13} 67^{13} 71^{13} 73^{13}}$
$\phantom{2^{2^{11}}}{}^{79^7 83^{11} 89^{11} 97^5 101^{11} 103^{17} 107^{13} 109^{13} 113^7 127^{11} 131^{17} 137^{13} 139^{13}}$
$\cdot 3^{2^{11} 3^5 5^{11} 7^{17} 11^{13} 13^{13} 17^7 19^{11} 23^{11} 29^{17} 31^{13} 37^7 41^{11} 43^{17} 47^{13} 53^{13}}$
$\cdot 5^{2^{11} 3^5 5^{11} 7^{17} 11^{13} 13^{13} 17^7 19^{11} 23^{17} 29^{13}}$

This number is truly gigantic, with its decimal places far surpassing our universe's estimated count of elementary particles. Therefore, we are well advised to keep the Gödel number in its factorized representation.

> natürliche Zahl zugeordnet. **Die dem Grundzeichen (bzw. der Grundzeichenreihe) a zugeordnete Zahl bezeichnen wir mit $\Phi(a)$.** Sei nun

> The number corresponding to the primitive symbol (or sequence of primitive symbols) a will be written $\Phi(a)$.

In modern terminology, the number Gödel refers to as $\Phi(a)$ is known as the *Gödel number* of a. It's worth noting that Gödel employs the symbol Φ for both primitive symbols and sequences of primitive symbols. Although it is generally clear from the context which of the two is meant, the notation occasionally leads to ambiguities. In $\Phi(0)$, for instance, Φ may refer to the symbol 0 or a string of length 1 containing 0 as its sole symbol. In the first case, $\Phi(0)$ equals 1; in the second case, $\Phi(0)$ equals $2^1 = 2$.

We will address this issue by restricting Gödel's notation $\Phi(x)$ to primitive symbols only. For the Gödel number of a formula φ, we will use the notation $\ulcorner \varphi \urcorner$, which is already familiar to us from the proof sketch. This notation is the one most commonly used today.

4.5 The Arithmetization of Syntax

> Note: In our notation,
> - $\Phi(x)$ denotes the Gödel number of a symbol and
> - $\ulcorner x \urcorner$ denotes the Gödel number of a string of symbols.

Next, Gödel points out that any relation between the syntactic objects of a formal system also has an arithmetic interpretation:

> zeichenreihe) a zugeordnete Zahl bezeichnen wir mit $\Phi(a)$. Sei nun irgend eine Klasse oder Relation $R(a_1, a_2 \ldots a_n)$ zwischen Grundzeichen oder Reihen von solchen gegeben. Wir ordnen ihr diejenige Klasse (Relation) $R'(x_1, x_2 \ldots x_n)$ zwischen natürlichen Zahlen zu, welche dann und nur dann zwischen $x_1, x_2 \ldots x_n$ besteht, wenn es solche $a_1, a_2 \ldots a_n$ gibt, daß $x_i = \Phi(a_i)$ $(i = 1, 2, \ldots n)$ und $R(a_1, a_2 \ldots a_n)$ gilt. Diejenigen Klassen und Relationen natürlicher

> Assume given now any class or relation $R(a_1, a_2 \ldots a_n)$ between primitive symbols or sequences of such symbols. We correlate to it that class (relation) $R'(x_1, x_2 \ldots x_n)$ of natural numbers which holds for $x_1, x_2 \ldots x_n$ when and only when there exist $a_1, a_2 \ldots a_n$ such that $x_i = \Phi(a_i)$ $(i = 1, 2, \ldots, n)$ and $R(a_1, a_2, \ldots, a_n)$ is true.

In Gödel's own account, R refers to a relation

$$R \subseteq (\Sigma^*)^n$$

that establishes a relationship between n strings, where Σ^* denotes the set of all finite strings formable with the primitive symbols contained in the set Σ. R has an isomorphic image in the domain of natural numbers, that is, the relation

$$R' \subseteq \mathbb{N}^n$$

with:

$$(\ulcorner \varphi_1 \urcorner, \ulcorner \varphi_2 \urcorner, \ldots, \ulcorner \varphi_n \urcorner) \in R' :\Leftrightarrow (\varphi_1, \varphi_2, \ldots, \varphi_n) \in R$$

More concisely, this relationship can be stated as:

$$R'(\ulcorner \varphi_1 \urcorner, \ulcorner \varphi_2 \urcorner, \ldots, \ulcorner \varphi_n \urcorner) :\Leftrightarrow R(\varphi_1, \varphi_2, \ldots, \varphi_n)$$

For instance, the unary relations (predicates)

$$R_1 := \{\varphi \in \Sigma^* \mid \varphi \text{ is a variable}\}$$
$$R_2 := \{\varphi \in \Sigma^* \mid \varphi \text{ is a formula}\}$$

$$R_3 := \{\varphi \in \Sigma^* \mid \varphi \text{ is a sentence formula}\}$$
$$R_4 := \{\varphi \in \Sigma^* \mid \varphi \text{ is an axiom}\}$$
$$R_5 := \{\varphi \in \Sigma^* \mid \varphi \text{ is a provable formula}\}$$

have the following isomorphic images:

$$R_1' := \{n \in \mathbb{N} \mid n \text{ is the Gödel number of a variable}\}$$
$$R_2' := \{n \in \mathbb{N} \mid n \text{ is the Gödel number of a formula}\}$$
$$R_3' := \{n \in \mathbb{N} \mid n \text{ is the Gödel number of a sentence formula}\}$$
$$R_4' := \{n \in \mathbb{N} \mid n \text{ is the Gödel number of an axiom}\}$$
$$R_5' := \{n \in \mathbb{N} \mid n \text{ is the Gödel number of a provable formula}\}$$

Consequently, every meta-statement about a formal system has a counterpart in arithmetic. For instance, at the syntax level – the level of symbol strings – the statement about the existence of undecidable formulas reads:

"There is a sentence formula φ, such that neither φ nor the negation of φ are provable formulas."

On the arithmetic level, the corresponding statement goes like this:

"There is a natural number n, which is the Gödel number of a sentence formula φ, such that neither n nor the number corresponding to the Gödel number of the negation of φ is the Gödel number of a provable formula."

This formulation sounds rather cumbersome. For this reason, Gödel introduces a distinctive italic notation that is used consistently throughout his subsequent work.

$R(a_1, a_2 \ldots a_n)$ gilt. Diejenigen Klassen und Relationen natürlicher Zahlen, welche auf diese Weise den bisher definierten metamathematischen Begriffen, z. B. „Variable", „Formel", „Satzformel", „Axiom", „beweisbare Formel" usw. zugeordnet sind, bezeichnen wir mit denselben Worten in Kursivschrift. Der Satz, daß es im System P unentscheidbare Probleme gibt, lautet z. B. folgendermaßen: Es gibt *Satzformeln a*, so daß weder *a* noch die *Negation* von *a beweisbare Formeln* sind.

Those classes and relations of natural numbers which correspond in this manner to the previously defined metamathematical concepts, e. g. "variable", "formula", "sentence",

4.5 The Arithmetization of Syntax

> "axiom", "provable formula", etc., are denoted by the same words in italics. For example, the proposition that there exist undecidable problems in the system *P* becomes: There exist *sentences a* such that neither *a* nor the *negation* of *a* is a *provable formula*.

Gödel associates the italic notation with a distinct semantic meaning. When he typesets a term upright, he refers to a relation at the syntax level. When he typesets a term in italics, he refers to the corresponding relation at the arithmetic level.

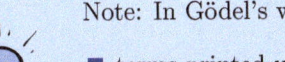
> Note: In Gödel's work,
> - terms printed upright refer to the syntactic level and
> - terms printed in italics refer to the arithmetic level.

At this juncture, our investigation of Gödel's system P comes to an end.

5 Primitive-Recursive Functions

"Number rules the universe."

Pythagoras [88]

After defining the formal system P, Gödel proceeds with an interim discussion spanning several pages. He introduces a class of number-theoretical functions, today referred to as *primitive-recursive functions*.

5.1 Definition and Properties

> Wir schalten nun eine Zwischenbetrachtung ein, die mit dem formalen System P vorderhand nichts zu tun hat, und geben zunächst folgende Definition: Eine zahlentheoretische Funktion [25] $\varphi(x_1, x_2 \ldots x_n)$ heißt **rekursiv definiert aus** den zahlentheoretischen Funktionen $\psi(x_1, x_2 \ldots x_{n-1})$ und $\mu(x_1, x_2 \ldots x_{n+1})$, wenn für alle $x_2 \ldots x_n, k$ [26] folgendes gilt:
>
> $$\varphi(0, x_2 \ldots x_n) = \psi(x_2 \ldots x_n)$$
> $$\varphi(k+1, x_2 \ldots x_n) = \mu(k, \varphi(k, x_2 \ldots x_n), x_2 \ldots x_n). \quad (2)$$
>
> ---
> [25] D. h. ihr Definitionsbereich ist die Klasse der nicht negativen ganzen Zahlen (bzw. der n-tupel von solchen) und ihre Werte sind nicht negative ganze Zahlen.
> [26] Kleine lateinische Buchstaben (ev. mit Indizes) sind im folgenden immer Variable für nicht negative ganze Zahlen (falls nicht ausdrücklich das Gegenteil bemerkt ist).

> We now introduce a digression which, for the moment, has nothing to do with the system P, and, first, we present the following definition: A number-theoretic function [25] $\varphi(x_1, x_2 \ldots x_n)$ is said to be **recursively defined from** the number-theoretic functions $\psi(x_1, x_2 \ldots x_{n-1})$ and $\mu(x_1, x_2 \ldots x_{n+1})$ if the following holds for all $x_2 \ldots x_n, k$ [26]:
>
> $$\varphi(0, x_2 \ldots x_n) = \psi(x_2 \ldots x_n)$$
> $$\varphi(k+1, x_2 \ldots x_n) = \mu(k, \varphi(k, x_2 \ldots x_n), x_2 \ldots x_n). \quad (2)$$
>
> ---
> [25] That is, its domain of definition is the class of non-negative integers (of n-tuples of such integers) and its values are non-negative integers.

[26] Small Roman letters (possibly with subscripts) are, in what follows, always variables for non-negative integers (in case nothing is expressly said to the contrary).

Gödel describes what is now known as the schema of *primitive recursion*:

 Definition 5.1 — Primitive recursion

Let $g : \mathbb{N}^{n-1} \to \mathbb{N}$ and $h : \mathbb{N}^{n+1} \to \mathbb{N}$ be two functions over the natural numbers. A function $f : \mathbb{N}^n \to \mathbb{N}$ is defined according to the schema of *primitive recursion* if it satisfies:

$$f(0, x_2, \ldots, x_n) = g(x_2, \ldots, x_n)$$
$$f(k+1, x_2, \ldots, x_n) = h(k, f(k, x_2, \ldots, x_n), x_2, \ldots, x_n)$$

Many number-theoretical functions can be defined recursively, including addition, multiplication, and exponentiation of natural numbers:

$$\operatorname{add}(0, x) = x$$
$$\operatorname{add}(k+1, x) = \operatorname{s}(\operatorname{add}(k, x))$$
$$\operatorname{mult}(0, x) = 0$$
$$\operatorname{mult}(k+1, x) = \operatorname{add}(\operatorname{mult}(k, x), x)$$
$$\operatorname{pow}(0, x) = 1$$
$$\operatorname{pow}(k+1, x) = \operatorname{mult}(\operatorname{pow}(k, x), x)$$

The first two schemata are already familiar to us. We have used them on page 191 in a slightly modified form to integrate addition and multiplication into system P.

Characterizing arithmetic functions this way is not Gödel's invention. Similar formation schemata were already employed in 1861 by Hermann Graßmann in his *Lehrbuch der Arithmetik für höhere Lehrveranstaltungen*. On pages 17 and 18, Graßmann defines multiplication as follows:

"Under $a \cdot 1$ (read a times one or a multiplied by one) one understands the size a itself, i.e. $a \cdot 1 = a$. [...] Multiplication with the other numbers (except 1) is determined by the following formulas: $a \cdot (\beta + 1) = a \cdot \beta + a$, where β is a positive number."

"Unter $a \cdot 1$ (gelesen a mal eins oder a multiplicirt mit eins) versteht man die Grösse a selbst, d. h. $a \cdot 1 = a$. [...] Die Multiplikation mit den übrigen Zahlen (ausser 1), wird durch folgende Formeln bestimmt: $a \cdot (\beta + 1) = a \cdot \beta + a$, wo β eine positive Zahl ist."

Hermann Graßmann, 1861 [38]

5.1 Definition and Properties

Figure 5.1

HERMANN GRASSMANN
1809 – 1877

On page 73, he similarly introduces exponentiation:

> "To exponentiate a number with a whole number means to link both numbers in such a way that, if the second is zero, the result is 1, and if the second increases by 1, the result multiplies with the first number, or under the power a^n, read a to the n-th, one understands the connection for which the formulas $a^0 = 1$ [and] $a^{n+1} = a^n a$ apply."

> "Eine Zahl mit einer ganzen Zahl potenziren heisst beide Zahlen so verknüpfen, dass, wenn die zweite null ist, das Resultat 1 wird, und wenn die zweite um 1 wächst, das Resultat sich mit der ersten Zahl multiplicirt, oder unter der Potenz a^n, gelesen a zur n-ten, versteht man diejenige Verknüpfung, für welche die Formeln $a^0 = 1$ [und] $a^{n+1} = a^n a$ gelten."

Hermann Graßmann, 1861 [38]

Another mathematician who employed the schema of primitive recursion to establish the fundamental arithmetic operations on natural numbers was Richard Dedekind. The corresponding definitions can be found on pages 44 to 49 of his renowned work *What are numbers and what should they be?*:

> Page 44: "[...] and call this number the sum which arises from the number m by the addition of the number n, or in short the sum of the numbers m, n. Therefore by (126) this sum is completely determined by the conditions
>
> II. $m + 1 = m'$
> III. $m + n' = (m + n)'$."

Figure 5.2

Rózsa Péter
1905 – 1977

Page 47: "[...] and call this number the product arising from the number m by multiplication by the number n, or, for short, the product of the numbers m, n. This therefore by (126) is completely determined by the conditions

  II. $m \cdot 1 = m$
  III. $m n' = m n + m$."

Page 49: "[...] and call this number a power of the base a, while n is called the exponent of this power of a. Hence this notion is completely determined by the conditions

  II. $a^1 = a$
  III. $a^{n'} = a \cdot a^n = a^n \cdot a$"

Richard Dedekind, 1888 [16]

Dedekind utilized the term n' to represent the number $n+1$. For many contemporary authors, this is still the preferred convention for naming the successor of a natural number.

Below, Gödel will employ the concept of primitive recursion to define the class of *primitive-recursive functions*. However, he will refer to them simply as *recursive functions* since the term *primitive-recursive* did not yet exist in 1931. This term was first coined in 1934 by the Hungarian mathematician Rózsa Péter, who was renowned for many significant contributions to recursion theory. Subsequently, it was adopted by David Hilbert and Paul Bernays in the first volume of their influential book *Foundations of Mathematics*, thus quickly becoming an integral part of the mathematical vocabulary [57].

5.1 Definition and Properties

The term *recursive function* is still used today but refers to something different now. Today, a *recursive function* is generally understood as a *computable function*. All primitive recursive functions are computable, but not vice versa.

> Eine zahlentheoretische Funktion φ heißt rekursiv, wenn es eine endliche Reihe von zahlentheor.Funktionen $\varphi_1, \varphi_2 \ldots \varphi_n$ gibt, welche mit φ endet und die Eigenschaft hat, daß jede Funktion φ_k der Reihe entweder aus zwei der vorhergehenden rekursiv definiert ist oder
>
> 180   Kurt Gödel,
>
> aus irgend welchen der vorhergehenden durch Einsetzung entsteht [27]) oder schließlich eine Konstante oder die Nachfolgerfunktion $x+1$ ist. Die Länge der kürzesten Reihe von φ_i, welche zu einer
>
> ---
> [27]) Genauer: durch Einsetzung gewisser der vorhergehenden Funktionen an die Leerstellen einer der vorhergehen- den, z. B. $\varphi_k(x_1, x_2) = \varphi_p[\varphi_q(x_1, x_2), \varphi_r(x_2)]$ $(p, q, r < k)$. Nicht alle Variable der linken Seite müssen auch rechts vorkommen (ebenso im Rekursionsschema (2)).

> A number-theoretic function φ is said to be recursive if there exists a finite sequence of number-theoretic functions $\varphi_1, \varphi_2 \ldots \varphi_n$ which ends with φ and has the property that each function φ_k of the sequence either is defined recursively from
>
> 180   Kurt Gödel,
>
> two of the preceding functions, or results [27]) from one of the preceding functions by substitution, or, finally, is a constant or the successor function $x+1$.
>
> ---
> [27]) More precisely: by substitution of some of the preceding functions for the arguments of one of the preceding functions, e. g. $\varphi_k(x_1, x_2) = \varphi_p[\varphi_q(x_1, x_2), \varphi_r(x_2)]$ $(p, q, r < k)$. Not all variables of the left side have to occur on the right (likewise in the recursion schema (2)).

In a more contemporary formulation, the definition reads like this:

 Definition 5.2 — Primitive-recursive function

The following functions are primitive recursive:

(PR1) The zero function   ☞ $\text{null}(x) := 0$

(PR2) The successor function   ☞ $s(x) := x + 1$

(PR3) The projection functions   ☞ $\pi_i^n(x_1, \ldots, x_n) := x_i$

Furthermore, two recursive construction rules apply:

(PR4) If $h : \mathbb{N}^k \to \mathbb{N}$ and $g_1, \ldots, g_k : \mathbb{N}^n \to \mathbb{N}$ are primitive-recursive, then so is the following function that arises from h by substitution:

$$f(x_1, \ldots, x_n) := h(g_1(x_1, \ldots, x_n), \ldots, g_k(x_1, \ldots, x_n))$$

(PR5) If $g : \mathbb{N}^{n-1} \to \mathbb{N}$ and $h : \mathbb{N}^{n+1} \to \mathbb{N}$ are primitive-recursive, then so is the following function that arises from g and h by primitive recursion:

$$f(0, x_2, \ldots, x_n) = g(x_2, \ldots, x_n)$$
$$f(k+1, x_2, \ldots, x_n) = h(k, f(k, x_2, \ldots, x_n), x_2, \ldots, x_n)$$

The modern definition differs from Gödel's version in two aspects:

- Instead of declaring all constant functions as primitive-recursive, only the zero function is declared as such. This suffices since all constant functions are constructible from the zero functions and the successor function by applying the substitution scheme.

- The modern definition declares all projections as primitive-recursive, whereas Gödel's definition nowhere mentions these functions. To understand why Gödel could do without them, we first convince ourselves that the identity function, mapping each natural number to itself, is also primitive recursive. The function can be easily derived using the schema of primitive recursion:

$$\text{id}(0) := 0 \qquad \text{(PR5)}$$
$$\text{id}(k+1) := s(\text{id}(k)) \qquad ☞ \text{id}(x) = x$$

Now, the projection functions π_i^n can be obtained by substitution:

$$\pi_i^n(x_1, \ldots, x_n) := \text{id}(\text{id}(x_i)) \qquad \text{(PR4)}$$
$$☞ \pi_i^n(x_1, \ldots, x_n) = x_i$$

Note that the construction relies on an additional freedom that Gödel grants himself in footnote 27. In particular, he permits that not all variables on the left-hand side must necessarily appear on the right. Indeed, the construction would not succeed otherwise, as not even the definition of the identical mapping precisely corresponds to the schema of primitive recursion. Strictly interpreted, the definition stipulates that the outer function must have at

5.1 Definition and Properties

least two argument positions. The successor function s, however, is a unary function.

From a mathematical perspective, Gödel's footnote 27 is rather informal, which is why most modern definitions take a slightly different route and declare all projections as primitive-recursive. As the following examples demonstrate, the projection functions prove versatile in flexibly combining functions with different signatures, rendering Gödel's informal rule obsolete.

The following series of primitive-recursive functions shows how addition, multiplication, and exponentiation can be defined by primitive recursion:

$f_1(x) := s(x)$ (PR2)
☞ $f_1(x) = x + 1$

$f_2(x_1, x_2, x_3) := \pi_2^3(x_1, x_2, x_3)$ (PR3)
☞ $f_2(x_1, x_2, x_3) = x_2$

$f_3(x_1, x_2, x_3) := f_1(f_2(x_1, x_2, x_3))$ (PR4)
☞ $f_3(x_1, x_2, x_3) = x_2 + 1$

$f_4(x) := \pi_1^1(x)$ (PR3)
☞ $f_4(x) = x$

$f_5(0, x) := f_4(x)$ (PR5)
$f_5(k+1, x) := f_3(k, f_5(k, x), x)$ ☞ $f_5(k, x) = x + k$

$f_6(x_1, x_2, x_3) := \pi_3^3(x_1, x_2, x_3)$ (PR3)
☞ $f_6(x_1, x_2, x_3) = x_3$

$f_7(x_1, x_2, x_3) := f_5(f_2(x_1, x_2, x_3), f_6(x_1, x_2, x_3))$ (PR4)
☞ $f_7(x_1, x_2, x_3) = x_2 + x_3$

$f_8(x) := \text{null}(x)$ (PR1)
☞ $f_8(x) = 0$

$f_9(0, x) := f_8(x)$ (PR5)
$f_9(k+1, x) := f_7(k, f_9(k, x), x)$ ☞ $f_9(k, x) = x \cdot k$

$f_{10}(x) := f_1(f_8(x))$ (PR4)
☞ $f_{10}(x) = 1$

$f_{11}(x_1, x_2, x_3) := f_9(f_2(x_1, x_2, x_3), f_6(x_1, x_2, x_3))$ (PR4)
☞ $f_{11}(x_1, x_2, x_3) = x_2 \cdot x_3$

$f_{12}(0, x) := f_{10}(x)$ (PR5)
$f_{12}(k+1, x) := f_{11}(k, f_{12}(k, x), x)$ ☞ $f_{12}(k, x) = x^k$

In this series, addition is in 5th place, multiplication is in 9th place, and exponentiation is in 12th place.

The next concept only plays a marginal role. Gödel defines the *level* of a primitive-recursive function as the length of the shortest series producing that function:

> ist. Die Länge der kürzesten Reihe von φ_i, welche zu einer rekursiven Funktion φ gehört, heißt ihre **Stufe**. Eine Relation

> The length of the shortest sequence of φ_i's belonging to a recursive function φ is called its rank.

Assuming that the series printed above is the shortest to define our example functions, their levels can be directly read off. Addition is a function of level 5, multiplication is a function of level 9, and exponentiation is a function of level 12. Gödel refers to the level whenever he proves a statement about primitive-recursive functions by induction. For such a proof to succeed, each function must belong to a certain level, but its specific value is usually irrelevant.

Next, we will convince ourselves that the predecessor function

$$\mathrm{p}(x) := \begin{cases} 0 & \text{if } x = 0 \\ x - 1 & \text{otherwise} \end{cases}$$

and the function computing the (saturated) difference

$$x \mathbin{\dot{-}} y := \begin{cases} x - y & \text{if } x > y \\ 0 & \text{otherwise} \end{cases}$$

are also primitive-recursive. For this purpose, it is sufficient to continue the above formula series as such:

$f_{13}(x_1, x_2) := \pi_1^2(x_1, x_2)$ \hfill (PR3)

☞ $f_{13}(x_1, x_2) = x_1$

$f_{14}(0, x) := f_8(x)$ \hfill (PR5)
$f_{14}(k+1, x) := f_{13}(k, f_{14}(k, x))$

☞ $f_{14}(k, x) = \mathrm{p}(k)$

$f_{15}(x) := f_{14}(f_4(x), f_4(x))$ \hfill (PR4)

☞ $f_{15}(x) = \mathrm{p}(x)$

$f_{16}(x_1, x_2) := \pi_2^2(x_1, x_2)$ \hfill (PR3)

☞ $f_{16}(x_1, x_2) = x_2$

5.1 Definition and Properties

$$f_{17}(x_1, x_2) := f_{15}(f_{16}(x_1, x_2)) \qquad \text{(PR4)}$$
☞ $f_{17}(x_1, x_2) = p(x_2)$ (PR5)

$$f_{18}(0, x) := f_4(x)$$
$$f_{18}(k+1, x) := f_{17}(k, f_{18}(k, x))$$
☞ $f_{18}(k, x) = x \dotminus k$

$$f_{19}(x_1, x_2) := f_{18}(f_{16}(x_1, x_2), f_{13}(x_1, x_2)) \qquad \text{(PR4)}$$
☞ $f_{19}(x_1, x_2) = x_1 \dotminus x_2$

Let's take a closer look at the last function. Besides its functional meaning, which is to calculate the saturated difference, it also has a relational meaning. Its function value is 0 exactly when the number x is less than or equal to y:

$$x \leq y \iff f_{19}(x, y) = 0$$

Thus, whether for two given numbers x and y the relationship $x \leq y$ holds can be decided by calculating the function value $f_{19}(x, y)$.

By associating the existence or non-existence of a relation with the function value in the way described, we can extend our vocabulary with the notion of *primitive-recursive relations*:

Definition 5.3   Primitive-recursive relation

A relation R between the natural numbers $x_1, \ldots, x_n$ is called *primitive-recursive* if a primitive-recursive function f with the following property exists:
$$R(x_1, \ldots, x_n) \iff f(x_1, \ldots, x_n) = 0$$
f is called the *characteristic function* of R.

As no fundamental difference exists between sets and unary relations, defining the concept of primitive-recursive sets is just a stone's throw away. It arises as a particular case from Definition 5.3, with the following wording:

Definition 5.4   Primitive-recursive set

A set $M \subseteq \mathbb{N}$ is called *primitive-recursive*, if a primitive-recursive function f with the following property exists:
$$x \in M \iff f(x) = 0$$
f is called the *characteristic function* of M.

Gödel's words read very familiar now:

> rekursiven Funktion φ gehört, heißt ihre Stufe. Eine Relation zwischen natürlichen Zahlen $R(x_1 \ldots x_n)$ heißt rekursiv [28]), wenn es eine rekursive Funktion $\varphi(x_1 \ldots x_n)$ gibt, so daß für alle $x_1, x_2 \ldots x_n$
> $$R(x_1 \ldots x_n) \sim [\varphi(x_1 \ldots x_n) = 0]\,{}^{29}).$$
>
> ---
>
> [28]) Klassen rechnen wir mit zu den Relationen (einstellige Relationen). Rekursive Relationen R haben natürlich die Eigenschaft, daß man für jedes spezielle Zahlen-n-tupel entscheiden kann, ob $R(x_1 \ldots x_n)$ gilt oder nicht.
> [29]) Für alle inhaltlichen (insbes. auch die metamathematischen) Überlegungen wird die Hilbertsche Symbolik verwendet. Vgl. Hilbert-Ackermann, Grundzüge der theoretischen Logik, Berlin 1928.

> A relation among natural numbers $R(x_1 \ldots x_n)$ is called recursive [28]) if there exists a recursive function $\varphi(x_1 \ldots x_n)$ such that, for all $x_1, x_2 \ldots x_n$
> $$R(x_1 \ldots x_n) \sim [\varphi(x_1 \ldots x_n) = 0]\,{}^{29}).$$
>
> ---
>
> [28]) We consider classes as relations (one-place relations). Naturally, recursive relations R have the property that, for every particular n-tuple of numbers, one can decide whether or not $R(x_1 \ldots x_n)$ holds.
> [29]) In all informal (in particular, metamathematical) considerations Hilbert's symbolism is employed. Cf. Hilbert-Ackermann, Grundzüge der theoretischen Logik, Berlin 1928.

In this context, Gödel uses the symbol '$\sim$' to express the equivalence between the left-hand and right-hand sides, spoken as if and only if ("genau dann, wenn"). It is synonymous with the symbol '$\Leftrightarrow$', which we use today for the same purpose.

Footnote 28 is significant. Gödel points out that all primitive-recursive relations are decidable, that is, it is always possible to answer the question of whether n given numbers $x_1, \ldots, x_n$ satisfy the relation R or not. All that is required is to calculate the function value $f(x_1, \ldots, x_n)$, with f being the primitive-recursive function associated with R. If the result equals 0, then $(x_1, \ldots, x_n)$ is in R; otherwise $(x_1, \ldots, x_n)$ is not in R.

Next, Gödel establishes four fundamental theorems regarding primitive-recursive functions and relations.

> Es gelten folgende Sätze:
> I. Jede aus rekursiven Funktionen (Relationen) durch Einsetzung rekursiver Funktionen an Stelle der Variablen entstehende Funktion (Relation) ist rekursiv; ebenso jede Funktion, die aus rekursiven Funktionen durch rekursive Definition nach dem Schema (2) entsteht.

5.1 Definition and Properties

> The following theorems hold:
> I. Every function (relation) resulting from recursive functions (relations) by substitution of recursive functions for variables is recursive; likewise, every function which arises from recursive functions by recursive definition according to schema (2) is recursive.

Let's look at a specific example. According to this theorem, the function

$$\begin{aligned}\operatorname{mult}(0,x) &:= \operatorname{null}(x) \\ \operatorname{mult}(k+1,x) &:= \operatorname{add}(\pi_2^3(k,\operatorname{mult}(k,x),x), \\ &\qquad \pi_3^3(k,\operatorname{mult}(k,x),x))\end{aligned} \quad (5.1)$$

is primitive-recursive, as it has been created from other primitive-recursive functions through the scheme of substitution and the scheme of primitive recursion. Note that Gödel includes this theorem primarily for the sake of completeness, as its correctness directly follows from the ability to translate a composite formula into a series of formulas, where the substitutions are performed step by step. We already know what this series looks like for the example formula (5.1). It is the series formed from the formulas $f_1, \ldots, f_9$ defined above.

Despite its unspectacular appearance in terms of content, Theorem I offers a high degree of comfort. It permits us to forgo the elaborate construction of formula series and represent primitive-recursive functions in the compact notation employed in (5.1).

> II. Wenn R und S rekursive Relationen sind, dann auch $\overline{R}$, $R \vee S$ (daher auch $R \,\&\, S$).

> II. If R and S are recursive relations, then so are $\overline{R}$ and v $R \vee S$ (hence also $R \,\&\, S$).

This theorem asserts that if R and S are n-ary primitive-recursive relations, then the following relations are also primitive-recursive:

$$\begin{aligned}\overline{R} &:= \{(x_1,\ldots,x_n) \mid (x_1,\ldots,x_n) \notin R\} \\ R \vee S &:= \{(x_1,\ldots,x_n) \mid (x_1,\ldots,x_n) \in R \text{ or } (x_1,\ldots,x_n) \in S\} \\ R \,\&\, S &:= \{(x_1,\ldots,x_n) \mid (x_1,\ldots,x_n) \in R \text{ and } (x_1,\ldots,x_n) \in S\}\end{aligned}$$

We will continue to use the notation $\overline{R}$ as it is still common today. Since relations are sets, we will employ the standard set notations $R \cup S$ and $R \cap S$ instead of $R \vee S$ and $R \,\&\, S$, respectively.

A few pen strokes suffice to complete the proof of Theorem II:

- **Complementary relation $\overline{R}$**

  If R is a primitive-recursive relation, then there exists a function f_R with

  $$(x_1, \ldots, x_n) \in R \Leftrightarrow f_R(x_1, \ldots, x_n) = 0$$

  Consequently,

  $$(x_1, \ldots, x_n) \in \overline{R} \Leftrightarrow \alpha(f_R(x_1, \ldots, x_n)) = 0$$

  with α being defined as follows:

  $$\alpha(x) := \begin{cases} 1 & \text{if } x = 0 \\ 0 & \text{otherwise} \end{cases}$$

  Function α is primitive-recursive, as it is easily derivable from saturated subtraction and the constant 1:

  $$\alpha(x) = 1 \dot{-} x$$

  The composition of α and f_R is also primitive-recursive according to the substitution scheme, and so is $\overline{R}$.

- **Union relation $R \cup S$**

  If R and S are primitive-recursive relations, then there exist functions f_R and f_S with

  $$(x_1, \ldots, x_n) \in R \Leftrightarrow f_R(x_1, \ldots, x_n) = 0$$
  $$(x_1, \ldots, x_n) \in S \Leftrightarrow f_S(x_1, \ldots, x_n) = 0$$

  Then

  $$(x_1, \ldots, x_n) \in R \cup S \Leftrightarrow \beta(f_R(x_1, \ldots, x_n), f_S(x_1, \ldots, x_n)) = 0 \quad (5.2)$$

  with β being defined as follows:

  $$\beta(x, y) = \begin{cases} 0 & \text{if } x = 0 \text{ or } y = 0 \\ 1 & \text{otherwise} \end{cases}$$

  Since β can be obtained from multiplication, saturated subtraction, and the constant 1, it is also primitive-recursive:

  $$\beta(x, y) = 1 \dot{-} (1 \dot{-} x \cdot y)$$

  Thus, the function on the right-hand side of (5.2) is also primitive-recursive, and so is $R \cup S$.

5.1 Definition and Properties

■ Intersection relation $R \cap S$

The assertion follows directly from the reducibility of the intersection to the complement and union:

$$R \cap S = \overline{\overline{R} \cup \overline{S}}$$ □

> III. Wenn die Funktionen $\varphi\,(\mathfrak{x}), \psi\,(\mathfrak{y})$ rekursiv sind, dann auch die Relation: $\varphi(\mathfrak{x}) = \psi(\mathfrak{y})$ [30]).
>
> ---
> [30]) Wir verwenden deutsche Buchstaben $\mathfrak{x}, \mathfrak{y}$ als abkürzende Bezeichnung für beliebige Variablen-n-tupel, z. B. $x_1\, x_2 \ldots x_n$.

> III. If the functions $\varphi\,(\mathfrak{x}), \psi\,(\mathfrak{y})$ are recursive, then so is the relation: $\varphi(\mathfrak{x}) = \psi(\mathfrak{y})$ [30]).
>
> ---
> [30]) We use German letters $\mathfrak{x}, \mathfrak{y}$ as abbreviations for arbitrary n-tuples of variables, e.g. $x_1\, x_2 \ldots x_n$.

For any n-ary function f and m-ary function g, both being primitive-recursive, Theorem III states that the relation $R \subseteq \mathbb{N}^{n+m}$ with

$$R := \{(x_1, \ldots, x_n, y_1, \ldots, y_m) \mid f(x_1, \ldots, x_n) = g(y_1, \ldots, y_m)\}$$

is also primitive-recursive.

This is also easy to see. If f and g are primitive-recursive, so is

$$\gamma(x_1, \ldots, x_n, y_1, \ldots, y_m) := \alpha(\alpha(c) \cdot \alpha(d)) \text{ with}$$
$$c := f(x_1, \ldots, x_n) \dot{-} g(y_1, \ldots, y_m)$$
$$d := g(y_1, \ldots, y_m) \dot{-} f(x_1, \ldots, x_n)$$

For this function, the following holds:

$$\gamma(x_1, \ldots, x_n, y_1, \ldots, y_m) = \begin{cases} 0 & \text{if } f(x_1, \ldots, x_n) = g(y_1, \ldots, y_m) \\ 1 & \text{otherwise} \end{cases}$$

Consequently,

$$(x_1, \ldots, x_n, y_1, \ldots, y_m) \in R \Leftrightarrow \gamma(x_1, \ldots, x_n, y_1, \ldots, y_m) = 0$$

which proves R to be primitive-recursive.

In footnote 30, Gödel explains the meaning of the small old German letters appearing in various places in his work. They serve as a compact notation for arbitrary tuples of variables:

$$\mathfrak{x} = (x_1, \ldots, x_n),$$
$$\mathfrak{y} = (y_1, \ldots, y_m), \text{ etc.}$$

The letters appear again in the very next theorem:

> IV. Wenn die Funktion $\varphi(\mathfrak{x})$ und die Relation $R(x, \mathfrak{y})$ rekursiv sind, dann auch die Relationen S, T
> $$S(\mathfrak{x},\mathfrak{y}) \sim (Ex)[x \leq \varphi(\mathfrak{x}) \,\&\, R(x,\mathfrak{y})]$$
> $$T(\mathfrak{x},\mathfrak{y}) \sim (x)[x \leq \varphi(\mathfrak{x}) \to R(x,\mathfrak{y})]$$
> sowie die Funktion ψ
> $$\psi(\mathfrak{x},\mathfrak{y}) = \varepsilon x[x \leq \varphi(\mathfrak{x}) \,\&\, R(x,\mathfrak{y})],$$
> wobei $\varepsilon x F(x)$ bedeutet: Die kleinste Zahl x, für welche $F(x)$ gilt und 0, falls es keine solche Zahl gibt.

> IV. If the function $\varphi(\mathfrak{x})$ and the relation $R(x, \mathfrak{y})$ are recursive, then so are the relations S and T
> $$S(\mathfrak{x},\mathfrak{y}) \sim (Ex)[x \leq \varphi(\mathfrak{x}) \,\&\, R(x,\mathfrak{y})]$$
> $$T(\mathfrak{x},\mathfrak{y}) \sim (x)[x \leq \varphi(\mathfrak{x}) \to R(x,\mathfrak{y})]$$
> as well as the function
> $$\psi(\mathfrak{x},\mathfrak{y}) = \varepsilon x[x \leq \varphi(\mathfrak{x}) \,\&\, R(x,\mathfrak{y})],$$
> where $\varepsilon x F(x)$ denotes: the smallest number x for which $F(x)$ holds, and 0 if there is no such number.

To facilitate the comprehension of Theorem IV, we confine our analysis to the scenario where $\mathfrak{x}$ represents an empty variable list. In this context, Gödel's words would read as follows:

> IV. If the constant φ and the relation $R(x, \mathfrak{y})$ are recursive, then so are the relations S and T
> $$S(\mathfrak{y}) \sim (Ex)[x \leq \varphi \,\&\, R(x,\mathfrak{y})]$$
> $$T(\mathfrak{y}) \sim (x)[x \leq \varphi \to R(x,\mathfrak{y})]$$
> as well as the function
> $$\psi(\mathfrak{y}) = \varepsilon x[x \leq \varphi \,\&\, R(x,\mathfrak{y})],$$
> where $\varepsilon x F(x)$ denotes: the smallest number x for which $F(x)$ holds, and 0 if there is no such number.

To dissect Gödel's words, let's start by translating the definitions of S, T, and ψ into a form that is more accessible to us:

$(y_1, \ldots, y_m) \in S :\Leftrightarrow$ there exists an x with $x \leq \varphi$ and $(x, y_1, \ldots, y_m) \in R$

5.1 Definition and Properties

$(y_1, \ldots, y_m) \in T \;:\Leftrightarrow\;$ for all $x \leq \varphi$, $(x, y_1, \ldots, y_m) \in R$
$\psi(y_1, \ldots, y_m) :=$ the smallest x with $x \leq \varphi$ and $(x, y_1, \ldots, y_m) \in R$,
or 0, if no such x exists

Let's start by considering the function ψ. Theorem IV assumes that R is a primitive-recursive relation, implying the existence of a function – later denoted by Gödel as ρ – with the following property:

$$(x, y_1, \ldots, y_m) \in R \;\Leftrightarrow\; \rho(x, y_1, \ldots, y_m) = 0$$

Using this function, we can rewrite the definition of ψ as follows:

$\psi(y_1, \ldots, y_m) =$ the smallest x with $x \leq \varphi$ and $\rho(x, y_1, \ldots, y_m) = 0$,
or 0, if no such x exists

Gödel proves function ψ to be primitive recursive with a smart move. First, he constructs a function $\chi(x, y_1, \ldots, y_m)$, which allows the value $\psi(y_1, \ldots, y_m)$ to be calculated elegantly. Then, he proves that χ is a primitive-recursive function and that this property transfers to ψ.

The abovementioned function χ is defined as follows:

$$\chi(0, y_1, \ldots, y_m) := 0$$

$$\chi(n+1, y_1, \ldots, y_m) := \begin{cases} n+1 & \text{if } \rho(n+1, y_1, \ldots, y_m) = 0 \text{ and} \\ & \chi(n, y_1, \ldots, y_m) = 0 \\ \chi(n, y_1, \ldots, y_m) & \text{otherwise} \end{cases} \quad (5.3)$$

To understand the meaning of χ, consider the two examples in Figure 5.3. The plotted values illustrate that χ remains 0 until ρ becomes 0 for the first time. From this point onward, the value of χ equals the smallest n satisfying $\rho(n, y_1, \ldots, y_m) = 0$, and this value persists indefinitely. Consequently, $\psi(y_1, \ldots, y_m)$ can be easily computed as

$$\psi(y_1, \ldots, y_m) = \chi(\varphi, y_1, \ldots, y_m). \quad (5.4)$$

To prove that ψ is primitive recursive, it suffices to demonstrate that χ is a primitive-recursive function, which is what we will do next.

We begin by slightly rewriting the definition of χ. It holds that

$$\chi(0, y_1, \ldots, y_m) = 0$$
$$\chi(n+1, y_1, \ldots, y_m) = (n+1) \cdot a + \chi(n, y_1, \ldots, y_m) \cdot (1 \dot{-} a)$$

with

$$a := \begin{cases} 1 & \text{if } \rho(n+1, y_1, \ldots, y_m) = 0 \text{ and } \chi(n, y_1, \ldots, y_m) = 0 \\ 0 & \text{otherwise} \end{cases}$$

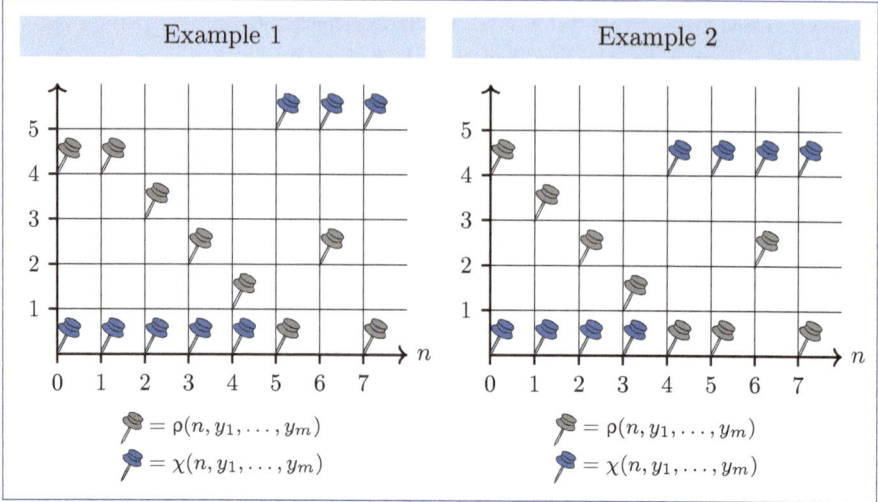

Figure 5.3: If $\rho(0, y_1, \ldots, y_m) \neq 0$, the function ψ can be calculated via the primitive-recursive function χ.

By utilizing the primitive-recursive function α defined above, we can easily calculate the value of a:

$$a = \alpha(\rho(n+1, y_1, \ldots, y_m)) \cdot \alpha(\chi(n, y_1, \ldots, y_m))$$

This concludes the proof of function χ being primitive recursive.

However, we are not quite finished yet. For the examples in Figure 5.3, the function χ works splendidly, but we have missed a critical case. If $\rho(n, y_1, \ldots, y_m)$ equals 0 for $n = 0$ and $n = 1$, then $\chi(1, y_1, \ldots, y_m)$ is equal to 1, but the function value should be 0.

For this reason, we must adjust Definition (5.3) to account for this particular case. The following modification fulfills our needs (Figure 5.4):

$$\chi(0, y_1, \ldots, y_m) := 0$$

$$\chi(n+1, y_1, \ldots, y_m) := \begin{cases} n+1 & \text{if } \begin{aligned} \rho(0, y_1, \ldots, y_m) &\neq 0 \text{ and} \\ \rho(n+1, y_1, \ldots, y_m) &= 0 \text{ and} \\ \chi(n, y_1, \ldots, y_m) &= 0 \end{aligned} \\ \chi(n, y_1, \ldots, y_m) & \text{otherwise} \end{cases} \quad (5.5)$$

The modification does not change the property of χ being primitive-recursive. It is

$$\chi(0, y_1, \ldots, y_m) = 0$$
$$\chi(n+1, y_1, \ldots, y_m) = (n+1) \cdot a + \chi(n, y_1, \ldots, y_m) \cdot (1 \dotminus a)$$

5.1 Definition and Properties

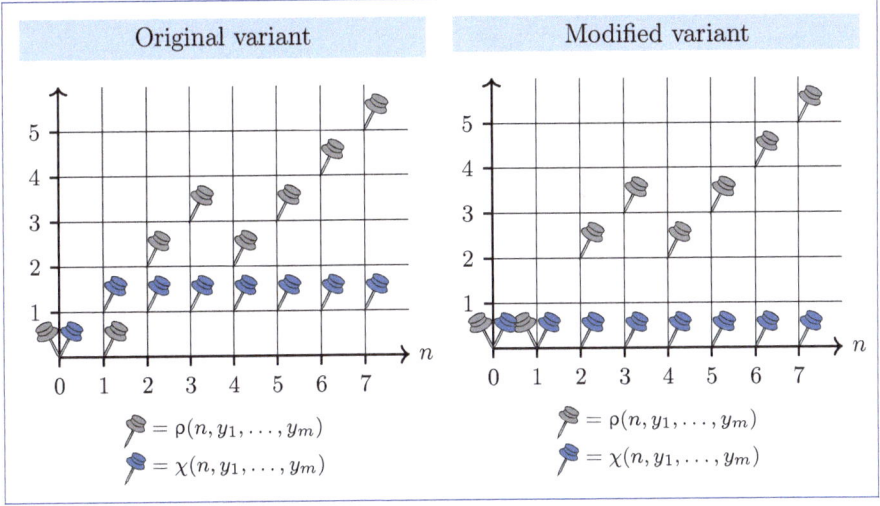

Figure 5.4: The modified variant is also correct for $\rho(0, y_1, \ldots, y_m) = 0$.

with

$$a = \alpha(\alpha(\rho(0, y_1, \ldots, y_m))) \cdot \alpha(\rho(n+1, y_1, \ldots, y_m)) \cdot \alpha(\chi(n, y_1, \ldots, y_m))$$

This completes the first part of the proof.

Next, we turn to the relations S and T. With ψ being primitive recursive, we can easily show that S is primitive recursive, too. It holds that:

$$(y_1, \ldots, y_m) \in S \Leftrightarrow (x, y_1, \ldots, y_m) \in R \text{ for some } x \text{ with } x \leq \varphi$$
$$\Leftrightarrow (\psi(y_1, \ldots, y_m), y_1, \ldots, y_m) \in R \quad (5.6)$$

The relation T is also primitive recursive, as it can be reduced analogously. First, the following applies to the complementary relation $\overline{T}$:

$$(y_1, \ldots, y_m) \in \overline{T} \Leftrightarrow \text{Not for all } x \text{ with } x \leq \varphi, (x, y_1, \ldots, y_m) \in R$$
$$\Leftrightarrow (x, y_1, \ldots, y_m) \in \overline{R} \text{ for some } x \text{ with } x \leq \varphi$$

Given the primitive recursiveness of $\overline{R}$, established by Theorem II, this property extends to $\overline{T}$ and subsequently to T, as has just been proven.

If you've been able to follow the explanations above, you'll find no further difficulties in understanding the proofs in Gödel's original words:

> Satz I folgt unmittelbar aus der Definition von „rekursiv". Satz II und III beruhen darauf, daß die den logischen Begriffen $\overline{}$, $\vee$, $=$ entsprechenden zahlentheoretischen Funktionen
> $$\alpha(x), \beta(x, y), \gamma(x, y)$$

nämlich:

$$\alpha(0) = 1; \ \alpha(x) = 0 \ \text{für} \ x \neq 0$$

$$\beta(0,x) = \beta(x,0) = 0; \ \beta(x,y) = 1, \text{ wenn } x,y \text{ beide } \neq 0 \text{ sind}$$

Über formal unentscheidbare Sätze der Principia Mathematica etc. 181

$$\gamma(x,y) = 0, \text{ wenn } x = y; \ \gamma(x,y) = 1, \text{ wenn } x \neq y$$

rekursiv sind, wie man sich leicht überzeugen kann. Der Beweis für Satz IV ist kurz der folgende: Nach der Voraussetzung gibt es ein rekursives $\rho(x,\mathfrak{y})$, so daß:

$$R(x,\mathfrak{y}) \sim [\rho(x,\mathfrak{y}) = 0].$$

Wir definieren nun nach dem Rekursionsschema (2) eine Funktion $\chi(x,\mathfrak{y})$ folgendermaßen:

$$\chi(0,\mathfrak{y}) = 0$$
$$\chi(n+1,\mathfrak{y}) = (n+1) \cdot a + \chi(n,\mathfrak{y}) \cdot \alpha(a) \ ^{31})$$

wobei $a = \alpha[\alpha(\rho(0,\mathfrak{y}))] \cdot \alpha[\rho(n+1,\mathfrak{y})] \cdot \alpha[\chi(n,\mathfrak{y})]$.

$\chi(n+1,\mathfrak{y})$ ist daher entweder $= n+1$ (wenn $a = 1$) oder $= \chi(n,\mathfrak{y})$ (wenn $a = 0$) $^{32})$. Der erste Fall tritt offenbar dann und nur dann ein, wenn sämtliche Faktoren von a 1 sind, d. h. wenn gilt:

$$\overline{R}(0,\mathfrak{y}) \ \& \ R(n+1,\mathfrak{y}) \ \& \ [\chi(n,\mathfrak{y}) = 0].$$

Daraus folgt, daß die Funktion $\chi(n,\mathfrak{y})$ (als Funktion von n betrachtet) 0 bleibt, bis zum kleinsten Wert von n, für den $R(n,\mathfrak{y})$ gilt, und von da ab gleich diesem Wert ist (falls schon $R(0,\mathfrak{y})$ gilt, ist dem entsprechend $\chi(n,\mathfrak{y})$ konstant und $= 0$). Demnach gilt:

$$\psi(\mathfrak{x},\mathfrak{y}) = \chi(\varphi(\mathfrak{x}),\mathfrak{y})$$
$$S(\mathfrak{x},\mathfrak{y}) \sim R[\psi(\mathfrak{x},\mathfrak{y}),\mathfrak{y}]$$

Die Relation T läßt sich durch Negation auf einen zu S analogen Fall zurückführen, womit Satz IV bewiesen ist.

---

[31]) Wir setzen als bekannt voraus, daß die Funktionen $x+y$ (Addition), $x \cdot y$ (Multiplikation) rekursiv sind.

[32]) Andere Werte als 0 und 1 kann a, wie aus der Definition für α ersichtlich ist, nicht annehmen.

Theorem I follows directly from the definition of "recursive". Theorems II and III depend upon the fact that the number-theoretic functions

$$\alpha(x), \ \beta(x,y), \ \gamma(x,y)$$

corresponding to the logical concepts $\overline{}$, $\vee$, $=$, namely:

$$\alpha(0) = 1; \ \alpha(x) = 0 \ \text{for} \ x \neq 0$$

5.1 Definition and Properties

$$\beta(0,x) = \beta(x,0) = 0; \ \beta(x,y) = 1 \text{ if } x \text{ and } y \text{ are both} \neq 0$$

On formally undecidable propositions of Principia Mathematica etc. 181

$$\gamma(x,y) = 0 \text{ if } x = y; \ \gamma(x,y) = 1 \text{ if } x \neq y$$

are recursive, as one can easily confirm. The proof of Theorem IV is briefly the following: By hypothesis, there exists a recursive $\rho(x,\mathfrak{y})$ such that

$$R(x,\mathfrak{y}) \sim [\rho(x,\mathfrak{y}) = 0].$$

We now define a function $\chi(x,\mathfrak{y})$, according to recursion schema (2), as follows:

$$\chi(0,\mathfrak{y}) = 0$$
$$\chi(n+1,\mathfrak{y}) = (n+1) \cdot a + \chi(n,\mathfrak{y}) \cdot \alpha(a)^{31})$$

where $a = \alpha[\alpha(\rho(0,\mathfrak{y}))] \cdot \alpha[\rho(n+1,\mathfrak{y})] \cdot \alpha[\chi(n,\mathfrak{y})]$.

$\chi(n+1,\mathfrak{y})$ is therefore either $= n+1$ (if $a=1$) or $= \chi(n,\mathfrak{y})$ (if $a=0$).[32]). Obviously the first case occurs when and only when all factors of a are 1, i.e., when

$$\overline{R}(0,\mathfrak{y}) \ \& \ R(n+1,\mathfrak{y}) \ \& \ [\chi(n,\mathfrak{y}) = 0].$$

holds.

From this it follows that the function $\chi(n,\mathfrak{y})$ (considered as a function of n) remains 0 until the least value of n for which $R(n,\mathfrak{y})$ holds, and, from there on, is equal to this value (if $R(0,\mathfrak{y})$ already holds, then the corresponding $\chi(n,\mathfrak{y})$ is constant and $=0$). Hence we have:

$$\psi(\mathfrak{x},\mathfrak{y}) = \chi(\varphi(\mathfrak{x}),\mathfrak{y})$$

☞ corresponds to (5.4)

$$S(\mathfrak{x},\mathfrak{y}) \sim R[\psi(\mathfrak{x},\mathfrak{y}),\mathfrak{y}]$$

☞ corresponds to (5.6)

The relation T can, by negation, be reduced to a case analogous to that of S, thus proving Theorem IV.

---

[31]) We assume known that the functions $x+y$ (addition), $x \cdot y$ (multiplication) are recursive.

[32]) As is apparent from the definition of α, a cannot assume values other than 0 and 1.

5.2 Primitive-Recursive Functions and Relations

> Die Funktionen $x+y$, $x \cdot y$, x^y, ferner die Relationen $x < y$, $x = y$ sind, wie man sich leicht überzeugt, rekursiv und wir definieren nun, von diesen Begriffen ausgehend, eine Reihe von Funktionen (Relationen) 1—45, deren jede aus den vorhergehenden mittels der in den Sätzen I bis IV genannten Verfahren definiert ist. Dabei sind meistens mehrere der nach Satz I bis IV erlaubten Definitionsschritte in einen zusammengefaßt. Jede der Funktionen (Relationen) 1—45, unter denen z. B. die Begriffe „Formel", „Axiom", „unmittelbare Folge" vorkommen, ist daher rekursiv.

> The functions $x+y$, $x \cdot y$, x^y and the relations $x < y$, $x = y$ are, as one can easily check, recursive, and we now define, starting from these concepts, a sequence of functions (relations) 1 – 45, of which each is defined from the preceding ones by the methods indicated in Theorems I-IV. In so doing, several of the definitional steps allowed by Theorems I-IV are often combined into one step. Each of the functions (relations) 1 – 45, among which occur, for example, the concepts *"formula"*, *"axiom"*, *"direct consequence"*, is therefore recursive.

In the final sentence, Gödel foreshadows his objectives with the functions and relations he will soon define. As the construction progresses, he will ensure that, for instance, the Gödel numbers of all formulas, the Gödel numbers of axioms, or the Gödel numbers of formulas generated by inference rules constitute primitive recursive sets. Similar examples have been utilized on page 205 to demonstrate the arithmetization of syntax.

Remember that all these considerations occur on the arithmetic level. Gödel's constructed relations don't directly link the formulas of a formal system; instead, they connect their Gödel numbers, aka their numerical representations. In his article, Gödel expresses this fact by italicizing the words "formula", "axiom", or "immediate consequence".

Tables 5.1 and 5.2 provide an overview of the 45 primitive-recursive functions and relations that Gödel will introduce.

5.2 Primitive-Recursive Functions and Relations

Table 5.1: Gödel's primitive-recursive functions and relations

1.	x/y	x is divisible by y
2.	$\mathrm{Prim}(x)$	x is a prime number
3.	$n \,\mathrm{Pr}\, x$	n-th prime number contained in x
4.	$n!$	Factorial of n
5.	$Pr(n)$	n-th prime number
6.	$n \,Gl\, x$	n-th member of the number series x
7.	$l(x)$	Length of the number series x
8.	$x * y$	Concatenation of x and y
9.	$R(x)$	Number series with x as the only element
10.	$E(x)$	Formula x in parentheses
11.	$n \,\mathrm{Var}\, x$	x is a variable of the n-th type
12.	$\mathrm{Var}(x)$	x is a variable
13.	$\mathrm{Neg}(x)$	Negation of x
14.	$x \,\mathrm{Dis}\, y$	Disjunction of x and y
15.	$x \,\mathrm{Gen}\, y$	Generalization of y with respect to x
16.	$n \,N\, x$	String x with n preceding f's
17.	$Z(n)$	String $\bar{n}$
18.	$\mathrm{Typ}_1'(x)$	x is a symbol of the first type (term)
19.	$\mathrm{Typ}_n(x)$	x is a symbol of the n-th type
20.	$Elf(x)$	x is an elementary formula (atomic formula)
21.	$Op(x,y,z)$	Auxiliary relation for $FR(x)$
22.	$FR(x)$	Auxiliary relation for $\mathrm{Form}(x)$
23.	$\mathrm{Form}(x)$	x is a formula
24.	$v \,\mathrm{Geb}\, n, x$	Variable v is bound at position n
25.	$v \,Fr\, n, x$	Variable v occurs freely at position n
26.	$v \,Fr\, x$	Variable v occurs freely in x at least once
27.	$Su\, x \binom{n}{y}$	Formula x, after inserting y at position n
28.	$k \,St\, v, x$	Auxiliary function for $A(v,x)$
29.	$A(v,x)$	Number of positions where v occurs freely in x
30.	$Sb_n\left(x^v_y\right)$	Auxiliary function for $Sb\left(x^v_y\right)$

5 Primitive-Recursive Functions

Table 5.2: Gödel's primitive-recursive functions and relations (continued)

31.	$Sb\,(x\,{}^v_y)$	Substitution of v by y
32.	$x\,\mathrm{Imp}\,y$	Definition of '$\rightarrow$'
	$x\,\mathrm{Con}\,y$	Definition of '$\wedge$'
	$x\,\mathrm{Aeq}\,y$	Definition of '$\leftrightarrow$'
	$v\,\mathrm{Ex}\,y$	Definition of '$\exists$'
33.	$n\,Th\,x$	n-th type elevation of x
34.	$Z\text{-}Ax(x)$	x is an instance of the axiom schema I.1, I.2, or I.3
35.	$A_i\text{-}Ax(x)$	x is an instance of axiom schema II.i
36.	$A\text{-}Ax(x)$	x is an axiom of axiom group II
37.	$Q(z,y,v)$	Auxiliary predicate to ensure collision-freeness
38.	$L_1\text{-}Ax(x)$	x is an instance of axiom schema III.1
39.	$L_2\text{-}Ax(x)$	x is an instance of axiom schema III.2
40.	$R\text{-}Ax(x)$	x is an instance of axiom schema IV.1
41.	$M\text{-}Ax(x)$	x is an instance of axiom schema V.1
42.	$Ax(x)$	x is an axiom
43.	$Fl(x,y,z)$	x can be derived from y and z
44.	$Bw(x)$	x is a formal proof chain of the system P
45.	xBy	x is a proof for the formula y

   1. x/y x is divisible by y

(Primitive-recursive relation)

---

182   Kurt Gödel,

1. $x/y \equiv (E z)\,[z \leqq x \,\&\, x = y \cdot z]$ [33])
   x ist teilbar durch y [34]).

---

[33]) Das Zeichen $\equiv$ wird im Sinne von „Definitionsgleichheit" verwendet, vertritt also bei Definitionen entweder = oder ∞ (im übrigen ist die Symbolik die Hilbertsche).

[34]) Überall, wo in den folgenden Definitionen eines der Zeichen (x), $(E x)$, εx auftritt, ist es von einer Abschätzung für x gefolgt. Diese Abschätzung dient lediglich dazu, um die rekursive Natur des definierten Begriffs (vgl. Satz IV) zu sichern. Dagegen würde sich der Umfang der definierten Begriffe durch Weglassung dieser Abschätzung meistens nicht ändern.

5.2 Primitive-Recursive Functions and Relations

---

182   Kurt Gödel,

1. $x/y \equiv (E z) [z \leq x \,\&\, x = y \cdot z]$ [33])
x is divisable by y [34]).

---

[33]) The symbol $\equiv$ will be used in the sense of "definitional equality", and therefore in definitions it represents either $=$ or $\sim$ (otherwise the symbolism is Hilbert's).

[34]) Everywhere in the following definitions where one of the expressions (x), $(E x)$, εx occurs it is followed by a bound for x. This bound serves only to assure the recursive nature of the defined concept (cf. Theorem IV). On the other hand the extension of the defined concept would, in most cases, not be changed by omission of this bound.

In footnote 33, Gödel explains how to interpret the symbols. When defining a relation, the symbol '$\equiv$' has the same meaning as '$\Leftrightarrow$'. Thus, the numbers on the left are related if and only if the right-hand side is a true statement. When defining a function, '$\equiv$' corresponds to the equality sign '$=$'. In this case, there is no statement on the right-hand side but a formula defining the function value.

Let's start by translating Gödel's definition into modern notation:

$$x/y \;:\Leftrightarrow\; \exists z \, (z \leq x \wedge x = y \cdot z) \tag{5.7}$$

Formally, (5.7) states that the relation $R \subseteq \mathbb{N}^2$ with

$$(x, y) \in R \;:\Leftrightarrow\; x \text{ is divisible by } y \text{ without remainder}$$

is primitive recursive. Theorem IV makes it evident that this assertion holds.

To fully grasp footnote 34, let us consider a simplified definition of divisibility:

$$x/y \;:\Leftrightarrow\; \exists z \; x = y \cdot z \tag{5.8}$$

The modified variant drops the constraint $z \leq x$, yet it still defines the notion of divisibility on the natural numbers. That is, relation (5.7) comprises the exact same elements as relation (5.8). This is what Gödel means by stating that "the extension of the defined concept would, in most cases, not be changed by omission of this bound".

The rationale behind including the estimate in Gödel's definition is simple. Its omission would fail to ensure that the defined relation remains primitive-recursive. For theorem IV to be applicable, the scope of the existential quantifier must be limited by a natural number calculatable through primitive recursion. This is the reason why these estimates consistently accompany the forthcoming definitions.

To succinctly denote the bound of an existential quantifier, we will frequently use the following two abbreviations:

$$\exists (z \leq c)\, \varphi \quad \text{or} \quad \exists (c_1 \leq z \leq c_2)\, \varphi$$

The first represents

$$\exists z\, (z \leq c \wedge \varphi) \tag{5.9}$$

whereas the second signifies

$$\exists z\, (c_1 \leq z \wedge z \leq c_2 \wedge \varphi). \tag{5.10}$$

Armed with this convenient notation, we can rewrite Gödel's definition as follows:

$$(x, y) \in R :\Leftrightarrow \exists (z \leq x)\, x = y \cdot z \tag{5.11}$$

> 2. $\mathrm{Prim}(x)$ x is a prime number
>
> (Primitive-recursive relation)

> 2. $\mathrm{Prim}\,(x) \equiv \overline{(E z)}\, [z \leqq x\, \&\, z \neq 1\, \&\, z \neq x\, \&\, x/z]\, \&\, x > 1$
> x ist Primzahl.

> 2. $\mathrm{Prim}\,(x) \equiv \overline{(E z)}\, [z \leqq x\, \&\, z \neq 1\, \&\, z \neq x\, \&\, x/z]\, \&\, x > 1$
> x is a prime number.

In modern notation, the definition reads like this:

$$\mathrm{Prim}(x) :\Leftrightarrow \neg\exists (z \leq x)\, (z \neq 1 \wedge z \neq x \wedge x/z) \wedge x > 1$$

The relation comprises precisely the prime numbers,

$$\mathrm{Prim} = \{2, 3, 5, 7, 11, 13, 17, 19, 23, \ldots\},$$

which is easy to see by examining the individual formula components more closely. A natural number is a prime number if

- it is greater than 1,   ☞ $x > 1$
- and cannot be factorized.   ☞ $\neg$

A number is factorizable if

- a natural number $z \leq x$ exists,   ☞ $\exists (z \leq x)$

5.2 Primitive-Recursive Functions and Relations

- which is not equal to 1 and not equal to x,   ☞ $z \neq 1 \wedge z \neq x$
- and divides x.   ☞ x/z

3. $n \, Pr \, x$ n-th prime number contained in x

(Primitive-recursive function)

> 3. $0 \, Pr \, x \equiv 0$
> $(n+1) \, Pr \, x \equiv \varepsilon y \, [y \leq x \, \& \, \text{Prim}(y) \, \& \, x/y \, \& \, y > n \, Pr \, x]$
> $n \, Pr \, x$ ist die n-te (der Größe nach) in x enthaltene Primzahl [34a].
>
> [34a] Für $0 < n \leq z$, wenn z die Anzahl der verschiedenen in x aufgehenden Primzahlen ist. Man beachte, daß für $n = z+1$ $n \, Pr \, x = 0$ ist!

> 3. $0 \, Pr \, x \equiv 0$
> $(n+1) \, Pr \, x \equiv \varepsilon y \, [y \leq x \, \& \, \text{Prim}(y) \, \& \, x/y \, \& \, y > n \, Pr \, x]$
> $n \, Pr \, x$ is the n-th prime factor of x (according to magnitude). [34a]
>
> [34a] For $0 < n \leq z$, where z is the number of distinct prime numbers dividing x. Observe that, for $n = z+1$, $n \, Pr \, x = 0$.

In modern notation:

$$0 \, Pr \, x := 0$$
$$(n+1) \, Pr \, x := \min\{y \leq x \mid \text{Prim}(y) \wedge x/y \wedge y > n \, Pr \, x\}$$

In colloquial terms, $(n+1) \, Pr \, x$ is

- the smallest number y less than or equal to x,   ☞ $\min\{y \leq x \mid \ldots\}$
- which is a prime factor of x, and   ☞ $\text{Prim}(y) \wedge x/y$
- greater than n other prime factors of x.   ☞ $y > n \, Pr \, x$

Simply put, $n \, Pr \, x$ calculates the n-th prime factors of x in ascending order. For instance, the following holds for $x = 45864$:

$0 \, Pr \, 45864 = 0$
$1 \, Pr \, 45864 = 2$
$2 \, Pr \, 45864 = 3$
$3 \, Pr \, 45864 = 7$
$4 \, Pr \, 45864 = 13$

$$\begin{array}{cc} 1 \, Pr \, x & 3 \, Pr \, x \\ \updownarrow & \updownarrow \\ 45864 = 2 \cdot 2 \cdot 2 \cdot 3 \cdot 3 \cdot 7 \cdot 7 \cdot 13 \\ \updownarrow & \updownarrow \\ 2 \, Pr \, x & 4 \, Pr \, x \end{array}$$

If the parameter n exceeds the number of distinct prime factors in x, then $n\, Pr\, x$ equals 0. The same applies to the numbers $x = 0$ and $x = 1$, which have no prime factors at all.

4. $n!$ — Factorial of n
(Primitive-recursive function)

4. $0! \equiv 1$
$(n+1)! \equiv (n+1) \cdot n!$

---

4. $0! \equiv 1$
$(n+1)! \equiv (n+1) \cdot n!$

The factorial function
$$n! := 1 \cdot 2 \cdot 3 \cdot \ldots \cdot n$$
is primitive recursive, as it can be defined in a straight-forward manner through the schema of primitive recursion:

$$\text{factorial}(0) := 1$$
$$\text{factorial}(n+1) := \text{mult}(\text{s}(n), \text{factorial}(n))$$

The factorial function is used in the next definition:

5. $Pr(n)$ — n-th prime number
(Primitive-recursive function)

5. $Pr(0) \equiv 0$
$Pr(n+1) \equiv \varepsilon y\, [y \leqq \{Pr(n)\}! + 1\, \&\, \text{Prim}(y)\, \&\, y > Pr(n)]$
$Pr(n)$ ist die n-te Primzahl (der Größe nach).

---

5. $Pr(0) \equiv 0$
$Pr(n+1) \equiv \varepsilon y\, [y \leqq \{Pr(n)\}! + 1\, \&\, \text{Prim}(y)\, \&\, y > Pr(n)]$
$Pr(n)$ is the n-th prime number (according to magnitude).

In modern notation:
$$Pr(0) := 0$$

5.2 Primitive-Recursive Functions and Relations

Figure 5.5

JOSEPH LOUIS FRANÇOIS BERTRAND
1822 – 1900

$$Pr(n+1) := \min\{y \leq (Pr(n))! + 1 \mid \text{Prim}(y) \land y > Pr(n)\}$$

$Pr(n+1)$ is the smallest prime number greater than $Pr(n)$. Thus, we have

$$Pr\,1 = 2,\ Pr\,2 = 3,\ Pr\,3 = 5,\ Pr\,4 = 7,\ Pr\,5 = 11,\ Pr\,6 = 13, \ldots$$

In short: $Pr(n)$ is the n-th prime number.

Like all primitive-recursive definitions that include a minimal element (in this case, the number y), the potential range of values must be constrained by a number that is computable by a primitive-recursive function. Hence, for the function $Pr(n)$ to accurately compute the n-th prime number, this prime must fall within the specified range. Gödel chose the limit $(Pr(n))! + 1$ based on Euclid's renowned discovery about the infinitude of prime numbers. In his proof, Euclid demonstrated that the product of the first n prime numbers

$$p_1 \cdot p_2 \cdot \ldots \cdot p_n + 1$$

is either a prime number itself or can be divided by a prime number greater than p_n. The estimation

$$p_1 \cdot p_2 \cdot \ldots \cdot p_n + 1 \leq p_n! + 1$$

ensures that the successor of the n-th prime is less than or equal to $p_n! + 1$, which is precisely the value Gödel had chosen to limit the existential quantifier.

In fact, Gödel could have further narrowed the search range by drawing upon a result from recent mathematical history:

Pafnuty Chebyshev (1821 – 1894)   Srinivasa Ramanujan (1887 – 1920)   Paul Erdős [64] (1913 – 1996)

Figure 5.6: Key contributors to the elucidation of Bertrand's postulate.

Theorem 5.5 — Bertrand's Postulate

For $n > 1$, there is always at least one prime number between n and $2n$.

The postulate was proposed in 1845 by the French mathematician Joseph Louis François Bertrand, offering a crucial insight into the density of prime numbers. Bertrand successfully demonstrated its correctness for numbers up to 3,000,000, but he couldn't provide a proof encompassing all numbers.

Bertrand's conjecture was confirmed by Pafnuty Lvovich Chebyshev (Figure 5.6). The Russian mathematician was the first to provide complete proof of the postulate in 1852. Later, the Indian mathematician Srinivasa Ramanujan and the Hungarian mathematician Paul Erdős made further simplifications. Today, most textbooks on number theory follow Erdős's line of proof.

Subsequently, Gödel defines several functions and relations that establish a direct link to the syntactic objects of a formal system. Once more, it's important to note that primitive-recursive functions and relations are arithmetic functions and relations. Thus, the objects of a formal system are never referenced directly but always indirectly through their Gödel numbers.

 6. $n \, Gl \, x$ — n-th member of the number series x

(Primitive-recursive function)

$$6. \; n \, Gl \, x \equiv \varepsilon y \, [y \leqq x \, \& \, x/(n \, Pr \, x)^y \, \& \, \overline{x/(n \, Pr \, x)^{y+1}}]$$

5.2 Primitive-Recursive Functions and Relations

> $n \, Gl \, x$ ist das n-te Glied der der Zahl x zugeordneten Zahlen-
> reihe (für $n > 0$ und n nicht größer als die Länge dieser Reihe).

> 6. $n \, Gl \, x \equiv \varepsilon y \, [y \leq x \, \& \, x/(n \, Pr \, x)^y \, \& \, \overline{x/(n \, Pr \, x)^{y+1}}]$
> $n \, Gl \, x$ is the n-th term of the sequence of numbers correspond-
> ing to the number x (for $n > 0$ and n not greater than the length
> of this sequence).

In modern notation:

$$n \, Gl \, x := \min \left\{ y \leq x \mid x/(n \, Pr \, x)^y \wedge \neg \left(x/(n \, Pr \, x)^{y+1} \right) \right\}$$

On page 200, we have discussed the computation of Gödel numbers. Let us walk through the following example as a reminder:

$$(x_2(x_1)) \vee (y_2(x_1))$$

In the first step, the formula is translated into a series of natural numbers, character by character:

$$
\begin{array}{ccccccc}
11 & 11 & 13 & 7 & 19^2 & 17 & 13 \\
\updownarrow & \updownarrow & \updownarrow & \updownarrow & \updownarrow & \updownarrow & \updownarrow \\
(\,x_2 & (\,x_1 &)\, &)\, \vee \, (\,y_2 & (\,x_1 &)\, &)\, \\
\updownarrow & \updownarrow & \updownarrow & \updownarrow & \updownarrow & \updownarrow \\
17^2 & 17 & 13 & 11 & 11 & 13
\end{array}
$$

In the second step, this series is merged into a single natural number by utilizing the computed numbers as the exponents of prime numbers and subsequently multiplying them together. For the formula provided above, the following product is obtained:

$$2^{11} \cdot 3^{17^2} \cdot 5^{11} \cdot 7^{17} \cdot 11^{13} \cdot 13^{13} \cdot 17^7 \cdot 19^{11} \cdot 23^{19^2} \cdot 29^{11} \cdot 31^{17} \cdot 37^{13} \cdot 41^{13} \quad (5.12)$$

Function $n \, Gl \, x$ operates in the opposite direction. Given a product x in the form of (5.12), it identifies the factor $p_n{}^y$, where p_n represents the n-th prime number, and returns the exponent y. In particular, this exponent is

- the smallest number $y \leq x$ with the property that ☞ $\min \{y \leq x \mid \ldots \}$
- x is divisible by the n-th prime number to the power of y, ☞ $x/(n \, Pr \, x)^y$
- but not by the next higher power. ☞ $\neg \left(x/(n \, Pr \, x)^{y+1} \right)$

Simply put, if x is the Gödel number of a formula, then $n \, Gl \, x$ is the n-th character of this formula. If n equals 0 or exceeds the length of the formula, then $n \, Gl \, x$ equals 0.

7. $l(x)$ — Length of the number series x
(Primitive-recursive function)

> 7. $l(x) \equiv \varepsilon y\,[y \leqq x \,\&\, y\,Pr\,x > 0 \,\&\, (y+1)\,Pr\,x = 0]$
> $l(x)$ ist die Länge der x zugeordneten Zahlenreihe.

> 7. $l(x) \equiv \varepsilon y\,[y \leqq x \,\&\, y\,Pr\,x > 0 \,\&\, (y+1)\,Pr\,x = 0]$
> $l(x)$ is the length of the sequence of numbers correlated with x

In modern notation:

$$l(x) := \min\{y \leq x \mid y\,Pr\,x > 0 \wedge (y+1)\,Pr\,x = 0\}$$

In colloquial terms, $l(x)$ is

- the smallest number $y \leq x$ with the property that ☞ $\min\{y \leq x \mid \ldots\}$
- there are y different prime factors in x, ☞ $y\,Pr\,x > 0$
- but not $y+1$. ☞ $(y+1)\,Pr\,x = 0$

For example,

$$l(2^{17}) = 1 \qquad (\text{☞ 1 prime factor})$$
$$l(2^{11} \cdot 3^{17} \cdot 5^{13}) = 3 \qquad (\text{☞ 3 prime factors})$$
$$l(2^{19^2} \cdot 3^{11} \cdot 5^{17} \cdot 7^{13}) = 4 \qquad (\text{☞ 4 prime factors})$$

Using the bracket notation ⌜...⌝, this can be rephrased as:

$$l(\ulcorner x_1 \urcorner) = 1$$
$$l(\ulcorner (x_1) \urcorner) = 3$$
$$l(\ulcorner y_2(x_1) \urcorner) = 4$$

Now, the meaning of $l(x)$ is evident. If x is the Gödel number of a formula φ, then $l(x)$ is the number of symbols in φ.

5.2 Primitive-Recursive Functions and Relations

 8. $x * y$ Concatenation of x and y

                                                                                                   (Primitive-recursive function)

8. $x * y \equiv \varepsilon z \{ z \leq [Pr\,(l\,(x) + l\,(y))]^{x+y}\,\&$
         $(n)\,[n \leq l\,(x) \rightarrow n\,Gl\,z = n\,Gl\,x]\,\&$
         $(n)\,[0 < n \leq l\,(y) \rightarrow (n + l\,(x))\,Gl\,z = n\,Gl\,y]\}$

$x * y$ entspricht der Operation des "Aneinanderfügens" zweier endlicher Zahlenreihen.

8. $x * y \equiv \varepsilon z \{ z \leq [Pr\,(l\,(x) + l\,(y))]^{x+y}\,\&$
         $(n)\,[n \leq l\,(x) \rightarrow n\,Gl\,z = n\,Gl\,x]\,\&$
         $(n)\,[0 < n \leq l\,(y) \rightarrow (n + l\,(x))\,Gl\,z = n\,Gl\,y]\}$

$x * y$ corresponds to the operation of juxtaposing two finite sequences of numbers.

In modern notation:

$$x * y := \min \left\{ z \leq c \;\middle|\; \begin{array}{l} \forall n\,(n \leq l(x) \rightarrow n\,Gl\,z = n\,Gl\,x)\,\wedge \\ \forall n\,(0 < n \leq l(y) \rightarrow (n + l(x))\,Gl\,z = n\,Gl\,y) \end{array} \right\}$$

with $c := (Pr(l(x) + l(y)))^{x+y}$

Since $0\,Gl\,x$ equals 0 for all x, this definition is equivalent to:

$$x * y := \min \left\{ z \leq c \;\middle|\; \begin{array}{l} \forall n\,(0 < n \leq l(x) \rightarrow n\,Gl\,z = n\,Gl\,x)\,\wedge \\ \forall n\,(0 < n \leq l(y) \rightarrow (n + l(x))\,Gl\,z = n\,Gl\,y) \end{array} \right\}$$

with $c := (Pr(l(x) + l(y)))^{x+y}$

The star operator '$*$' maps the Gödel numbers of two formulas φ and ψ to the Gödel number of the formula $\varphi\psi$, which is the formula obtained by appending ψ to φ. For example:

$$\ulcorner y_2 \urcorner * \ulcorner (x_1) \urcorner = \ulcorner y_2(x_1) \urcorner$$

After expressing the Gödel numbers in factorized form, the equation appears as follows:

$$2^{19^2} * 2^{11} \cdot 3^{17} \cdot 5^{13} = 2^{19^2} \cdot 3^{11} \cdot 5^{17} \cdot 7^{13}$$

This representation reveals the meaning of the definition. If x is the Gödel number of φ and y the Gödel number of ψ, then the Gödel number of $\varphi\psi$ is

- the smallest number z with the property that
                                    ☞ $\min\{z \leq (Pr(l(x) + l(y)))^{x+y} \mid \ldots\}$

- its leading prime factors encode φ and
                                    ☞ $\forall n\,(0 < n \leq l(x) \rightarrow n\,Gl\,z = n\,Gl\,x)$

■ its trailing prime factors encode ψ.

☞ $\forall n \ (0 < n \leq l(y) \to (n + l(x))\ Gl\ z = n\ Gl\ y)$

It remains to be shown that z is smaller than the chosen limit

$$c = Pr(l(x) + l(y))^{x+y}.$$

The Gödel number z is of the form

$$z = p_1^{k_1} \cdot p_2^{k_2} \cdot p_3^{k_3} \cdot \ldots \cdot p_{l(z)}^{k_{l(z)}}$$

with

$$l(z) = l(x) + l(y)$$

The sum of the exponents belonging to φ is less than x, and the sum of the exponents belonging to ψ is less than y. Thus, the sum $k_1 + k_2 + \ldots + k_{l(z)}$ of all exponents is less than $x + y$, and the value of z can be estimated as follows:

$$\begin{aligned}
z &\leq p_{l(z)}^{k_1} \cdot p_{l(z)}^{k_2} \cdot p_{l(z)}^{k_3} \cdot \ldots \cdot p_{l(z)}^{k_{l(z)}} \\
&= p_{l(z)}^{k_1+k_2+k_3+\ldots+k_{l(z)}} \\
&\leq p_{l(z)}^{x+y} \\
&= p_{l(x)+l(y)}^{x+y}
\end{aligned}$$

This is the number Gödel used to limit the scope of the existential quantifier.

---

9. $R(x)$ — Number series with x as the only element

(Primitive-recursive function)

> 9. $R(x) \equiv 2^x$
> $R(x)$ entspricht der nur aus der Zahl x bestehenden Zahlenreihe (für $x > 0$).

> 9. $R(x) \equiv 2^x$
> $R(x)$ corresponds to the sequence of numbers consisting of only the number x (for $x > 0$).

Function $R(x)$ maps the parameter x to the number 2^x. It can be utilized to generate the Gödel numbers of character strings containing a sole symbol. For instance:

$R(1) = 2^{\Phi(0)} = \ulcorner 0 \urcorner \qquad R(7) = 2^{\Phi(\vee)} = \ulcorner \vee \urcorner \qquad R(11) = 2^{\Phi('(')} = \ulcorner (\urcorner$

$R(3) = 2^{\Phi(f)} = \ulcorner f \urcorner \qquad R(9) = 2^{\Phi(\Pi)} = \ulcorner \Pi \urcorner \qquad R(13) = 2^{\Phi(')')} = \ulcorner) \urcorner$

$R(5) = 2^{\Phi(\sim)} = \ulcorner \sim \urcorner$

5.2 Primitive-Recursive Functions and Relations

In modern notation, '$\sim$' is written as '$\neg$', and 'Π' is written as '$\forall$'.

For the encoding of variables, Gödel utilizes the prime powers p^n with $p \geq 17$ and n encoding the type. For instance:

$$R(17^1) = 2^{\Phi(x_1)} = \ulcorner x_1 \urcorner \quad R(17^2) = 2^{\Phi(x_2)} = \ulcorner x_2 \urcorner \quad R(17^3) = 2^{\Phi(x_3)} = \ulcorner x_3 \urcorner$$
$$R(19^1) = 2^{\Phi(y_1)} = \ulcorner y_1 \urcorner \quad R(19^2) = 2^{\Phi(y_2)} = \ulcorner y_2 \urcorner \quad R(19^3) = 2^{\Phi(y_3)} = \ulcorner y_3 \urcorner$$
$$R(23^1) = 2^{\Phi(z_1)} = \ulcorner z_1 \urcorner \quad R(23^2) = 2^{\Phi(z_2)} = \ulcorner z_2 \urcorner \quad R(23^3) = 2^{\Phi(z_3)} = \ulcorner z_3 \urcorner$$

---

10. $E(x)$ — Formula x in parentheses
(Primitive-recursive function)

---

10. $E(x) \equiv R(11) * x * R(13)$
$E(x)$ entspricht der Operation des „Einklammerns" [11 und 13 sind den Grundzeichen „(" und „)" zugeordnet].

---

10. $E(x) \equiv R(11) * x * R(13)$
$E(x)$ corresponds to the operation of placing in parentheses (11 and 13 are correlated with the primitive symbols "(" and ")").

In modern notation:
$$E(x) := \ulcorner (\ulcorner * x * \ulcorner) \urcorner$$

If x is the Gödel number of a formula φ, then $E(x)$ is the Gödel number of the formula obtained by parenthesizing φ. For example:

$$E(\underbrace{131072}_{= 2^{17} \atop = \ulcorner x_1 \urcorner}) = \underbrace{322850407500000000000}_{= 2^{11} \cdot 3^{17} \cdot 5^{13} \atop = \ulcorner (x_1) \urcorner}$$

---

11. $n\,\text{Var}\,x$ — x is a variable of the n-th type
(Primitive-recursive relation)

---

11. $n\,\text{Var}\,x \equiv (Ez)\,[13 < z \leq x \,\&\, \text{Prim}(z) \,\&\, x = z^n] \,\&\, n \neq 0$
x ist eine *Variable n-ten Typs*.

> 11. $n \operatorname{Var} x \equiv (E z) \, [13 < z \leq x \,\&\, \operatorname{Prim}(z) \,\&\, x = z^n] \,\&\, n \neq 0$
> x is a *variable of the n-th type*.

In modern notation:

$$n \operatorname{Var} x \;:\Leftrightarrow\; n \neq 0 \wedge \exists\, (z \leq x)\, (z > 13 \wedge \operatorname{Prim}(z) \wedge x = z^n)$$

For any given value of n, $n \operatorname{Var} x$ is the unary relation containing the Gödel numbers of all variables of type n:

$$\begin{aligned}
1\operatorname{Var} &= \{17^1, 19^1, 23^1, \ldots\} = \{\Phi(x_1), \Phi(y_1), \Phi(z_1), \ldots\} \\
2\operatorname{Var} &= \{17^2, 19^2, 23^2, \ldots\} = \{\Phi(x_2), \Phi(y_2), \Phi(z_2), \ldots\} \\
3\operatorname{Var} &= \{17^3, 19^3, 23^3, \ldots\} = \{\Phi(x_3), \Phi(y_3), \Phi(z_3), \ldots\}
\end{aligned}$$

Observe the italic notation in Gödel's writing, which is crucial for accuracy. After all, x is a natural number, not a variable of any type. If $n \operatorname{Var} x$ holds, then x is the Gödel number of a variable of type n, which is precisely the meaning of italicized text in Gödel's work.

12. $\operatorname{Var}(x)$ — x is a variable

(Primitive-recursive relation)

> 12. $\operatorname{Var}(x) \equiv (E n) \, [n \leq x \,\&\, n \operatorname{Var} x]$
> x ist eine *Variable*.

> 12. $\operatorname{Var}(x) \equiv (E n) \, [n \leq x \,\&\, n \operatorname{Var} x]$
> x is a *variable*.

In modern notation:

$$x \in \operatorname{Var} \;:\Leftrightarrow\; \exists\, (n \leq x)\, n \operatorname{Var} x$$

The relation Var contains the Gödel numbers of all variables:

$$\begin{aligned}
\operatorname{Var} &= \bigcup_{n \in \mathbb{N}} n \operatorname{Var} \\
&= \{17^1, 19^1, 23^1, \ldots, 17^2, 19^2, 23^2, \ldots, 17^3, 19^3, 23^3, \ldots\} \\
&= \{\Phi(x_1), \Phi(y_1), \Phi(z_1), \ldots, \Phi(x_2), \Phi(y_2), \Phi(z_2), \ldots, \Phi(x_3), \Phi(y_3), \Phi(z_3), \ldots\}
\end{aligned}$$

Gödel's formula builds upon the idea that a variable with the Gödel number x must have a type less than x. Thus, the property of x being the Gödel number

5.2 Primitive-Recursive Functions and Relations

of a variable can be determined by checking whether x is a member of one of the sets 1 Var to x Var.

---

13. Neg(x) — Negation of x
(Primitive-recursive function)

13. Neg $(x) \equiv R\,(5) * E\,(x)$
Neg (x) ist die *Negation* von x.

13. Neg $(x) \equiv R\,(5) * E\,(x)$
Neg (x) is the *negation* of x.

In modern notation:

$$\mathrm{Neg}(x) := \ulcorner \infty \urcorner * E(x) = \ulcorner \infty \urcorner * \ulcorner (\urcorner * x * \ulcorner)\urcorner$$

The function Neg maps the Gödel number of a string φ to the Gödel number of the string $\infty\,(\varphi)$.

Let's consider the formula x₂(0) as an example. Because of

$$\ulcorner \mathsf{x_2(0)} \urcorner = 2^{17^2} \cdot 3^{11} \cdot 5^1 \cdot 7^{13}$$
$$= 85358558703482190127297085877476961847445387329503425822247990708594951927598812588177312402905487114240$$

the following holds:

$$\mathrm{Neg}(\ulcorner \mathsf{x_2(0)} \urcorner) = 2^5 \cdot 3^{11} \cdot 5^{17^2} \cdot 7^{11} \cdot 11^1 \cdot 13^{13} \cdot 17^{13}$$
$$= 37186535291129474224030921595627504500176940744077944442088782802856573615631770417779750185171257219696857962653208737811638051265145191193682635872751216634782366174295922704866472475453001082845431893124377165094074371154420077800750732421875000000$$
$$= \ulcorner \infty\,(\mathsf{x_2(0)}) \urcorner$$

In modern notation, $\infty\,(\mathsf{x_2(0)})$ is the formula $\neg(\mathsf{x_2(0)})$.

244　　　5 Primitive-Recursive Functions

 **14.** $x\,\mathrm{Dis}\,y$ — Disjunction of x and y

(Primitive-recursive function)

> Über formal unentscheidbare Sätze der Principia Mathematica etc. 183
>
> 14. $x\,\mathrm{Dis}\,y \equiv E(x) * R(7) * E(y)$
> $x\,\mathrm{Dis}\,y$ ist die *Disjunktion* aus x und y.

> On formally undecidable propositions of Principia Mathematica etc. 183
>
> 14. $x\,\mathrm{Dis}\,y \equiv E(x) * R(7) * E(y)$
> $x\,\mathrm{Dis}\,y$ is the *disjunction* of x and y.

In modern notation:

$$x\,\mathrm{Dis}\,y := E(x) * \ulcorner \vee \urcorner * E(y) = \ulcorner (\urcorner * x * \ulcorner) \urcorner * \ulcorner \vee \urcorner * \ulcorner (\urcorner * y * \ulcorner) \urcorner$$

The function maps the Gödel numbers of the strings φ and ψ to the Gödel number of the string $(\varphi) \vee (\psi)$. For example:

$$\ulcorner \mathsf{x_2(x_1)} \urcorner \,\mathrm{Dis}\, \ulcorner \mathsf{y_2(x_1)} \urcorner = \ulcorner (\mathsf{x_2(x_1)}) \vee (\mathsf{y_2(x_1)}) \urcorner$$

  15. $x\,\mathrm{Gen}\,y$ — Generalization of y with respect to x

(Primitive-recursive function)

> 15. $x\,\mathrm{Gen}\,y \equiv R(x) * R(9) * E(y)$
> $x\,\mathrm{Gen}\,y$ ist die *Generalisation* von y mittels der *Variablen* x (vorausgesetzt, daß x eine *Variable* ist).

> 15. $x\,\mathrm{Gen}\,y \equiv R(x) * R(9) * E(y)$
> $x\,\mathrm{Gen}\,y$ is the *generalization* of y by means of the *variable* x (assuming that x is a *variable*).

In modern notation:

$$x\,\mathrm{Gen}\,y := R(x) * \ulcorner \Pi \urcorner * E(y) = R(x) * \ulcorner \Pi (\urcorner * y * \ulcorner) \urcorner$$

5.2 Primitive-Recursive Functions and Relations

If x is the Gödel number of a symbol ξ and y the Gödel number of a string φ, that is, $x = \Phi(\xi)$ and $y = \ulcorner \varphi \urcorner$, then $x \operatorname{Gen} y$ is the Gödel number of the string $\xi \Pi (\varphi)$. For example:

$$\begin{aligned}
\Phi(\mathsf{x}_2) \operatorname{Gen} \ulcorner \mathsf{x}_2(0) \urcorner &= R(\Phi(\mathsf{x}_2)) * R(9) * E(\ulcorner \mathsf{x}_2(0) \urcorner) \\
&= \ulcorner \mathsf{x}_2 \urcorner * \ulcorner \Pi \urcorner * \ulcorner (\urcorner * \ulcorner \mathsf{x}_2(0) \urcorner * \ulcorner) \urcorner \\
&= \ulcorner \mathsf{x}_2 \Pi (\mathsf{x}_2(0)) \urcorner
\end{aligned}$$

In modern notation, $\mathsf{x}_2 \Pi (\mathsf{x}_2(0))$ is the formula $\forall \mathsf{x}_2\, \mathsf{x}_2(0)$.

---

16. $n\,N\,x$ — String x with n preceeding f's

(Primitive-recursive function)

---

16. $0\,N\,x \equiv x$
$(n+1)\,N\,x \equiv R\,(3) * n\,N\,x$
$n\,N\,x$ entspricht der Operation: "n-maliges Vorsetzen des Zeichens ‚f' vor x".

16. $0\,N\,x \equiv x$
$(n+1)\,N\,x \equiv R\,(3) * n\,N\,x$
$n\,N\,x$ corresponds to the n-fold prefexing of the symbol "f" in front of x.

In modern notation:

$$0\,N\,x := x$$
$$(n+1)\,N\,x := \ulcorner f \urcorner * n\,N\,x$$

The function N prefixes a string with n repetitions of the symbol 'f'. For example:

$$\begin{array}{ll}
0\,N\,\ulcorner 0 \urcorner = \ulcorner 0 \urcorner & 0\,N\,\ulcorner \mathsf{x}_1 \urcorner = \ulcorner \mathsf{x}_1 \urcorner \\
1\,N\,\ulcorner 0 \urcorner = \ulcorner f\,0 \urcorner & 1\,N\,\ulcorner \mathsf{x}_1 \urcorner = \ulcorner f\,\mathsf{x}_1 \urcorner \\
2\,N\,\ulcorner 0 \urcorner = \ulcorner f\,f\,0 \urcorner & 2\,N\,\ulcorner \mathsf{x}_1 \urcorner = \ulcorner f\,f\,\mathsf{x}_1 \urcorner \\
3\,N\,\ulcorner 0 \urcorner = \ulcorner f\,f\,f\,0 \urcorner & 3\,N\,\ulcorner \mathsf{x}_1 \urcorner = \ulcorner f\,f\,f\,\mathsf{x}_1 \urcorner
\end{array}$$

17. $Z(n)$ — String $\bar{n}$

(Primitive-recursive function)

> 17. $Z(n) \equiv n\,N\,[R(1)]$
> $Z(n)$ ist das *Zahlzeichen* für die Zahl n.

> 17. $Z(n) \equiv n\,N\,[R(1)]$
> $Z(n)$ is the *numeral* for the number n.

In modern notation:
$$Z(n) := n\,N\,\ulcorner 0 \urcorner$$

Z maps the natural number n to the Gödel number of the formula $\bar{n}$. For instance:

$$Z(0) = \ulcorner 0 \urcorner$$
$$Z(1) = \ulcorner f\ 0 \urcorner$$
$$Z(2) = \ulcorner f\ f\ 0 \urcorner$$
$$\cdots$$
$$Z(n) = \ulcorner \underbrace{f\ f\ \ldots f}_{n \text{ times}}\ 0 \urcorner$$

18. $\mathrm{Typ}_1'(x)$ — x is a symbol of the first type (term)

(Primitive-recursive relation)

> 18. $\mathrm{Typ}_1'(x) \equiv (E\,m,n)\{m,n \leqq x\,\&\,[m=1 \vee 1\,\mathrm{Var}\,m]$
> $\&\,x = n\,N\,[R(m)]\}$ $^{34b)}$
> x ist *Zeichen ersten Typs*.

$^{34b)}$ $m, n \leqq x$ steht für: $m \leqq x\,\&\,n \leqq x$ (ebenso für mehr als 2 Variable).

> 18. $\mathrm{Typ}_1'(x) \equiv (E\,m,n)\{m,n \leqq x\,\&\,[m=1 \vee 1\,\mathrm{Var}\,m]$
> $\&\,x = n\,N\,[R(m)]\}$ $^{34b)}$
> x is a *term of the first type*.

$^{34b)}$ $m, n \leqq x$ stands for: $m \leqq x\,\&\,n \leqq x$ (and similarly for more than two variables).

5.2 Primitive-Recursive Functions and Relations

In modern notation:

$$x \in \mathrm{Typ}_1' :\Leftrightarrow \exists\, (m, n \leq x)\, ((m = 1 \vee 1\,\mathrm{Var}\,m) \wedge x = n\,N\,R(m))$$

It is easy to see that the natural number x is contained in Typ_1' precisely if it is the Gödel number of a symbol of type 1. In Definition 4.2 on page 143, a symbol of type 1 was referred to as a *term*, which is an expression of the following form:

$$\underbrace{\mathrm{f\,f\,\ldots f}}_{n\ \text{times}}\, \sigma \quad \text{with } \sigma = 0 \text{ or } \sigma = \xi_1$$

Thus, x is the Gödel number of a term, if

- x is derived from m by prefixing a series of f's, and   ☞ $x = n\,N\,R(m)$
  - m is the Gödel number of 0, or   ☞ $m = 1\,\vee$
  - m is the Gödel number of a variable of type 1   ☞ $1\,\mathrm{Var}\,m$

---

19. $\mathrm{Typ}_n(x)$ *x is a symbol of the n-th type*

(Primitive-recursive relation)

19. $\mathrm{Typ}_n(x) \equiv [n = 1\,\&\,\mathrm{Typ}_1'(x)] \vee [n > 1\,\&\,(E v)\,\{v \leq x\,\&\,n\,\mathrm{Var}\,v\,\&\,x = R(v)\}]$

x ist *Zeichen n-ten Typs*.

19. $\mathrm{Typ}_n(x) \equiv [n = 1\,\&\,\mathrm{Typ}_1'(x)] \vee [n > 1\,\&\,(E v)\,\{v \leq x\,\&\,n\,\mathrm{Var}\,v\,\&\,x = R(v)\}]$

x is a *term of the n-th type*.

In modern notation:

$$x \in \mathrm{Typ}_n :\Leftrightarrow \begin{pmatrix} (n = 1 \wedge \mathrm{Typ}_1'(x))\ \vee \\ (n > 1 \wedge \exists\,(v \leq x)\,(n\,\mathrm{Var}\,v \wedge x = R(v))) \end{pmatrix}$$

Recall that a symbol of type 2 is the same as a variable of type 2, a symbol of type 3 is the same as a variable of type 3, and so on. Thus, x is the Gödel number of a symbol of type n, if

- n equals 1 and   ☞ $n = 1\,\wedge$
- x is the Gödel number of a term   ☞ $\mathrm{Typ}_1'(x)$

or

- n is greater than 1 and ☞ $n > 1 \wedge$
- x is the Gödel number of a variable of type n. ☞ $n \operatorname{Var} v \wedge x = R(v)$

20. $Elf(x)$ — x is an elementary formula (atomic formula)

(Primitive-recursive relation)

> 20. $Elf(x) \equiv (E\ y, z, n)\,[y, z, n \leq x\ \&\ \operatorname{Typ}_n(y)$
> $\&\ \operatorname{Typ}_{n+1}(z)\ \&\ x = z * E(y)]$
> x ist *Elementarformel.*

> 20. $Elf(x) \equiv (E\ y, z, n)\,[y, z, n \leq x\ \&\ \operatorname{Typ}_n(y)$
> $\&\ \operatorname{Typ}_{n+1}(z)\ \&\ x = z * E(y)]$
> x is an *elementary formula.*

In modern notation:

$$x \in Elf :\Leftrightarrow \exists (y, z, n \leq x)\,(\operatorname{Typ}_n(y) \wedge \operatorname{Typ}_{n+1}(z) \wedge x = z * E(y))$$

What Gödel calls elementary formulas was called *atomic formulas* on page 144. They take the form

$$\xi_2(\sigma) \text{ or } \xi_{n+1}(\zeta_n)$$

and can be easily described with the tools developed above. x is the Gödel number of an elementary formula, if

- y is the Gödel number of σ or ζ_n, ☞ $\operatorname{Typ}_n(y)$
- z is the Gödel number of ξ_2 or ξ_{n+1}, ☞ $\operatorname{Typ}_{n+1}(z)$
- x is the Gödel number of the formula,
  - which begins with ξ_2 or ξ_{n+1},
  - and is followed by σ or ζ_n in parentheses. ☞ $x = z * E(y)$

21. $Op(x, y, z)$ — Auxiliary relation for $FR(x)$

(Primitive-recursive relation)

> 21. $Op(x\ y\ z) \equiv x = \operatorname{Neg}(y) \vee x = y \operatorname{Dis} z \vee$
> $(E\ v)\,[v \leq x\ \&\ \operatorname{Var}(v)\ \&\ x = v \operatorname{Gen} y]$

5.2 Primitive-Recursive Functions and Relations

> 21. $Op\,(x\,y\,z) \equiv x = \text{Neg}\,(y) \lor x = y\,\text{Dis}\,z \lor$
> $(E\,v)\,[v \leqq x\,\&\,\text{Var}\,(v)\,\&\,x = v\,\text{Gen}\,y]$

In modern notation:

$$(x, y, z) \in Op \;:\Leftrightarrow\; \begin{pmatrix} x = \text{Neg}(y) \;\lor \\ x = y\,\text{Dis}\,z \;\lor \\ \exists\,(v \leq x)\,(\text{Var}(v) \land x = v\,\text{Gen}\,y) \end{pmatrix}$$

Let x, y, and z be the Gödel numbers of the symbol strings φ, ψ, and χ, respectively. The defined relation holds among these numbers precisely when

- φ is the string $\sim(\psi)$ or   ☞ $x = \text{Neg}(y)$

- φ is the string $(\psi) \lor (\chi)$ or   ☞ $x = y\,\text{Dis}\,z$

  - ξ is a variable and   ☞ $\exists\,(v \leq x)\,\text{Var}(v)$

  - φ is the string $\xi\,\Pi\,(\psi)$.   ☞ $x = v\,\text{Gen}\,y$

---

22. $FR(x)$ Auxiliary relation for $\text{Form}(x)$

(Primitive-recursive relation)

> 22. $FR\,(x) \equiv (n)\,\{0 < n \leqq l(x) \to Elf\,(n\,Gl\,x) \lor$
> $(E\,p,q)\,[0 < p,q < n\,\&\,Op\,(n\,Gl\,x, p\,Gl\,x, q\,Gl\,x)]\}$
> $\&\,l(x) > 0$
>
> x ist eine Reihe von *Formeln*, deren jede entweder *Elementarformel* ist oder aus den vorhergehenden durch die Operationen der *Negation, Disjunktion, Generalisation* hervorgeht.

> 22. $FR\,(x) \equiv (n)\,\{0 < n \leqq l(x) \to Elf\,(n\,Gl\,x) \lor$
> $(E\,p,q)\,[0 < p,q < n\,\&\,Op\,(n\,Gl\,x, p\,Gl\,x, q\,Gl\,x)]\}$
> $\&\,l(x) > 0$
>
> x is a sequence of *formulas* each one of which is either an *elementary formula* or comes from preceding ones by the operations of *negation, disjunction,* or *generalization*.

In modern notation:

$x \in FR :\Leftrightarrow$

$l(x) > 0 \wedge \forall (0 < n \leq l(x)) \begin{pmatrix} Elf(n \, Gl \, x) \vee \\ \exists (0 < p, q < n) \;\; Op(n \, Gl \, x, p \, Gl \, x, q \, Gl \, x) \end{pmatrix}$

The right-hand side is true exactly when

- x is the Gödel number of a non-empty number series, and   ☞ $l(x) > 0$
- every number in this series is the Gödel number   ☞ $\forall (0 < n \leq l(x))$
  - of an elementary formula or   ☞ $Elf(n \, Gl \, x)$
  - arises from previous formulas through   ☞ $\exists (0 < p, q < n)$
  - negation, disjunction, or generalization.   ☞ $Op(n \, Gl \, x, p \, Gl \, x, q \, Gl \, x)$

Negation, disjunction, and generalization are the three recursive construction schemata governing the construction of formulas in Definition 4.4. Any formula of the system P is obtained by starting with a series of elementary formulas and subsequently combining them into more complex structures using these schemata. For instance, the formula

$$x_2 \, \Pi \, ((x_2(x_1)) \vee (\neg(x_2(x_1))))$$

is constructed as follows:

1. $x_2(x_1)$ (Atomic formula)
2. $\neg(x_2(x_1))$ (Negation of 1.)
3. $(x_2(x_1)) \vee (\neg(x_2(x_1)))$ (Disjunction of 1. and 2.)
4. $x_2 \, \Pi \, ((x_2(x_1)) \vee (\neg(x_2(x_1))))$ (Generalization of 3.)

Arranged sequentially, we get the series

$$x_2(x_1), \neg(x_2(x_1)), (x_2(x_1)) \vee (\neg(x_2(x_1))), x_2 \, \Pi \, ((x_2(x_1)) \vee (\neg(x_2(x_1))))$$

which is also expressable on the arithmetic level:

$$\ulcorner x_2(x_1) \urcorner, \ulcorner \neg(x_2(x_1)) \urcorner, \ulcorner (x_2(x_1)) \vee (\neg(x_2(x_1))) \urcorner, \ulcorner x_2 \, \Pi \, ((x_2(x_1)) \vee (\neg(x_2(x_1)))) \urcorner$$

This series is what Gödel calls a series of *formulas*, that is, a series of numbers, each being the Gödel number of a formula constructed according to the syntactic rules of the formal system P.

5.2 Primitive-Recursive Functions and Relations

23. Form(x) x is a formula
(Primitive-recursive relation)

> 23. $\text{Form}(x) \equiv (En)\{n \leq (Pr\,[l\,(x)^2])^{x\,\cdot\,[l\,(x)]^2}$
> $\&\, FR\,(n)\, \&\, x = [l\,(n)]\, G\,l\,n\}$ [35]
>
> x ist *Formel* (d. h. letztes Glied einer *Formelreihe* n).
>
> [35] Die Abschätzung $n \leq (Pr\,[l\,(x)^2])^{x\,l\,(x)^2}$ erkennt man etwa so: Die Länge der kürzesten zu x gehörigen Formelreihe kann höchstens gleich der Anzahl der Teilformeln von x sein. Es gibt aber höchstens $l\,(x)$ Teilformeln der Länge 1, höchstens $l\,(x) - 1$ der Länge 2 usw., im ganzen also höchstens $\dfrac{l\,(x)\,[l\,(x) + 1]}{2} \leq l\,(x)^2$. Die Primzahlen aus n können also sämtlich kleiner als $Pr\{[l\,(x)]^2\}$ angenommen werden, ihre Anzahl $\leq l\,(x)^2$ und ihre Exponenten (welche Teilformeln von x sind) $\leq x$.

> 23. $\text{Form}(x) \equiv (En)\{n \leq (Pr\,[l\,(x)^2])^{x\,\cdot\,[l\,(x)]^2}$
> $\&\, FR\,(n)\, \&\, x = [l\,(n)]\, G\,l\,n\}$ [35]
>
> x is a *formula* (i.e. last term of a *sequence of formulas* n).
>
> [35] One finds the bound $n \leq (Pr\,[l\,(x)^2])^{x\,l\,(x)^2}$ as follows: the length of the shortest sequence of formulas belonging to x can be at most equal to the number of subformulas of x. There are, however, at most $l\,(x)$ subformulas of length 1, at most $l\,(x) - 1$ of length 2, etc., and, therefore, all together, at most $\dfrac{l\,(x)\,[l\,(x) + 1]}{2} \leq l\,(x)^2$. The prime divisors of n can therefore all be taken smaller than $Pr\{[l\,(x)]^2\}$, their number $\leq l\,(x)^2$ and their exponents (which are *subformulas* of x) $\leq x$.

In modern notation:

$$x \in \text{Form} :\Leftrightarrow \exists\,(n < c)\,(FR(n) \wedge x = l(n)\,Gl\,n)$$
$$\text{with } c = (Pr(l(x)^2))^{x \cdot l(x)^2}$$

The right-hand side is true precisely if

- there is a series of formulas, and   ☞ $\exists\,(n < c)\ FR(n)$

- x is the Gödel number of the last formula of this series.   ☞ $x = l(n)\,Gl\,n$

The natural number n is the Gödel number of this series. To ensure the primitive-recursive nature of this relation, the range of the existential quantifier must be restricted. Gödel has chosen

$$(Pr(l(x)^2))^{x \cdot l(x)^2}$$

as a suitable bound. In footnote 35, he elaborates on how he reached this value. To get a better sense of Gödel's idea, let's consider the following formula, which has already been used as an example above:

$$\varphi = \mathsf{x}_2 \, \Pi \, ((\mathsf{x}_2(\mathsf{x}_1)) \vee (\neg(\mathsf{x}_2(\mathsf{x}_1))))$$

Let x be the Gödel number of this formula:

$$x = \ulcorner \varphi \urcorner = \ulcorner \mathsf{x}_2 \, \Pi \, ((\mathsf{x}_2(\mathsf{x}_1)) \vee (\neg(\mathsf{x}_2(\mathsf{x}_1)))) \urcorner$$

Furthermore, let n be the Gödel number of the shortest formula series that generates φ. This formula series is already familiar to us. It looks like this:

$$\mathsf{x}_2(\mathsf{x}_1), \; \neg(\mathsf{x}_2(\mathsf{x}_1)), \; (\mathsf{x}_2(\mathsf{x}_1)) \vee (\neg(\mathsf{x}_2(\mathsf{x}_1))), \; \mathsf{x}_2 \, \Pi \, ((\mathsf{x}_2(\mathsf{x}_1)) \vee (\neg(\mathsf{x}_2(\mathsf{x}_1))))$$

Thus, the following holds true in our example:

$$n = 2^{\ulcorner \varphi_1 \urcorner} \cdot 3^{\ulcorner \varphi_2 \urcorner} \cdot 5^{\ulcorner \varphi_3 \urcorner} \cdot 7^{\ulcorner \varphi_4 \urcorner} \quad \text{with}$$

$$\varphi_1 = \mathsf{x}_2(\mathsf{x}_1)$$
$$\varphi_2 = \neg(\mathsf{x}_2(\mathsf{x}_1))$$
$$\varphi_3 = (\mathsf{x}_2(\mathsf{x}_1)) \vee (\neg(\mathsf{x}_2(\mathsf{x}_1)))$$
$$\varphi_4 = \mathsf{x}_2 \, \Pi \, ((\mathsf{x}_2(\mathsf{x}_1)) \vee (\neg(\mathsf{x}_2(\mathsf{x}_1))))$$

For other formulas, the construction adheres to the same scheme. If p_i denotes the i-th prime number, n has the following general form:

$$n = p_1^{\ulcorner \varphi_1 \urcorner} \cdot p_2^{\ulcorner \varphi_2 \urcorner} \cdot p_3^{\ulcorner \varphi_3 \urcorner} \cdot \ldots \cdot p_{l(n)}^{\ulcorner \varphi_{l(n)} \urcorner}$$

Since the formulas $\varphi_1, \varphi_2, \ldots, \varphi_{l(n)}$ are all subformulas of φ, the estimate

$$\ulcorner \varphi_i \urcorner \leq \ulcorner \varphi \urcorner = x$$

holds, which can be used to estimate n as follows:

$$n \leq p_1^x \cdot p_2^x \cdot p_3^x \cdot \ldots \cdot p_{l(n)}^x \tag{5.13}$$

Next, let's look closer at $l(n)$, the length of the formula series. For this purpose, we first examine how many subformulas each series member contains. For the above example formula, the following holds:

- φ_1 contains 1 subformula ☞ φ_1 itself
- φ_2 contains 2 subformulas ☞ φ_1 and φ_2
- φ_3 contains 4 subformulas ☞ $2 \times \varphi_1$, φ_2, and φ_3
- φ_4 contains 5 subformulas ☞ $2 \times \varphi_1$, φ_2, φ_3, and φ_4

In our example, the number of subformulas of the last series member exceeds the number of sequence elements $l(n)$. This comes as no surprise, as the number

5.2 Primitive-Recursive Functions and Relations

of subformulas increases by at least one from one sequence element to the next. In general, we have:

$$l(n) \leq \text{Number of subformulas of } \varphi \qquad (5.14)$$

Next, let us consider how many subformulas of a certain length can be contained in φ. For our example, we get:

- φ contains 1 subformula of length 4   ☞ Formula φ_1
- φ contains 1 subformula of length 7   ☞ Formula φ_2
- φ contains 1 subformula of length 16   ☞ Formula φ_3
- φ contains 1 subformula of length 20   ☞ Formula φ_4

In general, the following holds:

- φ contains at most 1 subformula of length $l(x)$
- φ contains at most 2 subformulas of length $l(x) - 1$
- φ contains at most 3 subformulas of length $l(x) - 2$
- $\ldots$
- φ contains at most $l(x)$ subformulas of length 1

Thus,

$$\begin{aligned}
&\text{Number of subformulas of } \varphi \\
&= \sum_{i=1}^{l(x)} \text{Number of subformulas of length } i \\
&\leq \sum_{i=1}^{l(x)} i = \frac{l(x)(l(x)+1)}{2} = \frac{l(x)^2}{2} + \frac{l(x)}{2} \leq \frac{l(x)^2}{2} + \frac{l(x)^2}{2} = l(x)^2
\end{aligned}$$

In combination with (5.14), this implies:

$$l(n) \leq l(x)^2 \qquad (5.15)$$

Consequently, the formula series producing φ can have no more than $l(x)^2$ elements. Combining this result with (5.13) yields the following estimate:

$$\begin{aligned}
n &\leq p_1{}^x \cdot p_2{}^x \cdot p_3{}^x \cdot \ldots \cdot p_{l(x)^2}{}^x \\
&\leq p_{l(x)^2}{}^x \cdot p_{l(x)^2}{}^x \cdot p_{l(x)^2}{}^x \cdot \ldots \cdot p_{l(x)^2}{}^x \\
&= p_{l(x)^2}{}^{x \cdot l(x)^2}
\end{aligned}$$

This value matches Gödel's bound on the existential quantifier.

24. $v\,\mathrm{Geb}\,n, x$ — Variable v is bound at position n

(Primitive-recursive relation)

> 24. $v\,\mathrm{Geb}\,n, x \equiv \mathrm{Var}\,(v)\,\&\,\mathrm{Form}\,(x)\,\&$
> $(E\,a, b, c)\,[a, b, c \leq x\,\&\,x = a * (v\,\mathrm{Gen}\,b) * c$
> $\&\,\mathrm{Form}\,(b)\,\&\,l\,(a) + 1 \leq n \leq l\,(a) + l\,(v\,\mathrm{Gen}\,b)]$
> Die *Variable* v ist in x an n-ter Stelle *gebunden*.

> 24. $v\,\mathrm{Geb}\,n, x \equiv \mathrm{Var}\,(v)\,\&\,\mathrm{Form}\,(x)\,\&$
> $(E\,a, b, c)\,[a, b, c \leq x\,\&\,x = a * (v\,\mathrm{Gen}\,b) * c$
> $\&\,\mathrm{Form}\,(b)\,\&\,l\,(a) + 1 \leq n \leq l\,(a) + l\,(v\,\mathrm{Gen}\,b)]$
> The *variable* v is *bound* at the n-th place in x.

In modern notation:

$v\,\mathrm{Geb}\,n, x \;:\Leftrightarrow$

$\mathrm{Var}(v) \wedge \mathrm{Form}(x) \wedge \exists\,(a, b, c \leq x) \begin{pmatrix} x = a * (v\,\mathrm{Gen}\,b) * c \wedge \mathrm{Form}(b) \wedge \\ l(a) + 1 \leq n \leq l(a) + l(v\,\mathrm{Gen}\,b) \end{pmatrix}$

Let v be the Gödel number of a variable ξ and x the Gödel number of a formula φ. Then, $v\,\mathrm{Geb}\,n, x$ holds true precisely when the n-th character of the formula φ is embedded in a subformula of the form $\forall \xi\,(\ldots)$:

$$\Phi(\xi) = v$$
$$\updownarrow$$
$$x = \ulcorner \varphi \urcorner = \ulcorner (\;\ldots\;\underbrace{\forall \xi\,(\;\ldots\;)}\;\ldots\;) \urcorner$$
$$\updownarrow$$
$$\text{Possible locations for } n$$

This is the case if and only if

■ v is the Gödel number of a variable ξ, and ☞ $\mathrm{Var}(v)$

■ x is the Gödel number of a formula ☞ $\mathrm{Form}(x)$

■ that can be divided into three parts, ☞ $\exists\,(a, b, c \leq x)$

- such that the middle part is a formula that binds ξ
  ☞ $x = a * (v\,\mathrm{Gen}\,b) * c \wedge \mathrm{Form}(b)$

- and the n-th character is within the middle part.
  ☞ $l(a) + 1 \leq n \leq l(a) + l(v\,\mathrm{Gen}\,b)$

5.2 Primitive-Recursive Functions and Relations

Let us consider the formula

$$\varphi = \underbrace{(\ \mathsf{x}_1\ \Pi\ (\ \mathsf{x}_2\ (\ \mathsf{x}_1\ )\ )\ )\ \vee\ (\ \mathsf{x}_2\ \Pi\ (\ \mathsf{x}_2\ (\ \mathsf{x}_1\ )\ )\ )}_{1\ \ 2\ \ \ 3\ \ 4\ \ 5\ \ 6\ \ 7\ \ \ 8\ 9\ 10\ 11\ 12\ 13\ \ 14\ 15\ 16\ 17\ 18\ 19\ 20\ 21}.$$

In this example, the following holds:

$17\,\text{Geb}\,n, \ulcorner\varphi\urcorner \Leftrightarrow n \in \{2,3,4,5,6,7,8,9\}$ (☞ $17 = \Phi(\mathsf{x}_1)$)

$17^2\,\text{Geb}\,n, \ulcorner\varphi\urcorner \Leftrightarrow n \in \{13,14,15,16,17,18,19,20\}$ (☞ $17^2 = \Phi(\mathsf{x}_2)$)

---

25. $v\,Fr\,n,x$ Variable v occurs freely at position n

(Primitive-recursive relation)

---

184   Kurt Gödel,

25. $v\,Fr\,n,x \equiv \text{Var}\,(v)\,\&\,\text{Form}\,(x)\,\&\,v = n\,Gl\,x\,\&$
$n \leq l(x)\,\&\,\overline{v\,\text{Geb}\,n,x}$

Die *Variable* v ist in x an n-ter Stelle *frei*.

---

184   Kurt Gödel,

25. $v\,Fr\,n,x \equiv \text{Var}\,(v)\,\&\,\text{Form}\,(x)\,\&\,v = n\,Gl\,x\,\&$
$n \leq l(x)\,\&\,\overline{v\,\text{Geb}\,n,x}$

The *variable* v is *free* at the n-th place in x.

In modern notation:

$$v\,Fr\,n,x :\Leftrightarrow \text{Var}(v) \wedge \text{Form}(x) \wedge v = n\,Gl\,x \wedge n \leq l(x) \wedge \neg(v\,\text{Geb}\,n,x)$$

The right-hand side holds true precisely when the variable with the Gödel number v appears freely at the n-th position in the formula with the Gödel number x. This is the case if and only if

- v is the Gödel number of a variable ξ,   ☞ $\text{Var}(v)$
- x is the Gödel number of a formula, and   ☞ $\text{Form}(x)$
- ξ appears at the n-th position   ☞ $v = n\,Gl\,x$
  - within the formula and   ☞ $n \leq l(x)$
  - is not bound at that position.   ☞ $\neg(v\,\text{Geb}\,n,x)$

As an example, let's revisit

$$\varphi = \underbrace{(}_{1} \underbrace{x_1}_{2} \underbrace{\Pi}_{3} \underbrace{(}_{4} \underbrace{x_2}_{5} \underbrace{(}_{6} \underbrace{x_1}_{7} \underbrace{)}_{8} \underbrace{)}_{9} \underbrace{)}_{10} \underbrace{\vee}_{11} \underbrace{(}_{12} \underbrace{x_2}_{13} \underbrace{\Pi}_{14} \underbrace{(}_{15} \underbrace{x_2}_{16} \underbrace{(}_{17} \underbrace{x_1}_{18} \underbrace{)}_{19} \underbrace{)}_{20} \underbrace{)}_{21}.$$

For this formula, the following holds:

$$17 \, Fr \, n, \ulcorner\varphi\urcorner \Leftrightarrow n \in \{18\} \qquad (\text{☞ } 17 = \Phi(\mathsf{x}_1))$$
$$17^2 \, Fr \, n, \ulcorner\varphi\urcorner \Leftrightarrow n \in \{5\} \qquad (\text{☞ } 17^2 = \Phi(\mathsf{x}_2))$$

 **26.** $v \, Fr \, x$ Variable v occurs freely in x at least once

(Primitive-recursive relation)

26. $v \, Fr \, x \equiv (E\,n)\,[n \leq l\,(x)\, \& \, v\, Fr\, n, x]$
v kommt in x als *freie Variable* vor.

26. $v \, Fr \, x \equiv (E\,n)\,[n \leq l\,(x)\, \& \, v\, Fr\, n, x]$
v occurs in x as a *free variable*.

In modern notation:

$$v \, Fr \, x \; :\Leftrightarrow \; \exists\,(n \leq l(x)) \; v \, Fr \, n, x$$

The right-hand side holds true precisely when the variable with the Gödel number v appears freely at any position in the formula with the Gödel number x.

Let us consider the following example:

$$(\mathsf{x}_1 \, \Pi \, (\mathsf{x}_2(\mathsf{x}_1))) \vee (\mathsf{x}_2 \, \Pi \, (\mathsf{x}_2(\mathsf{x}_1))).$$

Listed below are instances where the relationship is true (✔) and cases where it is not (✘):

$17 \, Fr \, \ulcorner\varphi\urcorner$	✔	$17^2 \, Fr \, \ulcorner\varphi\urcorner$	✔	$17^3 \, Fr \, \ulcorner\varphi\urcorner$	✘
$19 \, Fr \, \ulcorner\varphi\urcorner$	✘	$19^2 \, Fr \, \ulcorner\varphi\urcorner$	✘	$19^3 \, Fr \, \ulcorner\varphi\urcorner$	✘

For each n, 17^n is the Gödel number of variable x_n, and 19^n is the Gödel number of variable y_n.

5.2 Primitive-Recursive Functions and Relations

> 27. $Su\,x\,\binom{n}{y}$ Formula x, after inserting y at position n
>
> (Primitive-recursive function)

> 27. $Su\,x\,\binom{n}{y} \equiv \varepsilon z\,\{z \leq [Pr(l(x) + l(y))]^{x+y}\,\&\,[(E\,u,v)\,u,v \leq x\,\&$
> $x = u * R\,(n\,Gl\,x) * v\,\&\,z = u * y * v\,\&\,n = l(u) + 1]\}$
> $Su\,x\,\binom{n}{y}$ entsteht aus x, wenn man an Stelle des n-ten Gliedes
> von x y einsetzt (vorausgesetzt, daß $0 < n \leq l(x)$).

> 27. $Su\,x\,\binom{n}{y} \equiv \varepsilon z\,\{z \leq [Pr(l(x) + l(y))]^{x+y}\,\&\,[(E\,u,v)\,u,v \leq x\,\&$
> $x = u * R\,(n\,Gl\,x) * v\,\&\,z = u * y * v\,\&\,n = l(u) + 1]\}$
> $Su\,x\,\binom{n}{y}$ arises from x by substituting y in place of the n-th term
> of x (assuming that $0 < n \leq l(x)$).

In modern notation:

$$Su\,x\,\binom{n}{y} := \min\left\{z \leq c\ \middle|\ \exists\,(u,v \leq x)\ \binom{x = u * R(n\,Gl\,x) * v\,\wedge}{z = u * y * v \wedge n = l(u) + 1}\right\}$$

with $c := (Pr(l(x) + l(y)))^{x+y}$

To get a firm grip on this definition, let us again consider the formula

$$\varphi = (\,\mathsf{x_1}\ \Pi\ (\,\mathsf{x_2}\ (\,\mathsf{x_1}\,)\,)\,)\ \vee\ (\,\mathsf{x_2}\ \Pi\ (\,\mathsf{x_2}\ (\,\mathsf{x_1}\,)\,)\,)\,.$$
$$1\ \ 2\ \ \ 3\ \ 4\ \ 5\ \ 6\ \ 7\ \ \ 8\ 9\ 10\ 11\ 12\ 13\ \ 14\ 15\ 16\ 17\ 18\ 19\ 20\ 21$$

Assuming variable $\mathsf{x_2}$ is to be replaced with variable $\mathsf{y_2}$ at position 5, we need to

- split the formula φ ☞ $x = u * R(n\,Gl\,x) * v$
- at position $n = 5$ ☞ $n = l(u) + 1$

$$\underbrace{\ulcorner(\mathsf{x_1}\ \Pi\ (\ulcorner}_{u} * \ulcorner\mathsf{x_2}\urcorner * \underbrace{\ulcorner(\mathsf{x_1}))) \vee (\mathsf{x_2}\ \Pi\ (\mathsf{x_2}(\mathsf{x_1})))\urcorner}_{v}$$

- and recombine the parts with $y = \ulcorner\mathsf{y_2}\urcorner$.   ☞ $z = u * y * v$

$$\underbrace{\ulcorner(\mathsf{x_1}\ \Pi\ (\ulcorner}_{u} * \underbrace{\ulcorner\mathsf{y_2}\urcorner}_{y} * \underbrace{\ulcorner(\mathsf{x_1}))) \vee (\mathsf{x_2}\ \Pi\ (\mathsf{x_2}(\mathsf{x_1})))\urcorner}_{v}$$

The justification for z being smaller than the chosen limit

$$(Pr(l(x) + l(y)))^{x+y}$$

aligns with the argument previously employed on page 240 concerning the star operator '∗'. The resulting formula has the form

$$z = p_1{}^{k_1} \cdot p_2{}^{k_2} \cdot p_3{}^{k_3} \cdot \ldots \cdot p_{l(z)}{}^{k_{l(z)}}$$

with

$$l(z) = l(x) + l(y) - 1.$$

Thus, the value of z can be estimated as follows:

$$\begin{aligned} z &\leq p_{l(z)}{}^{k_1} \cdot p_{l(z)}{}^{k_2} \cdot p_{l(z)}{}^{k_3} \cdot \ldots \cdot p_{l(z)}{}^{k_{l(z)}} \\ &= p_{l(z)}{}^{k_1+k_2+k_3+\ldots+k_{l(z)}} \\ &\leq p_{l(z)}{}^{x+y} \\ &\leq p_{l(x)+l(y)}{}^{x+y} \end{aligned}$$

 28. $k\,St\,v,x$ Auxiliary function for $A(v,x)$

(Primitive-recursive function)

28. $0\,St\,v,x \equiv \varepsilon n\,\{n \leq l\,(x)\,\&\,v\,Fr\,n,x$
        $\&\,\overline{(E\,p)}\,[n < p \leq l\,(x)\,\&\,v\,Fr\,p,x]\}$
$(k+1)\,St\,v,x \equiv \varepsilon n\,\{n < k\,St\,v,x\,\&\,v\,Fr\,n,x$
        $\&\,\overline{(E\,p)}\,[n < p < k\,St\,v,x\,\&\,v\,Fr\,p,x]\}$

$k\,St\,v,x$ ist die $k+1$-te Stelle in x (vom Ende der *Formel* x an gezählt), an der v in x *frei* ist (und 0, falls es keine solche Stelle gibt).

28. $0\,St\,v,x \equiv \varepsilon n\,\{n \leq l\,(x)\,\&\,v\,Fr\,n,x$
        $\&\,\overline{(E\,p)}\,[n < p \leq l\,(x)\,\&\,v\,Fr\,p,x]\}$
$(k+1)\,St\,v,x \equiv \varepsilon n\,\{n < k\,St\,v,x\,\&\,v\,Fr\,n,x$
        $\&\,\overline{(E\,p)}\,[n < p < k\,St\,v,x\,\&\,v\,Fr\,p,x]\}$

$k\,St\,v,x$ is the $(k+1)$st place in x (counting from the end of the *formula* x) at which v is *free* in x (and 0, in case there is no such place)

In modern notation:

$$0\,St\,v,x := \min\left\{n \leq l(x)\,\middle|\,\begin{array}{c} v\,Fr\,n,x\,\wedge \\ \neg \exists\,(p \leq l(x))\,(n < p \wedge v\,Fr\,p,x) \end{array}\right\}$$

$$k+1\,St\,v,x := \min\left\{n < k\,St\,v,x\,\middle|\,\begin{array}{c} v\,Fr\,n,x\,\wedge \\ \neg \exists\,(p < k\,St\,v,x)\,(n < p \wedge v\,Fr\,p,x) \end{array}\right\}$$

5.2 Primitive-Recursive Functions and Relations

Let v be the Gödel number of a variable ξ and x the Gödel number of a formula φ. The function $k\,St\,v, x$ calculates the position in φ, where the variable ξ, counted from the right, appears freely for the $(k+1)$-th time. For instance, formula

$$\varphi = \underbrace{(\;\mathsf{x}_2\;\Pi\;(\;\mathsf{x}_2\;(\;\mathsf{x}_1\;)\;)\;)\;\vee\;(\;\mathsf{y}_2\;(\;\mathsf{x}_1\;)\;)}_{1\;\;2\;\;\;3\;\;4\;\;5\;\;6\;\;7\;\;\;8\;9\;10\;11\;12\;13\;\;14\;15\;16\;17},$$

satisfies the following:

$0\,St\,\Phi(\mathsf{x}_1), \ulcorner\varphi\urcorner = 15$	$0\,St\,\Phi(\mathsf{x}_2), \ulcorner\varphi\urcorner = 0$	$0\,St\,\Phi(\mathsf{y}_2), \ulcorner\varphi\urcorner = 13$
$1\,St\,\Phi(\mathsf{x}_1), \ulcorner\varphi\urcorner = 7$	$1\,St\,\Phi(\mathsf{x}_2), \ulcorner\varphi\urcorner = 0$	$1\,St\,\Phi(\mathsf{y}_2), \ulcorner\varphi\urcorner = 0$
$2\,St\,\Phi(\mathsf{x}_1), \ulcorner\varphi\urcorner = 0$	$2\,St\,\Phi(\mathsf{x}_2), \ulcorner\varphi\urcorner = 0$	$2\,St\,\Phi(\mathsf{y}_2), \ulcorner\varphi\urcorner = 0$

The definition arises from the following observation: The natural number n is the first position from the right where ξ appears free in φ, if

- ξ appears free at position n and ☞ $v\,Fr\,n, x$
- is nowhere free between n and the end of the formula. ☞ $\neg\exists\,(p \leq l(x))\,(n < p \wedge v\,Fr\,p, x)$

The natural number n is the $(k+2)$-th position from the right $(k \geq 0)$ where ξ appears free in φ, if

- ξ appears free at position n and ☞ $v\,Fr\,n, x$
- is nowhere free between n and the $(k+1)$-th free appearance. ☞ $\neg\exists\,(p < k\,St\,v, x)\,(n < p \wedge v\,Fr\,p, x)$

 **29.** $A(v, x)$ Number of positions where v occurs freely in x

(Primitive-recursive function)

---

29. $A(v, x) \equiv \varepsilon n\,\{n \leq l(x)\;\&\;n\,St\,v, x = 0\}$
$A(v, x)$ ist die Anzahl der Stellen, an denen v in x *frei* ist.

---

29. $A(v, x) \equiv \varepsilon n\,\{n \leq l(x)\;\&\;n\,St\,v, x = 0\}$
$A(v, x)$ is the number of places at which v is *free* in x.

In modern notation:

$$A(v, x) := \min\{n \leq l(x) \mid n\,St\,v, x = 0\}$$

According to this definition, $A(v,x)$ is the smallest natural number n for which $n \, St \, v, x$ yields 0. For example, the formula

$$\varphi = (x_2 \, \Pi \, (x_2(x_1))) \vee (y_2(x_1))$$

satisfies the following:

$$A(\Phi(x_1), \ulcorner \varphi \urcorner) = 2 \qquad A(\Phi(x_2), \ulcorner \varphi \urcorner) = 0 \qquad A(\Phi(y_2), \ulcorner \varphi \urcorner) = 1$$

The meaning of $A(v,x)$ is thus exposed: The function calculates how often the variable with the Gödel number v appears freely in the formula with the Gödel number x.

---

30. $Sb_n \left(x \, {}^v_y \right)$ — Auxiliary function for $Sb \left(x \, {}^v_y \right)$

(Primitive-recursive function)

$$30. \quad Sb_0 \left(x \, {}^v_y \right) \equiv x$$
$$Sb_{k+1} \left(x \, {}^v_y \right) \equiv Su \left[Sb_k \left(x \, {}^v_y \right) \right] \left({}^{k \, St \, v}_{\quad y}, x \right)$$

$$30. \quad Sb_0 \left(x \, {}^v_y \right) \equiv x$$
$$Sb_{k+1} \left(x \, {}^v_y \right) \equiv Su \left[Sb_k \left(x \, {}^v_y \right) \right] \left({}^{k \, St \, v}_{\quad y}, x \right)$$

In modern notation:

$$Sb_0 \left(x \, {}^v_y \right) := x$$
$$Sb_{k+1} \left(x \, {}^v_y \right) := Su \left(Sb_k \left(x \, {}^v_y \right) \right) \left({}^{k \, St \, v, x}_{\quad y} \right)$$

The function $Sb_k \left(x \, {}^v_y \right)$ replaces k free occurrences of the variable with the Gödel number v by the string with the Gödel number y. The replacement happens from right to left. The definition builds upon the following idea: If $k = 0$, there is nothing to do; the function returns x. If $k > 0$, the replacement can be carried out recursively by

- replacing $k - 1$ occurrences   ☞ $Sb_k \left(x \, {}^v_y \right)$
- and then replacing another occurrence.   ☞ $Su(Sb_k(x\,{}^v_y)) \left({}^{k \, St \, v, x}_{\quad y} \right)$

The following examples show the function in action:

$$Sb_0 \left(\ulcorner (x_2(x_1) \vee y_2(x_1)) \vee z_2(x_1) \urcorner \, {}^{17}_{24} \right) = \ulcorner (x_2(x_1) \vee y_2(x_1)) \vee z_2(x_1) \urcorner$$
$$Sb_1 \left(\ulcorner (x_2(x_1) \vee y_2(x_1)) \vee z_2(x_1) \urcorner \, {}^{17}_{24} \right) = \ulcorner (x_2(x_1) \vee y_2(x_1)) \vee z_2(f \, 0) \urcorner$$

5.2 Primitive-Recursive Functions and Relations

$$Sb_2\left(\ulcorner(x_2(x_1) \vee y_2(x_1)) \vee z_2(x_1)\urcorner\,{}^{17}_{24}\right) = \ulcorner(x_2(x_1) \vee y_2(f\,0)) \vee z_2(f\,0)\urcorner$$
$$Sb_3\left(\ulcorner(x_2(x_1) \vee y_2(x_1)) \vee z_2(x_1)\urcorner\,{}^{17}_{24}\right) = \ulcorner(x_2(f\,0) \vee y_2(f\,0)) \vee z_2(f\,0)\urcorner$$

Note that 17 is the Gödel number of x_1, and 24 is the Gödel number of f 0, as can be easily verified:

$$\ulcorner f\,0\urcorner = 2^{\Phi(f)} \cdot 3^{\Phi(0)} = 2^3 \cdot 3^1 = 24$$

---

31. $Sb\left(x\,{}^{v}_{y}\right)$ Substitution of v by y

(Primitive-recursive function)

---

31. $Sb\left(x\,{}^{v}_{y}\right) \equiv Sb_{A\,(v,\,x)}\left(x\,{}^{v}_{y}\right)$ [36])
$Sb\left(x\,{}^{v}_{y}\right)$ ist der oben definierte Begriff *Subst a* $\left({}^{v}_{b}\right)$ [37]).

---

[36]) Falls v keine *Variable* oder x keine *Formel* ist, ist $Sb\left(x\,{}^{v}_{y}\right) = x$.
[37]) Statt $Sb\left[Sb\left(x\,{}^{v}_{y}\right)\,{}^{w}_{z}\right]$ schreiben wir: $Sb\left(x\,{}^{v}_{y}\,{}^{w}_{z}\right)$ (analog für mehr als zwei *Variable*).

---

31. $Sb\left(x\,{}^{v}_{y}\right) \equiv Sb_{A\,(v,\,x)}\left(x\,{}^{v}_{y}\right)$ [36])
$Sb\left(x\,{}^{v}_{y}\right)$ is the concept *Subst a* $\left({}^{v}_{b}\right)$ defined above. [37])

---

[36]) In case v is not a *variable* or x is not a *formula*, then $Sb\left(x\,{}^{v}_{y}\right) = x$.
[37]) Instead of $Sb\left[Sb\left(x\,{}^{v}_{y}\right)\,{}^{w}_{z}\right]$ we write $Sb\left(x\,{}^{v}_{y}\,{}^{w}_{z}\right)$ (and similarly for more than two *variables*).

In modern notation:

$$Sb\left(x\,{}^{v}_{y}\right) := Sb_{A(v,x)}\left(x\,{}^{v}_{y}\right)$$

Let x be the Gödel number of a formula φ, v the Gödel number of a variable ξ, and y the Gödel number of a symbol string σ. The primitive-recursive function $A(v, x)$ calculates how often ξ appears freely in φ. Consequently, $Sb_{A(v,x)}\left(x\,{}^{v}_{y}\right)$ is the Gödel number of the formula φ, after all free occurrences of ξ have been replaced by σ:

$$Sb\left(x\,{}^{v}_{y}\right) = \ulcorner\varphi[\xi \leftarrow \sigma]\urcorner$$

For instance:

$$Sb\left(\ulcorner(x_2(x_1) \vee y_2(x_1)) \vee z_2(x_1)\urcorner\,{}^{17}_{24}\right) = \ulcorner(x_2(f\,0) \vee y_2(f\,0)) \vee z_2(f\,0)\urcorner$$

The example reveals that Sb is the substitution function discussed in Section 4.1.2. In footnote 36, Gödel explicitly points out that $Sb\left(x\,{}^{v}_{y}\right)$ changes the number x only when v is the Gödel number of a variable and x the Gödel number of a formula. To understand the reason, let us revisit the definition of

Fr from page 255:

$$v \, Fr \, n, x :\Leftrightarrow \text{Var}(v) \wedge \text{Form}(x) \wedge v = n \, Gl \, x \wedge n \leq l(x) \wedge \neg(v \, \text{Geb} \, n, x)$$

If v is not the Gödel number of a variable or x is not the Gödel number of a formula, the right-hand side is false, implying that the numbers v and x do not satisfy the relation Fr for any n. Consequently, $n \, St \, v, x = 0$ (Definition 28) and thus also $A(v, x) = 0$ (Definition 29). Now, the assertion of footnote 36 follows immediately from Definition 30 and Definition 31:

$$Sb \left(x \, {}^v_y \right) = Sb_{A(v,x)} \left(x \, {}^v_y \right) = Sb_0 \left(x \, {}^v_y \right) = x$$

In footnote 37, Gödel introduces a concise notation for the repeated execution of function Sb. He will use this notation extensively later on.

---

32. $x \, \text{Imp} \, y$, $x \, \text{Con} \, y$, $x \, \text{Aeq} \, y$, $v \, \text{Ex} \, y$ Definition of '$\rightarrow$', '$\wedge$', '$\leftrightarrow$', '$\exists$'

(Primitive-recursive function)

$$
\begin{aligned}
32. \quad & x \, \text{Imp} \, y \equiv [\text{Neg}(x)] \, \text{Dis} \, y \\
& x \, \text{Con} \, y \equiv \text{Neg} \, \{[\text{Neg}(x)] \, \text{Dis} \, [\text{Neg}(y)]\} \\
& x \, \text{Aeq} \, y \equiv (x \, \text{Imp} \, y) \, \text{Con} \, (y \, \text{Imp} \, x) \\
& v \, \text{Ex} \, y \equiv \text{Neg} \, \{v \, \text{Gen} \, [\text{Neg}(y)]\}
\end{aligned}
$$

$$
\begin{aligned}
32. \quad & x \, \text{Imp} \, y \equiv [\text{Neg}(x)] \, \text{Dis} \, y \\
& x \, \text{Con} \, y \equiv \text{Neg} \, \{[\text{Neg}(x)] \, \text{Dis} \, [\text{Neg}(y)]\} \\
& x \, \text{Aeq} \, y \equiv (x \, \text{Imp} \, y) \, \text{Con} \, (y \, \text{Imp} \, x) \\
& v \, \text{Ex} \, y \equiv \text{Neg} \, \{v \, \text{Gen} \, [\text{Neg}(y)]\}
\end{aligned}
$$

These functions are provided merely for convenience. They define implication, conjunction, logical equivalence, and existential quantification in terms of the native language constructs of system P.

---

33. $n \, Th \, x$ n-th type elevation of x

(Primitive-recursive function)

$$
\begin{aligned}
33. \quad n \, Th \, x \equiv \, & \varepsilon y \, \{y \leq x^{(x^n)} \, \& \, (k) \, [k \leq l(x) \rightarrow \\
& (k \, Gl \, x \leq 13 \, \& \, k \, Gl \, y = k \, Gl \, x) \vee \\
& (k \, Gl \, x > 13 \, \& \, k \, Gl \, y = k \, Gl \, x . \, [1 \, Pr \, (k \, Gl \, x)]^n)]\} \\
& n \, Th \, x \text{ ist die } n\text{-te Typenerhöhung von } x \text{ (falls } x \text{ und } n \, Th \, x \\
& \textit{Formeln sind)}.
\end{aligned}
$$

5.2 Primitive-Recursive Functions and Relations

> 33. $n\ Th\ x \equiv \varepsilon y\ \{y \leq x^{(x^n)}\ \&\ (k)\ [k \leq l(x) \rightarrow$
> $(k\ Gl\ x \leq 13\ \&\ k\ Gl\ y = k\ Gl\ x) \lor$
> $(k\ Gl\ x > 13\ \&\ k\ Gl\ y = k\ Gl\ x.[1\ Pr\ (k\ Gl\ x)]^n)]\}$
> $n\ Th\ x$ is the n-th type elevation of x (in case x and $n\ Th\ x$ are formulas).

In modern notation:

$$n\ Th\ x := \min\left\{y \leq x^{x^n}\ \middle|\ \forall(k \leq l(x))\begin{pmatrix}(k\ Gl\ x \leq 13 \land k\ Gl\ y = k\ Gl\ x) \lor \\ (k\ Gl\ x > 13\ \land \\ k\ Gl\ y = k\ Gl\ x \cdot (1\ Pr(k\ Gl\ x))^n)\end{pmatrix}\right\}$$

To get a handle on this definition, let's take a closer look at the following example:

$$2^{17^2} \cdot 3^9 \cdot 5^{11} \cdot 7^5 \cdot 11^{11} \cdot 13^{17^2} \cdot 17^{11} \cdot 19^{19} \cdot 23^{13} \cdot 29^{13} \cdot 31^{13}$$

$$\underbrace{\mathsf{x_2\ \Pi\ (\sim (x_2(y_1)))}}$$

type elevation

$$\underbrace{\mathsf{x_{2+n}\ \Pi\ (\sim (x_{2+n}(y_{1+n})))}}$$

$$2^{17^2+n} \cdot 3^9 \cdot 5^{11} \cdot 7^5 \cdot 11^{11} \cdot 13^{17^2+n} \cdot 17^{11} \cdot 19^{19^{1+n}} \cdot 23^{13} \cdot 29^{13} \cdot 31^{13}$$

From this example, the following procedure emerges:

- ■ If the k-th exponent is ≤ 13,   ☞ $k\ Gl\ x \leq 13$

$$3^9 \cdot 5^{11} \cdot 7^5 \cdot 11^{11} \qquad \cdot 17^{11} \qquad \cdot 23^{13} \cdot 29^{13} \cdot 31^{13}$$

type elevation

$$3^9 \cdot 5^{11} \cdot 7^5 \cdot 11^{11} \qquad \cdot 17^{11} \qquad \cdot 23^{13} \cdot 29^{13} \cdot 31^{13}$$

then keep the number series as it is.   ☞ $k\ Gl\ y = k\ Gl\ x$

- ■ If the k-th exponent is > 13,   ☞ $k\ Gl\ x > 13$

$$2^{17^2} \qquad \cdot 13^{17^2} \qquad \cdot 19^{19}$$

type elevation

$$2^{17^2+n} \qquad \cdot 13^{17^2+n} \qquad \cdot 19^{19^{1+n}}$$

- • then extract the exponent (which is of the form e^i),   ☞ $k\ Gl\ x$

- determine the number e
- and calculate e^{i+n}.
- The result is the k-th factor of y.

☞ $1\,Pr(k\,Gl\,x)$
☞ $k\,Gl\,x \cdot (1\,Pr(k\,Gl\,x))^n$
☞ $k\,Gl\,y = k\,Gl\,x \cdot (1\,Pr(k\,Gl\,x))^n$

It remains to be shown that the estimate $y \leq x^{x^n}$ always holds. Starting from the equation

$$x = p_1^{e_1^{i_1}} \cdot p_2^{e_2^{i_2}} \cdot \ldots \cdot p_k^{e_k^{i_k}},$$

the value of y can be estimated as follows:

$$\begin{aligned}
y &\leq p_1^{e_1^{i_1+n}} \cdot p_2^{e_2^{i_2+n}} \cdot \ldots \cdot p_k^{e_k^{i_k+n}} \\
&= p_1^{e_1^{i_1} \cdot e_1^n} \cdot p_2^{e_2^{i_2} \cdot e_2^n} \cdot \ldots \cdot p_k^{e_k^{i_k} \cdot e_k^n} \\
&= \left(p_1^{e_1^{i_1}}\right)^{e_1^n} \cdot \left(p_2^{e_2^{i_2}}\right)^{e_2^n} \cdot \ldots \cdot \left(p_k^{e_k^{i_k}}\right)^{e_k^n}
\end{aligned}$$

Let j be the index between 1 and k that maximizes $p_j^{e_j^{i_j}}$. Then:

$$\begin{aligned}
y &\leq \left(p_j^{e_j^{i_j}}\right)^{e_1^n} \cdot \left(p_j^{e_j^{i_j}}\right)^{e_2^n} \cdot \ldots \cdot \left(p_j^{e_j^{i_j}}\right)^{e_k^n} \\
&= \left(p_j^{e_j^{i_j}}\right)^{e_1^n + e_2^n + \ldots + e_k^n} \\
&\leq \left(p_j^{e_j^{i_j}}\right)^{(e_1 + e_2 + \ldots + e_k)^n} \\
&\leq \left(p_j^{e_j^{i_j}}\right)^{x^n} \\
&\leq x^{x^n}
\end{aligned}$$

---

34. $Z\text{-}Ax(x)$ x is an instance of the axiom schema I.1, I.2, or I.3

(Primitive-recursive relation)

Den Axiomen I, 1 bis 3 entsprechen drei bestimmte Zahlen, die wir mit z_1, z_2, z_3 bezeichnen, und wir definieren:
34. $Z\text{-}A\,x\,(x) \equiv (x = z_1 \lor x = z_2 \lor x = z_3)$

To the Axioms I, 1–3 correspond three definite numbers, which we denote by z_1, z_2, z_3, and we define:
34. $Z\text{-}A\,x\,(x) \equiv (x = z_1 \lor x = z_2 \lor x = z_3)$

In a more detailed spelling:

$$x \in Z\text{-}Ax :\Leftrightarrow (x = z_1 \lor x = z_2 \lor x = z_3) \text{ with}$$

5.2 Primitive-Recursive Functions and Relations

$$z_1 = \ulcorner \sim (f\, x_1 = 0) \urcorner$$
$$z_2 = \ulcorner f\, x_1 = f\, y_1 \supset x_1 = y_1 \urcorner$$
$$z_3 = \ulcorner x_2(0)\,.\,x_1\, \Pi\, (x_2(x_1) \supset x_2(f\, x_1)) \supset x_1\, \Pi\, (x_2(x_1)) \urcorner$$

The right-hand side holds true precisely when x is the Gödel number of an axiom obtained from the axiom schema I.1, I.2, or I.3. The axioms are left here in Gödel's original notation.

Note that the formulas contain operators that are not part of the language of the system P. To calculate the Gödel numbers accurately, these operators must initially be substituted with their respective definitions.

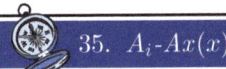

| 35. A_i-$Ax(x)$ | x is an instance of axiom schema II.i |

(Primitive-recursive relation)

Über formal unentscheidbare Sätze der Principia Mathematica etc. 185

35. A_1-$A\,x\,(x) \equiv (E\,y)\,[y \leq x\, \&\, \text{Form}\,(y)\,\&$
$\qquad x = (y\, \text{Dis}\, y)\, \text{Imp}\, y]$

x ist eine durch Einsetzung in das Axiomenschema II, 1 entstehende *Formel*. Analog werden A_2-$A\,x$, A_3-$A\,x$, A_4-$A\,x$ entsprechend den Axiomen II, 2 bis 4 definiert.

On formally undecidable propositions of Principia Mathematica etc. 185

35. A_1-$A\,x\,(x) \equiv (E\,y)\,[y \leq x\, \&\, \text{Form}\,(y)\,\&$
$\qquad x = (y\, \text{Dis}\, y)\, \text{Imp}\, y]$

x is a *formula* arising from a substitution in Axiom schema II, 1. Similarly, A_2-$A\,x$, A_3-$A\,x$, and A_4-$A\,x$, corresponding to axioms II, 2–4, are defined.

In modern notation:

$$x \in A_1\text{-}Ax \;:\Leftrightarrow\; \exists\,(y \leq x)\,(\text{Form}(y) \wedge x = (y\, \text{Dis}\, y)\, \text{Imp}\, y)$$

The right-hand side holds true precisely when x is the Gödel number of an instance of axiom schema II.1:

$$x \in A_1\text{-}Ax \;\Leftrightarrow\; x = \ulcorner \varphi \vee \varphi \supset \varphi \urcorner \text{ for some formula } \varphi$$

The axiom schemata II.2 to II.4 can be described in a similar way:

- Axiom schema II.2: $\varphi \supset \varphi \vee \psi$

$$x \in A_2\text{-}Ax :\Leftrightarrow \exists (y, z \leq x) \, (\text{Form}(y) \wedge \text{Form}(z) \wedge x = y \, \text{Imp}(y \, \text{Dis} \, z))$$

- Axiom schema II.3: $\varphi \vee \psi \supset \psi \vee \varphi$

$$x \in A_3\text{-}Ax :\Leftrightarrow \exists (y, z \leq x) \, (\text{Form}(y) \wedge \text{Form}(z) \wedge \\ x = (y \, \text{Dis} \, z) \, \text{Imp}(z \, \text{Dis} \, y))$$

- Axiom schema II.4: $(\varphi \supset \psi) \supset (\chi \vee \varphi \supset \chi \vee \psi)$

$$x \in A_4\text{-}Ax :\Leftrightarrow \exists (y, z, w \leq x) \, (\text{Form}(y) \wedge \text{Form}(z) \wedge \text{Form}(w) \wedge \\ x = (y \, \text{Imp} \, z) \, \text{Imp}((w \, \text{Dis} \, y) \, \text{Imp}(w \, \text{Dis} \, z)))$$

---

36. $A\text{-}Ax(x)$ — x is an axiom of axiom group II

(Primitive-recursive relation)

---

36. $A\text{-}A\,x\,(x) \equiv A_1\text{-}A\,x\,(x) \vee A_2\text{-}A\,x\,(x) \vee A_3\text{-}A\,x\,(x) \vee$
$\vee A_4\text{-}A\,x\,(x)$

x ist eine durch Einsetzung in ein Aussagenaxiom entstehende *Formel*.

---

36. $A\text{-}A\,x\,(x) \equiv A_1\text{-}A\,x\,(x) \vee A_2\text{-}A\,x\,(x) \vee A_3\text{-}A\,x\,(x) \vee$
$\vee A_4\text{-}A\,x\,(x)$

x is a *formula* resulting from substitution in a sentential axiom.

In modern notation:

$$x \in A\text{-}Ax :\Leftrightarrow A_1\text{-}Ax(x) \vee A_2\text{-}Ax(x) \vee A_3\text{-}Ax(x) \vee A_4\text{-}Ax(x)$$

The right-hand side holds true precisely when x is the Gödel number of an axiom derived from one of the axiom schemata II.1 to II.4.

---

37. $Q(z, y, v)$ — Auxiliary predicate to ensure collision-freeness

(Primitive-recursive relation)

---

37. $Q(z, y, v) \equiv \overline{(En, m, w)} \, [n \leq l(y) \, \& \, m \leq l(z) \, \& \, w \leq z \, \& \\ w = m \, Gl \, z \, \& \, w \, \text{Geb} \, n, y \, \& \, v \, Fr \, n, y]$

5.2 Primitive-Recursive Functions and Relations

> z enthält keine *Variable*, die in y an einer Stelle *gebunden* ist,
> an der v *frei* ist.

> 37. $Q(z,y,v) \equiv \overline{(En,m,w)}\,[n \leq l(y)\,\&\,m \leq l(z)\,\&\,w \leq z\,\&$
> $w = m\,Gl\,z\,\&\,w\,\text{Geb}\,n,y\,\&\,v\,Fr\,n,y]$
> z contains no *variable* which is *bound* at a place in y at which v is *free*.

In modern notation:

$$(z,y,v) \in Q \;:\Leftrightarrow\; \neg\exists\,(n \leq l(y))\,\exists\,(m \leq l(z))\,\exists\,(w \leq z)$$
$$(w = m\,Gl\,z \wedge w\,\text{Geb}\,n,y \wedge v\,Fr\,n,y)$$

In Gödel's words, the right-hand side states that z

- does not contain any   ☞ $\neg\exists$
- variable (with the Gödel number w)   ☞ $w = m\,Gl\,z$
- that is *bound* in y at a position n ☞ $w\,\text{Geb}\,n,y$
- where v is *free*.   ☞ $v\,Fr\,n,y$

To understand the full meaning of this predicate, let's consider the triple (z,y,v) with

$$z := \ulcorner\varphi\urcorner := \ulcorner\mathsf{y_1}\urcorner$$
$$y := \ulcorner\psi\urcorner := \ulcorner\exists\mathsf{y_1}\,(\mathsf{x_2(x_1)} \wedge \mathsf{y_2(y_1)})\urcorner$$
$$v := \Phi(\mathsf{x_1})$$

Variable $\mathsf{x_1}$ occurs freely at the marked position in ψ. Furthermore, the marked position falls within the scope of a quantifier that binds $\mathsf{y_1}$. Consequently, φ contains a variable, $\mathsf{y_1}$, which is bound in ψ at a position where the variable represented by the Gödel number v occurs freely. Thus, we have

$$(z,y,v) \notin Q.$$

On the other hand, consider the triple (z',y',v') with

$$z' := \ulcorner\varphi\urcorner := \ulcorner\mathsf{z_1}\urcorner$$
$$y' := \ulcorner\psi\urcorner := \ulcorner\exists\mathsf{y_1}\,(\mathsf{x_2(x_1)} \wedge \mathsf{y_2(y_1)})\urcorner$$
$$v' := \Phi(\mathsf{x_1})$$

The situation is different here because φ contains no variable that is bound in ψ at a location where the variable with the Gödel number v' occurs freely. Thus, we have

$$(z', y', v') \in Q.$$

Gödel will use this predicate to ensure that an applied substitution is collision-free, which is required to derive valid instances from axiom schema III.1.

 38. $L_1\text{-}Ax(x)$ x is an instance of axiom schema III.1

(Primitive-recursive relation)

38. $L_1\text{-}A\,x\,(x) \equiv (E\,v, y, z, n)\,\{v, y, z, n \leq x\,\&\,n\,\text{Var}\,v\,\&$
$\text{Typ}_n(z)\,\&\,\text{Form}(y)\,\&\,Q(z, y, v)\,\&$
$x = (v\,\text{Gen}\,y)\,\text{Imp}\,[S\,b\,(y\,{}^v_z)]\}$

x ist eine aus dem Axiomenschema III, 1 durch Einsetzung entstehende *Formel*.

---

38. $L_1\text{-}A\,x\,(x) \equiv (E\,v, y, z, n)\,\{v, y, z, n \leq x\,\&\,n\,\text{Var}\,v\,\&$
$\text{Typ}_n(z)\,\&\,\text{Form}(y)\,\&\,Q(z, y, v)\,\&$
$x = (v\,\text{Gen}\,y)\,\text{Imp}\,[S\,b\,(y\,{}^v_z)]\}$

x is a *formula* arising from axiom schema III, 1 by substitution.

In modern notation:

$$x \in L_1\text{-}Ax \;:\Leftrightarrow\; \exists\,(v, y, z, n \leq x) \begin{pmatrix} n\,\text{Var}\,v \wedge \text{Typ}_n(z) \wedge \text{Form}(y) \wedge \\ Q(z, y, v) \wedge \\ x = (v\,\text{Gen}\,y)\,\text{Imp}(Sb(y\,{}^v_z)) \end{pmatrix}$$

The right-hand side holds true precisely when x is the Gödel number of a formula that was obtained from axiom schema III.1. Let us recall: Axiom schema III.1 allows,

- for a formula φ (with the Gödel number y),   ☞ $\text{Form}(y)$
- a type-n variable ξ (with the Gödel number v), and   ☞ $n\,\text{Var}\,v$
- a symbol σ (with the Gödel number z)

to obtain the formula

- $\xi\,\Pi\,\varphi \supset \varphi[\xi \leftarrow \sigma]$ ☞ $x = (v\,\text{Gen}\,y)\,\text{Imp}(Sb(y\,{}^v_z))$

provided that

5.2 Primitive-Recursive Functions and Relations

- σ has the type of ξ ☞ $\mathrm{Typ}_n(z)$

- and the substitution is collision-free.   ☞ $Q(z,y,v)$

In modern notation, $\xi \,\Pi\, \varphi \supset \varphi[\xi \leftarrow \sigma]$ is the formula

$$\forall \xi \, \varphi \to \varphi[\xi \leftarrow \sigma].$$

---

39. $L_2\text{-}Ax(x)$ x is an instance of axiom schema III.2

(Primitive-recursive relation)

---

39. $L_2\text{-}A\,x\,(x) \equiv (E\,v, q, p)\, \{v, q, p \leq x\,\&\, \mathrm{Var}\,(v)\,\&\,\mathrm{Form}\,(p)$
   $\&\,\overline{v\,Fr\,p}\,\&\,\mathrm{Form}\,(q)\,\&$
   $x = [v\,\mathrm{Gen}\,(p\,\mathrm{Dis}\,q)]\,\mathrm{Imp}\,[p\,\mathrm{Dis}\,(v\,\mathrm{Gen}\,q)]\}$

x ist eine aus dem Axiomenschema III, 2 durch Einsetzung entstehende *Formel*.

---

39. $L_2\text{-}A\,x\,(x) \equiv (E\,v, q, p)\, \{v, q, p \leq x\,\&\, \mathrm{Var}\,(v)\,\&\,\mathrm{Form}\,(p)$
   $\&\,\overline{v\,Fr\,p}\,\&\,\mathrm{Form}\,(q)\,\&$
   $x = [v\,\mathrm{Gen}\,(p\,\mathrm{Dis}\,q)]\,\mathrm{Imp}\,[p\,\mathrm{Dis}\,(v\,\mathrm{Gen}\,q)]\}$

x is a *formula* arising from axiom schema III, 2 by substitution.

In modern notation:

$$x \in L_2\text{-}Ax \;:\Leftrightarrow\; \exists\,(v, q, p \leq x)\, \begin{pmatrix} \mathrm{Var}(v) \wedge \mathrm{Form}(p) \wedge \neg(v\,Fr\,p) \wedge \mathrm{Form}(q) \wedge \\ x = (v\,\mathrm{Gen}(p\,\mathrm{Dis}\,q))\,\mathrm{Imp}(p\,\mathrm{Dis}(v\,\mathrm{Gen}\,q)) \end{pmatrix}$$

The right-hand side holds true precisely when x is the Gödel number of a formula that was obtained from axiom schema III.2. We remember: Axiom schema III.2 allows us,

- for a formula φ (with the Gödel number p),   ☞ $\mathrm{Form}(p)$

- a formula ψ (with the Gödel number q) and   ☞ $\mathrm{Form}(q)$

- a variable ξ (with the Gödel number v)   ☞ $\mathrm{Var}(v)$

to obtain the formula

- $\xi\,\Pi\,(\varphi \vee \psi) \supset \varphi \vee \xi\,\Pi\,(\psi)$ ☞ $x = (v\,\mathrm{Gen}(p\,\mathrm{Dis}\,q))\,\mathrm{Imp}(p\,\mathrm{Dis}(v\,\mathrm{Gen}\,q))$

provided that

- ξ does not occur freely in φ. ☞ $\neg(v\,Fr\,p)$

In modern notation, $\xi\Pi(\varphi\vee\psi)\supset\varphi\vee\xi\Pi(\psi)$ is the formula

$$\forall\xi\,(\varphi\vee\psi)\to\varphi\vee\forall\xi\,\psi$$

40. $R\text{-}Ax(x)$ x is an instance of axiom schema IV.1

(Primitive-recursive relation)

40. $R\text{-}A\,x\,(x)\equiv(E\,u,v,y,n)\,[u,v,y,n\leqq x\ \&\ n\,\mathrm{Var}\,v\ \&$
$(n+1)\,\mathrm{Var}\,u\ \&\ \overline{u\,Fr\,y}\ \&\ \mathrm{Form}\,(y)\ \&$
$x=u\,\mathrm{Ex}\,\{v\,\mathrm{Gen}\,[[R\,(u)*E\,(R\,(v))]\,\mathrm{Aeq}\,y]\}]$

x ist eine aus dem Axiomenschema IV, 1 durch Einsetzung entstehende *Formel*.

40. $R\text{-}A\,x\,(x)\equiv(E\,u,v,y,n)\,[u,v,y,n\leqq x\ \&\ n\,\mathrm{Var}\,v\ \&$
$(n+1)\,\mathrm{Var}\,u\ \&\ \overline{u\,Fr\,y}\ \&\ \mathrm{Form}\,(y)\ \&$
$x=u\,\mathrm{Ex}\,\{v\,\mathrm{Gen}\,[[R\,(u)*E\,(R\,(v))]\,\mathrm{Aeq}\,y]\}]$

x is a *formula* arising from axiom schema IV, 1 by substitution.

In modern notation:

$$x\in R\text{-}Ax\ :\Leftrightarrow\ \exists(u,v,y,n\leq x)\begin{pmatrix}n\,\mathrm{Var}\,v\wedge(n+1)\,\mathrm{Var}\,u\wedge\\ \neg(u\,Fr\,y)\wedge\mathrm{Form}(y)\wedge\\ x=u\,\mathrm{Ex}(v\,\mathrm{Gen}((R(u)*E(R(v)))\,\mathrm{Aeq}\,y))\end{pmatrix}$$

The right-hand side holds true precisely when x is the Gödel number of a formula derived from axiom schema IV.1. Let us recall: Axiom schema IV.1 allows us,

- for a formula φ (with the Gödel number y), ☞ $\mathrm{Form}(y)$
- a type-n variable ξ (with the Gödel number v), and ☞ $n\,\mathrm{Var}\,v$
- a type-$(n+1)$ variable ζ (with the Gödel number u) ☞ $(n+1)\,\mathrm{Var}\,u$

to obtain the formula

- $(E\,\zeta)(\xi\,\Pi\,(\zeta(\xi)\equiv\varphi))$ ☞ $x=u\,\mathrm{Ex}(v\,\mathrm{Gen}((R(u)*E(R(v)))\,\mathrm{Aeq}\,y))$

5.2 Primitive-Recursive Functions and Relations

provided that

- ζ does not occur freely in φ.   ☞ $\neg(u \, Fr \, y)$

In modern notation, $(E \, \zeta)(\xi \, \Pi \, (\zeta(\xi) \equiv \varphi))$ is the formula

$$\exists \zeta \, \forall \xi \, (\zeta(\xi) \leftrightarrow \varphi).$$

---

41. $M\text{-}Ax(x)$ x is an instance of axiom schema V.1

(Primitive-recursive relation)

Dem Axiom V, 1 entspricht eine bestimmte Zahl z_4 und wir definieren:

41. $M\text{-}A \, x \, (x) \equiv (E \, n) \, [n \leq x \, \& \, x = n \, Th \, z_4].$

To axiom V, 1 corresponds a definite number z_4, and we define

41. $M\text{-}A \, x \, (x) \equiv (E \, n) \, [n \leq x \, \& \, x = n \, Th \, z_4].$

In modern notation:

$$x \in M\text{-}Ax \; :\Leftrightarrow \; \exists \, (n \leq x) \, x = n \, Th \, z_4 \quad \text{with}$$
$$z_4 := \ulcorner \mathsf{x}_1 \, \Pi \, (\mathsf{x}_2(\mathsf{x}_1) \equiv \mathsf{y}_2(\mathsf{x}_1)) \supset \mathsf{x}_2 = \mathsf{y}_2 \urcorner$$

The right-hand side holds true precisely when x is the Gödel number of a formula that arises from the principle of type elevation from the comprehension axiom.

---

42. $Ax(x)$ x is an axiom

(Primitive-recursive relation)

---

42. $A \, x \, (x) \equiv Z\text{-}A \, x \, (x) \vee A\text{-}A \, x \, (x) \vee L_1\text{-}A \, x \, (x)$
$ \vee L_2\text{-}A \, x \, (x) \vee R\text{-}A \, x \, (x) \vee M\text{-}A \, x \, (x)$

x ist ein *Axiom*.

---

42. $A \, x \, (x) \equiv Z\text{-}A \, x \, (x) \vee A\text{-}A \, x \, (x) \vee L_1\text{-}A \, x \, (x)$
$ \vee L_2\text{-}A \, x \, (x) \vee R\text{-}A \, x \, (x) \vee M\text{-}A \, x \, (x)$

x is an *axiom*.

The relation Ax summarizes the various axiom schemata:

$$Ax := \{n \in \mathbb{N} \mid n \text{ is the Gödel number of an axiom of system P}\}$$

> **43.** $Fl(x, y, z)$ — x can be derived from y and z
> (Primitive-recursive relation)

> 43. $Fl\ (x\ y\ z) \equiv y = z \operatorname{Imp} x \vee$
> $(E v)\ [v \leq x\ \&\ \operatorname{Var}(v)\ \&\ x = v \operatorname{Gen} y]$
> x ist *unmittelbare Folge* aus y und z.

> 43. $Fl\ (x\ y\ z) \equiv y = z \operatorname{Imp} x \vee$
> $(E v)\ [v \leq x\ \&\ \operatorname{Var}(v)\ \&\ x = v \operatorname{Gen} y]$
> x is an *immediate consequence* of y and z.

In modern notation:

$$(x, y, z) \in Fl \;:\Leftrightarrow\; y = z \operatorname{Imp} x \vee \exists\, (v \leq x)\, (\operatorname{Var}(v) \wedge x = v \operatorname{Gen} y)$$

If x, y, and z represent the Gödel numbers of the formulas φ, ψ, and χ, respectively, then the right-hand side holds true if and only if one of the following cases applies:

- ψ has the form $\chi \supset \varphi$ ☞ $y = z \operatorname{Imp} x$

  In this scenario, φ can be derived from ψ and χ by modus ponens.

- For some variable ξ, φ has the form $\xi \Pi \psi$ ☞ $\operatorname{Var}(v) \wedge x = v \operatorname{Gen} y$

  In this scenario, φ can be derived from ψ by generalization.

Thus, the relation Fl applies to the Gödel numbers $\ulcorner\varphi\urcorner$, $\ulcorner\psi\urcorner$, $\ulcorner\chi\urcorner$ if and only if the formula φ can be derived from the formulas ψ and χ by applying an inference rule of the formal system P.

5.2 Primitive-Recursive Functions and Relations

 44. $Bw(x)$ x is a formal proof chain of the system P
(Primitive-recursive relation)

186   Kurt Gödel,

44. $Bw(x) \equiv (n) \{0 < n \leqq l(x) \rightarrow Ax(n\,Gl\,x) \vee (Ep,q)\,[0 < p,q < n\,\&\,Fl(n\,Gl\,x, p\,Gl\,x, q\,Gl\,x)]\}$
$\&\,l(x) > 0$

x ist eine *Beweisfigur* (eine endliche Folge von *Formeln*, deren jede entweder *Axiom* oder *unmittelbare Folge* aus zwei der vorhergehenden ist).

186   Kurt Gödel,

44. $Bw(x) \equiv (n) \{0 < n \leqq l(x) \rightarrow Ax(n\,Gl\,x) \vee (Ep,q)\,[0 < p,q < n\,\&\,Fl(n\,Gl\,x, p\,Gl\,x, q\,Gl\,x)]\}$
$\&\,l(x) > 0$

x is a *proof figure* (a finite sequence of *formulas* each of which is either an *axiom* or an *immediate consequence* of two preceding ones).

In modern notation:

$$x \in Bw :\Leftrightarrow l(x) > 0 \wedge \forall\,(0 < n \leq l(x)) \begin{pmatrix} Ax(n\,Gl\,x)\,\vee \\ \exists\,(0 < p,q < n) \\ Fl(n\,Gl\,x, p\,Gl\,x, q\,Gl\,x) \end{pmatrix}$$

The right-hand side holds true precisely when

- x is a non-empty number series, and   ☞ $l(x) > 0$
- the n-th member   ☞ $\forall\,(0 < n \leq l(x))$
    - is an axiom, or   ☞ $Ax(n\,Gl\,x)$
    - is derived from preceding formulas   ☞ $\exists\,(0 < p,q < n)$
    - by an inference rule.   ☞ $Fl(n\,Gl\,x, p\,Gl\,x, q\,Gl\,x)$

Therefore, $Bw(x)$ is true if and only if x is the Gödel number of a formal proof. As a set, we can write the relation Bw as follows:

$$Bw = \{n \in \mathbb{N} \mid n \text{ is the Gödel number of a proof of the system P}\}$$

| 45. $x B y$ | x is a proof for the formula y |

(Primitive-recursive relation)

> 45. $x\,B\,y \equiv B w\,(x)\, \&\, [l\,(x)]\, Gl\,x = y$
> x ist ein *Beweis* für die *Formel* y.

> 45. $x\,B\,y \equiv B w\,(x)\, \&\, [l\,(x)]\, Gl\,x = y$
> x is a *proof* of the *formula* y.

In modern notation:

$$x\,B\,y \;:\Leftrightarrow\; Bw(x) \wedge l(x)\,Gl\,x = y$$

The right-hand side holds true precisely when

- x is the Gödel number of a proof chain, and   ☞ $Bw(x)$
- y is the Gödel number of the last formula.   ☞ $(l(x)\,Gl\,x) = y$

Thus, $x\,B\,y$ is true if and only if x is the Gödel number of a formal proof for the formula with the Gödel number y.

At this point, we have achieved an important result:

Theorem 5.6

The relation

$$B := \{(\ulcorner\varphi_1,\ldots,\varphi_n\urcorner,\ulcorner\varphi\urcorner) \in \mathbb{N}^2 \mid \varphi_1,\ldots,\varphi_n \text{ is a proof for } \varphi \text{ in P}\}$$

is primitive recursive.

5.3 Decision Procedures

Gödel concludes his long list of primitive recursive functions and relations with relation 46. It deserves our special attention because it plays a crucial role in what is yet to come:

5.3 Decision Procedures

> 46. Bew x x is provable within the system P

> 46. Bew $(x) \equiv (E y)\, y\, B\, x$
>
> x ist eine *beweisbare Formel.* [Bew (x) ist der einzige unter den Begriffen 1—46, von dem nicht behauptet werden kann, er sei rekursiv.]

> 46. Bew $(x) \equiv (E y)\, y\, B\, x$
>
> x is a *provable formula.* [Bew (x) is the only one of the concepts 1–46 which cannot be asserted to be recursive.]

In modern notation:

$$x \in \text{Bew} :\Leftrightarrow \exists y\, y\, B\, x \tag{5.16}$$

The right-hand side holds true precisely when x is the Gödel number of a formula provable in P.

Pay attention to the comment in square brackets! It is the first time Gödel avoids making a clear statement. He answers whether this relation is primitive recursive with the vague statement that it "cannot be asserted". A closer look at the right-hand side of (5.16) reveals why. Among the relations and functions presented so far, this relation is the first to be defined with an unrestricted existential quantifier, thus failing to meet the requirement of Theorem IV. This does not necessarily mean that Bew is *not* primitive recursive, as we would have to rule out the existence of another formula that captures Bew while simultaneously satisfying all the requirements of a primitive-recursive definition.

Before we lift the veil and reveal whether or not Bew is a primitive-recursive relation, let us consider the consequences. If Bew were a primitive-recursive function, the syntactic variant of the so-called *decision problem* would be solvable. The following definition clarifies the exact meaning of this term:

Definition 5.7   Decision problem

The syntactic variant of the *decision problem* is defined as follows:

- Given: A formal system and a formula φ
- Asked: Does $\vdash \varphi$ hold?

The semantic variant of the *decision problem* is defined as follows:

- Given: A formula φ
- Asked: Does $\models \varphi$ hold?

This definition is a generalization of the historical formulation by David Hilbert and Wilhelm Ackermann. The original formulation originates from [52] and refers to a question concerning first-order predicate logic.

> "The decision problem is solved if one knows a procedure that allows a decision about the validity or satisfiability of a given logical expression through a finite number of operations. The solution of the decision problem is of fundamental importance for the theory of all areas whose sentences are capable of logical development from a finite number of axioms."
>
> "Das Entscheidungsproblem ist gelöst, wenn man ein Verfahren kennt, das bei einem vorgelegten logischen Ausdruck durch endlich viele Operationen die Entscheidung über die Allgemeingültigkeit bzw. Erfüllbarkeit erlaubt. Die Lösung des Entscheidungsproblems ist für die Theorie aller Gebiete, deren Sätze überhaupt einer logischen Entwickelbarkeit aus endlich vielen Axiomen fähig sind, von grundsätzlicher Wichtigkeit."
>
> David Hilbert, Wilhelm Ackermann [52]

In a later edition of the cited textbook, the two become even more explicit:

> "The decision problem must be considered the main problem of mathematical logic."
>
> "Das Entscheidungsproblem muss als das Hauptproblem der mathematischen Logik bezeichnet werden."
>
> David Hilbert, Wilhelm Ackermann [53]

Hilbert and Ackermann did not distinguish between a syntactic and a semantic variant – and they didn't need to. In first-order predicate logic, a formula is universally valid ($\models \varphi$) if and only if it is provable ($\vdash \varphi$), making both formulations equivalent. However, by generalizing the concept to arbitrary formal systems, as done in Definition 5.7, the model relation '$\models$' and the provability relation '$\vdash$' no longer coincide, making the syntactic and semantic variants two concepts that must not be confused.

We will now prove a theorem that establishes a crucial relationship between the syntactic decision problem and the negation completeness of a formal system.

5.3 Decision Procedures

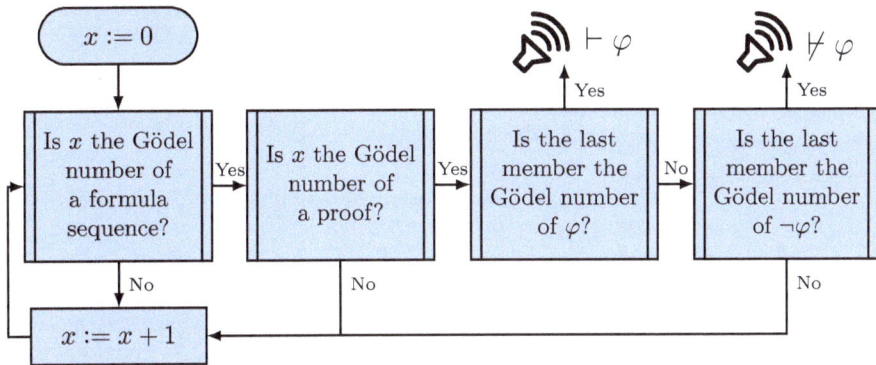

Figure 5.7: In a formal system being both consistent and negation complete, it is possible to decide for any given formula φ whether or not it is provable. In other words, the syntactic decision problem has a solution.

Theorem 5.8

For every consistent formal system K, the following holds:

K is negation complete $\Rightarrow$ K has a decision procedure

Figure 5.7 depicts a possible solution of the decision problem for consistent, negation-complete formal systems. It proceeds as follows:

- All symbol strings constructible in the formal system's artificial language are enumerated in order, e. g., by first enumerating the strings of length 1, then the strings of length 2, and so on. This way, every string will eventually appear in the enumeration.

- All strings that do not constitute a formula sequence are discarded. The same applies to formula sequences that do not constitute a proof. Due to the systematic enumeration, every proof will appear at some point.

- For each string that constitutes a proof, it is checked whether the final formula equals φ or $\neg\varphi$. In the first case, φ is a theorem. In the second case, it is not.

Two important aspects are to be taken into account:

- For the algorithm to effectively solve the syntactic decision problem, it must be guaranteed to stop for any input. This is ensured by the property of negation completeness, stating that among the two formulas φ and $\neg\varphi$, at least one is provable. Thus, the algorithm terminates for every input after a finite number of steps.

> ON COMPUTABLE NUMBERS, WITH AN APPLICATION TO
> THE ENTSCHEIDUNGSPROBLEM
>
> By A. M. TURING.
>
> [Received 28 May, 1936.—Read 12 November, 1936.]
>
> The "computable" numbers may be described briefly as the real numbers whose expressions as a decimal are calculable by finite means. Although the subject of this paper is ostensibly the computable *numbers*, it is almost equally easy to define and investigate computable functions of an integral variable or a real or computable variable, computable predicates, and so forth. The fundamental problems involved are, however, the same in each case, and I have chosen the computable numbers for explicit treatment as involving the least cumbrous technique. I hope shortly to give an account of the relations of the computable numbers, functions, and so forth to one another. This will include a development of the theory of functions of a real variable expressed in terms of computable numbers. According to my definition, a number is computable

Figure 5.8: In 1936, the British mathematician Alan Turing succeeded in answering one of the fundamental questions of mathematical logic. From the undecidability of the halting problem for Turing machines, he inferred that no decision procedure for first-order predicate logic can exist.

- The algorithm relies on the fact that the derivability of $\neg\varphi$ implies the non-derivability of φ ($\vdash \neg\varphi$ implies $\nvdash \varphi$). This is ensured by the property of consistency, which is an explicit requirement of Theorem 5.8. In a potentially contradictory system, however, $\neg\varphi$ could be provable, and yet a proof for φ could also appear at a later point in time. In this case, φ would also be a theorem.

Is the syntactic decision problem solvable for Gödel's system P? Theorem 5.8 would affirm this question if P were consistent and negation complete. However, Gödel will demonstrate that P cannot be consistent and negation complete simultaneously, thus thwarting this conclusion.

If the proof relation Bew were primitive recursive, the answer would again be yes! In this case, we could decide the question $\ulcorner\varphi\urcorner \in$ Bew simply by calculating the corresponding primitive-recursive function. The reverse conclusion is no less remarkable. If there is no decision procedure for P, we would be confident that Bew cannot be a primitive-recursive relation.

The hope of Hilbert and Ackermann to find a solution for the decision problem of predicate logic was shattered by Alan Turing (Figure 5.8). In his famous work "*On Computable Numbers with an Application to the Entscheidungsprob-*

lem" [97, 80], the British mathematician convincingly demonstrated the unsolvability of the decision problem for first-order predicate logic, and the method of proof can be easily transferred to the system P. Turing's result implies that the proof relation Bew cannot be primitive-recursive. However, Gödel was unaware of this fact at the time of writing. Turing made his stunning discovery in 1936, about five years after the publication of the incompleteness theorems.

5.4 Theorem V

Our journey through Gödel's work has led us to Theorem V, one of the most significant auxiliary theorems in the proof of the first incompleteness theorem. Metaphorically speaking, this theorem bridges two distinct realms. On one side resides Gödel's system P, extensively discussed in a separate chapter. On the other side reside the primitive-recursive functions and relations, which are also familiar to us by now. At its core, Theorem V states that the computation of a primitive-recursive function is replicable within the formal system P, that is, for any given primitive-recursive function f and any given number combination $y, x_1, \ldots, x_n$, it can be proved within P whether or not y is equal to the function value $f(x_1, \ldots, x_n)$.

Further down, it will become evident that the proof of Theorem V follows a simple line of argument, yet it involves intricate technicalities. For this reason, Gödel merely sketches the proof, leaving out many details. Due to the lack of complete formal justification, Theorem V may seem like the Achilles' heel of Gödel's work, but this fear is unfounded. A few years after the incompleteness theorems were published, David Hilbert and Paul Bernays had fully worked out the proof and published it in their two-volume *Foundations of Mathematics* [55, 56].

To convey the core idea as clearly as possible, we first derive the assertion of proposition V for a specific example before presenting its general form. For our investigation, we will employ the slightly extended formal system P' from Section 4.4.5, which significantly simplifies the formulation of statements involving addition and multiplication compared to Gödel's original system P.

Let's start with the following formula:

$$\varphi(\mathsf{x}_1) := \exists \mathsf{z}_1 \, (\mathsf{x}_1 = \mathsf{z}_1 \times \overline{2}) \tag{5.17}$$

With x_1, the formula contains a single free individual variable, which makes it a *class expression* according to Gödel's terminology. Replacing x_1 with a term of the form $\bar{n}$ yields a closed formula that must either be substantively true or substantively false. A closer look at the right-hand side of (5.17) reveals that

$\varphi(\overline{n})$ is substantively true if and only if n is an even natural number:

$$\models \exists z_1\ 0 = z_1 \times \overline{2} \qquad (\text{☞} \models \varphi(\overline{0}))$$
$$\models \exists z_1\ f\ f\ 0 = z_1 \times \overline{2} \qquad (\text{☞} \models \varphi(\overline{2}))$$
$$\models \exists z_1\ f\ f\ f\ f\ 0 = z_1 \times \overline{2} \qquad (\text{☞} \models \varphi(\overline{4}))$$
$$\ldots$$
$$\models \neg\exists z_1\ f\ 0 = z_1 \times \overline{2} \qquad (\text{☞} \models \neg\varphi(\overline{1}))$$
$$\models \neg\exists z_1\ f\ f\ f\ 0 = z_1 \times \overline{2} \qquad (\text{☞} \models \neg\varphi(\overline{3}))$$
$$\models \neg\exists z_1\ f\ f\ f\ f\ f\ 0 = z_1 \times \overline{2} \qquad (\text{☞} \models \neg\varphi(\overline{5}))$$
$$\ldots$$

By denoting the set of even natural numbers with

$$R := \{0, 2, 4, 6, 8, 10, \ldots\},$$

this relationship can be written as such:

$$n \in R \;\Rightarrow\; \models\ \varphi(\overline{n})$$
$$n \notin R \;\Rightarrow\; \models\ \neg\varphi(\overline{n})$$

We say the formula φ *semantically represents* the relation R, and generalize this concept straightforwardly:

Definition 5.9   Semantically representable relations

Let $R \subseteq \mathbb{N}^n$ be a relation and φ a formula with n free variables. R is *semantically represented* by φ if the following holds:

$$(x_1, \ldots, x_n) \in R \;\Rightarrow\; \models\ \varphi(\overline{x_1}, \ldots, \overline{x_n})$$
$$(x_1, \ldots, x_n) \notin R \;\Rightarrow\; \models\ \neg\varphi(\overline{x_1}, \ldots, \overline{x_n})$$

Functions can be represented in a similar fashion, as every n-ary function can be considered a relation with an arity of $n+1$:

Definition 5.10   Semantically representable functions

Let $f : \mathbb{N}^n \to \mathbb{N}$ be a function and φ a formula with $n+1$ free variables. f is *semantically represented* by φ if the following holds:

$$y = f(x_1, \ldots, x_n) \;\Rightarrow\; \models\ \varphi(\overline{y}, \overline{x_1}, \ldots, \overline{x_n})$$
$$y \neq f(x_1, \ldots, x_n) \;\Rightarrow\; \models\ \neg\varphi(\overline{y}, \overline{x_1}, \ldots, \overline{x_n})$$

5.4 Theorem V

Let's turn back to our example formula φ. For every even natural number n, we can notice that the instance $\varphi(\overline{n})$ is not only true, but also provable within P'. The following derivation shows how:

$\exists z_1\ 0 = z_1 \times \overline{2}$ $\varphi(\overline{0})$

1. $\vdash\ 0 = 0 + 0$ (M.3.2f)
2. $\vdash\ 0 \times \overline{2} = 0 + 0$ (M.3.5c)
3. $\vdash\ 0 = 0 \times \overline{2}$ (GL, 1,2)
4. $\vdash\ 0 = 0 \times \overline{2} \to \exists z_1\ 0 = z_1 \times \overline{2}$ (P.3)
5. $\vdash\ \exists z_1\ 0 = z_1 \times \overline{2}$ (MP, 3,4)

$\exists z_1\ f\ f\ 0 = z_1 \times \overline{2}$ $\varphi(\overline{2})$

1. $\vdash\ \exists z_1\ 0 = z_1 \times \overline{2}$ $(\varphi(\overline{0}))$
2. $\vdash\ \exists z_1\ 0 = z_1 \times \overline{2} \to \exists z_1\ f\ f\ 0 = z_1 \times \overline{2}$ (N.7)
3. $\vdash\ \exists z_1\ f\ f\ 0 = z_1 \times \overline{2}$ (MP, 1,2)

$\exists z_1\ f\ f\ f\ f\ 0 = z_1 \times \overline{2}$ $\varphi(\overline{4})$

1. $\vdash\ \exists z_1\ f\ f\ 0 = z_1 \times \overline{2}$ $(\varphi(\overline{2}))$
2. $\vdash\ \exists z_1\ f\ f\ 0 = z_1 \times \overline{2} \to \exists z_1\ f\ f\ f\ f\ 0 = z_1 \times \overline{2}$ (N.7)
3. $\vdash\ \exists z_1\ f\ f\ f\ f\ 0 = z_1 \times \overline{2}$ (MP, 1,2)

A similar relationship holds for the odd natural numbers. If n is odd, that is $n \notin R$, then the instance $\neg\varphi(\overline{n})$ is provable:

$\neg\exists z_1\ f\ 0 = z_1 \times \overline{2}$ $\neg\varphi(\overline{1})$

1. $\vdash\ \overline{1} \neq z_1 \times \overline{2}$ (N.4)
2. $\vdash\ \forall z_1\ \overline{1} \neq z_1 \times \overline{2}$ (G, 1)
3. $\vdash\ \forall z_1\ \overline{1} \neq z_1 \times \overline{2} \to \neg\neg\forall z_1\ \overline{1} \neq z_1 \times \overline{2}$ (H.4)
4. $\vdash\ \neg\neg\forall z_1\ \overline{1} \neq z_1 \times \overline{2}$ (MP, 2,3)
5. $\vdash\ \neg\exists z_1\ f\ 0 = z_1 \times \overline{2}$ (Def, 4)

$\neg\exists z_1\ f\ f\ f\ 0 = z_1 \times \overline{2}$ $\neg\varphi(\overline{3})$

1. $\vdash\ \neg\exists z_1\ f\ 0 = z_1 \times \overline{2}$ $(\neg\varphi(\overline{1}))$

2. ⊢ ∃z₁ f f f 0 = z₁ × 2̄ → ∃z₁ f 0 = z₁ × 2̄   (N.8)
3. ⊢ ¬∃z₁ f 0 = z₁ × 2̄ → ¬∃z₁ f f f 0 = z₁ × 2̄   (INV, 2)
4. ⊢ ¬∃z₁ f f f 0 = z₁ × 2̄   (MP, 1,3)

 ¬∃z₁ f f f f f 0 = z₁ × 2̄   ¬φ(5̄)

1. ⊢ ¬∃z₁ f f f 0 = z₁ × 2̄   (¬φ(3̄))
2. ⊢ ∃z₁ f f f f f 0 = z₁ × 2̄ → ∃z₁ f f f 0 = z₁ × 2̄ (N.8)
3. ⊢ ¬∃z₁ f f f 0 = z₁ × 2̄ → ¬∃z₁ f f f f f 0 = z₁ × 2̄ (INV, 2)
4. ⊢ ¬∃z₁ f f f f f 0 = z₁ × 2̄   (MP, 1,3)

For each natural number n, either the formula instance $\varphi(\bar{n})$ or the formula instance $\neg\varphi(\bar{n})$ can be derived in the demonstrated way. In particular, the first instance is provable when n is even and the second when n is odd. In symbolic terms:

$$n \in R \Rightarrow \vdash \varphi(\bar{n})$$
$$n \notin R \Rightarrow \vdash \neg\varphi(\bar{n})$$

We say the formula φ *syntactically represents* the relation R, and generalize this concept similar to what we did above:

Definition 5.11   Syntactically representable relations

Let $R \subseteq \mathbb{N}^n$ be a relation and φ a formula with n free variables. R is *syntactically represented* by φ, if the following holds:

$$(x_1, \ldots, x_n) \in R \Rightarrow \vdash \varphi(\bar{x_1}, \ldots, \bar{x_n})$$
$$(x_1, \ldots, x_n) \notin R \Rightarrow \vdash \neg\varphi(\bar{x_1}, \ldots, \bar{x_n})$$

Once more, the notion can be extended to functions straightforwardly:

Definition 5.12   Syntactically representable functions

Let $f : \mathbb{N}^n \to \mathbb{N}$ be a function and φ a formula with $n+1$ free variables. f is *syntactically represented* by φ if the following holds:

$$y = f(x_1, \ldots, x_n) \Rightarrow \vdash \varphi(\bar{y}, \bar{x_1}, \ldots, \bar{x_n})$$
$$y \neq f(x_1, \ldots, x_n) \Rightarrow \vdash \neg\varphi(\bar{y}, \bar{x_1}, \ldots, \bar{x_n})$$

With the acquired knowledge, Gödel's original formulation of Theorem V is easy to grasp:

5.4 Theorem V

> Die Tatsache, die man vage so formulieren kann: Jede rekursive Relation ist innerhalb des Systems P (dieses inhaltlich gedeutet) definierbar, wird, ohne auf eine inhaltliche Deutung der Formeln aus P Bezug zu nehmen, durch folgenden Satz exakt ausgedrückt:
>
> Satz V: Zu jeder rekursiven Relation $R(x_1 \ldots x_n)$ gibt es ein n-stelliges *Relationszeichen* r (mit den *freien Variablen*[38] $u_1, u_2 \ldots u_n$), so daß für alle Zahlen-n-tupel $(x_1 \ldots x_n)$ gilt:
>
> $$R(x_1 \ldots x_n) \rightarrow \text{Bew}\left[Sb\left(r \begin{array}{ccc} u_1 & \ldots & u_n \\ Z(x_1) & \ldots & Z(x_n) \end{array}\right)\right] \quad (3)$$
>
> $$\overline{R}(x_1 \ldots x_n) \rightarrow \text{Bew}\left[\text{Neg } Sb\left(r \begin{array}{ccc} u_1 & \ldots & u_n \\ Z(x_1) & \ldots & Z(x_n) \end{array}\right)\right] \quad (4)$$
>
> ---
>
> [38]) Die *Variablen* $u_1 \ldots u_n$ können willkürlich vorgegeben werden. Es gibt z. B. immer ein r mit den *freien Variablen* 17, 19, 23 ... usw., für welches (3) und (4) gilt.

> The fact which can be vaguely formulated as the assertion that every recursive relation is definable within the system P (under its intuitive interpretation), is rigorously expressed by the following theorem, without reference to the intuitive meaning of the formulas of P:
>
> Theorem V: For every recursive relation $R(x_1 \ldots x_n)$, there is an n-ary *predicate* r (with the *free variables*[38]) $u_1, u_2 \ldots u_n$) such that, for all n-tuples of numbers $(x_1 \ldots x_n)$, we have:
>
> $$R(x_1 \ldots x_n) \rightarrow \text{Bew}\left[Sb\left(r \begin{array}{ccc} u_1 & \ldots & u_n \\ Z(x_1) & \ldots & Z(x_n) \end{array}\right)\right] \quad (3)$$
>
> $$\overline{R}(x_1 \ldots x_n) \rightarrow \text{Bew}\left[\text{Neg } Sb\left(r \begin{array}{ccc} u_1 & \ldots & u_n \\ Z(x_1) & \ldots & Z(x_n) \end{array}\right)\right] \quad (4)$$
>
> ---
>
> [38]) The *variables* $u_1 \ldots u_n$ can be arbitrarily prescribed. There always exists, e.g. some r with the *free variables* 17, 19, 23, etc., for which (3) and (4) hold.

If φ is the formula with the Gödel number r, formulas (3) and (4) can be rewritten as follows:

$$R(x_1, \ldots, x_n) \Rightarrow \text{Bew } \ulcorner \varphi(\overline{x_1}, \ldots, \overline{x_n}) \urcorner \quad (5.18)$$

$$\overline{R}(x_1, \ldots, x_n) \Rightarrow \text{Bew } \ulcorner \neg\varphi(\overline{x_1}, \ldots, \overline{x_n}) \urcorner \quad (5.19)$$

Bew x substantively states that the formula with the Gödel number x is provable within P. Thus, (5.18) and (5.19) are the same as:

$$(x_1, \ldots, x_n) \in R \Rightarrow \vdash \varphi(\overline{x_1}, \ldots, \overline{x_n})$$
$$(x_1, \ldots, x_n) \notin R \Rightarrow \vdash \neg\varphi(\overline{x_1}, \ldots, \overline{x_n})$$

Another look at Definition 5.11 reveals the meaning of Theorem V:

Theorem 5.13 — Gödel's Theorem V

Every primitive-recursive relation is syntactically representable in P.

As said above, formally proving Theorem V is tedious, which is why Gödel confines himself to presenting a rough sketch of the proof:

> Wir begnügen uns hier damit, den Beweis dieses Satzes, da er keine prinzipiellen Schwierigkeiten bietet und ziemlich umständlich ist, in Umrissen anzudeuten [39]). Wir beweisen den Satz für alle Relationen $R(x_1 \ldots x_n)$ der Form: $x_1 = \varphi(x_2 \ldots x_n)$ [40]) (wo φ eine rekursive Funktion ist) und wenden vollständige Induktion nach
>
> ---
> [39]) Satz V beruht natürlich darauf, daß bei einer rekursiven Relation R für jedes n-tupel von Zahlen aus den Axiomen des Systems P entscheidbar ist, ob die Relation R besteht oder nicht.
> [40]) Daraus folgt sofort seine Geltung für jede rekursive Relation, da eine solche gleichbedeutend ist mit $0 = \varphi(x_1 \ldots x_n)$, wo φ rekursiv ist.

> We shall be content here to indicate the outline of the proof of this theorem, since if offers no theoretical difficulties and is fairly tedious.[39]) We shall prove the theorem for all relations $R(x_1 \ldots x_n)$ of the form $x_1 = \varphi(x_2 \ldots x_n)$ [40]) (where φ is a recursive function)
>
> ---
> [39]) Theorem V depends of course upon the fact that, for a recursive relation R, it is decidable on the basis of the axioms of the system P whether or not R holds for any given n-tuple of numbers.
> [40]) From this, its validity follows immediately for every recursive relation, since such a relation is equivalent to $0 = \varphi(x_1 \ldots x_n)$, where φ is recursive.

Gödel will establish the assertion not for primitive-recursive relations, but for primitive-recursive functions. The following theorem, briefly mentioned in footnote 40, demonstrates the sufficiency of this approach:

Theorem 5.14

If every primitive-recursive function is syntactically representable, then so is every primitive-recursive relation.

Proof: Let R be a primitive-recursive relation. Then there exists a primitive-recursive function f with the following property:

$$(x_1, \ldots, x_n) \in R \Leftrightarrow f(x_1, \ldots, x_n) = 0$$

5.4 Theorem V

If all primitive-recursive functions are syntactically representable, then so is f. This implies the existence of a formula φ satisfying

$$y = f(x_1, \ldots, x_n) \;\Rightarrow\; \vdash \varphi(\overline{y}, \overline{x_1}, \ldots, \overline{x_n})$$
$$y \neq f(x_1, \ldots, x_n) \;\Rightarrow\; \vdash \neg\varphi(\overline{y}, \overline{x_1}, \ldots, \overline{x_n})$$

A fortiori, this implies

$$0 = f(x_1, \ldots, x_n) \;\Rightarrow\; \vdash \varphi(0, \overline{x_1}, \ldots, \overline{x_n})$$
$$0 \neq f(x_1, \ldots, x_n) \;\Rightarrow\; \vdash \neg\varphi(0, \overline{x_1}, \ldots, \overline{x_n})$$

which is equivalent to

$$(x_1, \ldots, x_n) \in R \;\Rightarrow\; \vdash \varphi(0, \overline{x_1}, \ldots, \overline{x_n})$$
$$(x_1, \ldots, x_n) \notin R \;\Rightarrow\; \vdash \neg\varphi(0, \overline{x_1}, \ldots, \overline{x_n})$$

At this point, we are done: Substituting y with 0 turns φ into a formula that syntactically represents R. □

Next, Gödel outlines the inductive argument:

> eine rekursive Funktion ist) und wenden vollständige Induktion nach der Stufe von φ an. Für Funktionen erster Stufe (d. h. Konstante und die Funktion $x+1$) ist der Satz trivial. Habe also φ die m-te Stufe. Es entsteht aus Funktionen niedrigerer Stufe $\varphi_1 \ldots \varphi_k$ durch die Operationen der Einsetzung oder der rekursiven Definition. Da für $\varphi_1 \ldots \varphi_k$ nach induktiver Annahme bereits alles bewiesen ist, gibt es zugehörige *Relationszeichen* $r_1 \ldots r_k$, so daß (3), (4) gilt. Die Definitionsprozesse, durch die φ aus $\varphi_1 \ldots \varphi_k$ entsteht (Einsetzung und rekursive Definition), können sämtlich im System P formal nachgebildet werden. Tut man dies, so erhält man aus $r_1 \ldots r_k$ ein neues *Relationszeichen* r[41]), für welches man die Geltung von (3), (4) unter Verwendung der induktiven Annahme ohne Schwierigkeit beweisen kann. Ein *Rela-*
>
> ---
> [41]) Bei der genauen Durchführung dieses Beweises wird natürlich r nicht auf dem Umweg über die inhaltliche Deutung, sondern durch seine rein formale Beschaffenheit definiert.

> and we shall use complete induction on the rank of φ. For functions of rank one (i.e. constants and the function $x+1$) the theorem is trivial. Therefore let φ have rank m. It results from functions of lower rank $\varphi_1 \ldots \varphi_k$ by the operations

of substitution or recursive definition. Since, by inductive hypothesis, everything is already proved for $\varphi_1 \ldots \varphi_k$, there exist corresponding *predicates* $r_1 \ldots r_k$ for which (3) and (4) hold. The definitional procedures by which φ arises from $\varphi_1 \ldots \varphi_k$ (substitution and recursive definition) can both be formally imitated in the system P. If one does

On formally undecidable propositions of Principia Mathematica etc. 187

this, then one obtains from $r_1 \ldots r_k$ a new *predicate* r[41]) for which one can prove without difficulty the validity of (3) and (4) by using the inductive hypothesis.

[41]) When this proof is rigorously carried out, r will naturally not be defined by this shortcut through the intuitive interpretation, but rather by its purely formal structure.

To navigate some nasty technical difficulties, we will prove a weakened variant of Theorem V, which replaces syntactic representability with semantic representability.

Theorem 5.15 — Gödel's Theorem V, semantic variant

Every primitive-recursive relation is semantically representable in P.

Proof: We closely follow Gödel's line of reasoning and prove the theorem for all primitive-recursive functions f by induction over their *level*, a concept defined on page 216.

At the lowest level are the zero function, the successor function, and the projection functions. We will show that these functions are semantically representable:

- The zero function $\text{null}(x) = 0$ is represented by the formula (PR1)

$$\varphi_{\text{null}}(\mathsf{y}_1, \mathsf{x}_1) := (\mathsf{y}_1 = 0)$$

as this formula satisfies:

$$y = \text{null}(x) \;\Rightarrow\; \models \varphi_{\text{null}}(\overline{y}, \overline{x})$$
$$y \neq \text{null}(x) \;\Rightarrow\; \models \neg\varphi_{\text{null}}(\overline{y}, \overline{x})$$

- The successor function $\text{s}(x) = x+1$ is represented by the formula (PR2)

$$\varphi_{\text{s}}(\mathsf{y}_1, \mathsf{x}_1) := (\mathsf{y}_1 = \mathsf{f}\,\mathsf{x}_1)$$

5.4 Theorem V

as this formula satisfies:

$$y = x+1 \Rightarrow \models \varphi_s(\overline{y}, \overline{x})$$
$$y \neq x+1 \Rightarrow \models \neg\varphi_s(\overline{y}, \overline{x})$$

- The projection $\pi_i^n(x_1, \ldots, x_n) = x_i$ is represented by the formula (PR3)

$$\varphi_\pi(y_1, x_1, \ldots, x_n) := (y_1 = x_i)$$

as this formula satisfies:

$$y = \pi_i^n(x_1, \ldots, x_n) \Rightarrow \models \varphi_\pi(\overline{y}, \overline{x_1}, \ldots, \overline{x_n})$$
$$y \neq \pi_i^n(x_1, \ldots, x_n) \Rightarrow \models \neg\varphi_\pi(\overline{y}, \overline{x_1}, \ldots, \overline{x_n})$$

This concludes the induction start. In the induction step, we must demonstrate that the theorem's proposition extends from the primitive-recursive functions at a certain level to those at the subsequent level. Two cases need to be distinguished:

- f has been created by substitution: (PR4)

$$f(x_1, \ldots, x_n) = h(g_1(x_1, \ldots, x_n), \ldots, g_m(x_1, \ldots, x_n)) \quad (5.20)$$

To maintain clarity, we confine ourselves to the cases $n=1$ and $m=1$; the proof proceeds accordingly for other values. Equation (5.20) then takes this form:

$$f(x) = h(g(x)) \quad (5.21)$$

In a first attempt, we transform (5.21) into the following formula:

$$\varphi_f(y_1, x_1) := \forall u_1 \, (u_1 = g(x_1) \rightarrow y_1 = h(u_1)) \quad (5.22)$$

We haven't quite reached our goal yet, as the two function symbols g and h are unavailable in P or P'. However, the induction hypothesis guarantees that both g and h are semantically representable by two formulas φ_g and φ_h, respectively, allowing us to rewrite (5.22) as follows:

$$\varphi_f(y_1, x_1) := \forall u_1 \, (\varphi_g(u_1, x_1) \rightarrow \varphi_h(y_1, u_1))$$

Now, let us consider an arbitrary instance of the form $\varphi_f(\overline{y_1}, \overline{x_1})$:

$$\varphi_f(\overline{y_1}, \overline{x_1}) := \forall u_1 \, (\varphi_g(u_1, \overline{x_1}) \rightarrow \varphi_h(\overline{y_1}, u_1))$$

The subformula $\varphi_g(u_1, \overline{x_1})$ is true precisely when u_1 is interpreted as the number $g(x_1)$. In this case, the subformula $\varphi_h(\overline{y_1}, u_1)$ is true exactly when y_1 is the value $h(g(x_1))$, yielding the following relationship:

$$y_1 = h(g(x_1)) \Rightarrow \models \varphi_f(\overline{y_1}, \overline{x_1})$$

$$y_1 \neq h(g(x_1)) \Rightarrow \ \models \neg \varphi_f(\overline{y_1}, \overline{x_1})$$

Thus, the function f obtained from substitution is represented by φ_f.

■ f is derived from the schema of primitive recursion: (PR5)

Once again, we restrict ourselves to the case $n = 1$ for clarity.

$$f(0, x) := g(x)$$
$$f(k+1, x) := h(k, f(k, x), x)$$

Let us attempt to construct a formula $\psi(\mathsf{x}_2)$ that holds true precisely when x_2 is interpreted as the function f. As a starting point, consider:

$$\psi := \forall \mathsf{x}_1 \, \forall \mathsf{y}_1 \, \forall \mathsf{k}_1 \, ((\mathsf{x}_2(\mathsf{y}_1, 0, \mathsf{x}_1) \leftrightarrow \mathsf{y}_1 = g(\mathsf{x}_1)) \wedge$$
$$(\mathsf{x}_2(\mathsf{y}_1, \mathsf{f}\,\mathsf{k}_1, \mathsf{x}_1) \leftrightarrow (\forall \mathsf{u}_1 \, (\mathsf{x}_2(\mathsf{u}_1, \mathsf{k}_1, \mathsf{x}_1) \rightarrow \mathsf{y}_1 = h(\mathsf{k}_1, \mathsf{u}_1, \mathsf{x}_1)))))$$

As before, the function symbols can be substituted by formulas that are guaranteed to exist by the induction hypothesis:

$$\psi := \forall \mathsf{x}_1 \, \forall \mathsf{y}_1 \, \forall \mathsf{k}_1 \, ((\mathsf{x}_2(\mathsf{y}_1, 0, \mathsf{x}_1) \leftrightarrow \varphi_g(\mathsf{y}_1, \mathsf{x}_1)) \wedge$$
$$(\mathsf{x}_2(\mathsf{y}_1, \mathsf{f}\,\mathsf{k}_1, \mathsf{x}_1) \leftrightarrow (\forall \mathsf{u}_1 \, (\mathsf{x}_2(\mathsf{u}_1, \mathsf{k}_1, \mathsf{x}_1) \rightarrow \varphi_h(\mathsf{y}_1, \mathsf{k}_1, \mathsf{u}_1, \mathsf{x}_1)))))$$

The auxiliary function $\psi(\mathsf{x}_2)$ enables us to semantically represent f with less effort. The formula

$$\varphi_f(\mathsf{y}_1, \mathsf{k}_1, \mathsf{x}_1) := \forall \mathsf{x}_2 \, (\psi(\mathsf{x}_2) \rightarrow \mathsf{x}_2(\mathsf{y}_1, \mathsf{k}_1, \mathsf{x}_1))$$

serves this purpose, as the following holds:

$$y = f(k, x) \Rightarrow \ \models \ \varphi_f(\overline{y}, \overline{k}, \overline{x})$$
$$y \neq f(k, x) \Rightarrow \ \models \neg \varphi_f(\overline{y}, \overline{k}, \overline{x})$$

A minor technical hurdle remains: x_2 has three parameters, a construct not available in Gödel's system P, where variables of higher types always have a single parameter. However, on page 141, Gödel had already indicated that this is not an issue in the proper sense. Variables with multiple parameters can be simulated within P by variables with a single parameter, that is, each formula can be rewritten such that only variables with a single parameter are present. Nevertheless, the result would be so confusing that we are well advised to leave the formulas in their current form.

This completes the proof sketch for the weakened version of Theorem V. □

We will further illustrate the meaning of Theorem V by constructing the formulas for two specific primitive-recursive functions we know well by now, addition

5.4 Theorem V

and multiplication:

$$\begin{aligned}\text{add}(0,x) &= x\\ \text{add}(k+1,x) &= s(\text{add}(k,x))\\ \text{mult}(0,x) &= 0\\ \text{mult}(k+1,x) &= \text{add}(\text{mult}(k,x),x)\end{aligned}$$

Addition is semantically represented by the formula

$$\varphi_{\text{add}}(y_1,k_1,x_1) := \forall x_2\,(\psi_{\text{add}}(x_2) \to x_2(y_1,k_1,x_1))$$

where ψ_{add} is a placeholder for:

$$\begin{aligned}\psi_{\text{add}} := \forall x_1\,\forall y_1\,\forall k_1\,&((x_2(y_1,0,x_1) \leftrightarrow y_1 = x_1) \wedge\\ &(x_2(y_1,\mathsf{f}\,k_1,x_1) \leftrightarrow (\forall u_1\,(x_2(u_1,k_1,x_1) \to y_1 = \mathsf{f}\,u_1))))\end{aligned}$$

With the formula φ_{add} at hand, multiplication can easily be represented, too, by the formula

$$\varphi_{\text{mult}}(y_1,k_1,x_1) := \forall x_2\,(\psi_{\text{mult}}(x_2) \to x_2(y_1,k_1,x_1))$$

with

$$\begin{aligned}\psi_{\text{mult}} := \forall x_1\,\forall y_1\,\forall k_1\,&((x_2(y_1,0,x_1) \leftrightarrow y_1 = 0) \wedge\\ &(x_2(y_1,\mathsf{f}\,k_1,x_1) \leftrightarrow (\forall u_1\,(x_2(u_1,k_1,x_1) \to \varphi_{\text{add}}(y_1,u_1,x_1)))))\end{aligned}$$

For φ_{mult}, the following applies:

$$y = x \cdot z \Rightarrow \;\models\; \varphi_{\text{mult}}(\overline{y},\overline{x},\overline{z}) \tag{5.23}$$
$$y \neq x \cdot z \Rightarrow \;\models\; \neg\varphi_{\text{mult}}(\overline{y},\overline{x},\overline{z}) \tag{5.24}$$

The variables x, y, and z may be substituted by arbitrary numbers. For instance, the following holds:

$$\models\; \varphi_{\text{mult}}(\overline{4},\overline{2},\overline{2}) \tag{5.25}$$
$$\models\; \neg\varphi_{\text{mult}}(\overline{5},\overline{2},\overline{2}) \tag{5.26}$$

Gödel's Theorem V is quite similar to the weakened variant we just proved. The only difference is that the notion of *truth* ('$\models$') is replaced by the notion of *provability* ('$\vdash$'). (5.23) and (5.24) then become:

$$y = x \cdot z \Rightarrow \;\vdash\; \varphi_{\text{mult}}(\overline{y},\overline{x},\overline{z})$$
$$y \neq x \cdot z \Rightarrow \;\vdash\; \neg\varphi_{\text{mult}}(\overline{y},\overline{x},\overline{z})$$

Furthermore, the two instances (5.25) and (5.26) change to:

$$\vdash \varphi_{\text{mult}}(\overline{4}, \overline{2}, \overline{2}) \tag{5.27}$$
$$\vdash \neg \varphi_{\text{mult}}(\overline{5}, \overline{2}, \overline{2}) \tag{5.28}$$

Hence, considerably more effort is required to prove Theorem V in its original form. Regarding our specific example, it has to be demonstrated that the formula instances (5.27) and (5.28) are derivable within the formal system P. This also clarifies footnote 39, where Gödel states:

> [39] Theorem V depends of course upon the fact that, for a recursive relation R, it is decidable on the basis of the axioms of the system P whether or not R holds for any given n-tuple of numbers.

We want to sketch how to construct a corresponding proof using the formula instances (5.27) and (5.28) as examples. For this task, we exploit that every primitive-recursive function is readily computable. For instance:

$$\begin{aligned}
\text{mult}(2,2) &= \text{add}(\text{mult}(1,2),2) \\
&= \text{add}(\text{add}(\text{mult}(0,2),2),2) \\
&= \text{add}(\text{add}(0,2),2) \\
&= \text{add}(2,2) \\
&= \text{s}(\text{add}(1,2)) \\
&= \text{s}(\text{s}(\text{add}(0,2))) \\
&= \text{s}(\text{s}(2)) \\
&= \text{s}(3) \\
&= 4
\end{aligned}$$

Note that the calculation did not rely on the semantic meaning of addition or multiplication; the correct result was obtained by strictly adhering to the primitive-recursive definition of these functions. The proof of Theorem V builds upon the observation that all steps are replicable within the formal system P in reverse order. In detail, this means that the following theorems are derivable within P, one after the other:

$\varphi_{\text{mult}}(\overline{4}, \overline{2}, \overline{2})$

1. $\vdash \varphi_{\text{s}}(\overline{4}, \overline{3})$ ☞ $4 = \text{s}(3)$

  ⋮

2. $\vdash \exists u_1 \, (\varphi_{\text{s}}(\overline{4}, u_1) \land \varphi_{\text{s}}(u_1, \overline{2}))$ ☞ $4 = \text{s}(\text{s}(2))$

  ⋮

3. $\vdash \exists u_1 \, \exists v_1 \, (\varphi_{\text{s}}(\overline{4}, u_1) \land \varphi_{\text{s}}(u_1, v_1) \land \varphi_{\text{add}}(v_1, 0, \overline{2}))$ ☞ $4 = \text{s}(\text{s}(\text{add}(0, 2)))$

  ⋮

4. $\vdash \exists u_1 \, (\varphi_{\text{s}}(\overline{4}, u_1) \land \varphi_{\text{add}}(u_1, \overline{1}, \overline{2}))$ ☞ $4 = \text{s}(\text{add}(1, 2))$

5.4 Theorem V

$\vdots$

5. $\vdash \varphi_{\text{add}}(\overline{4}, \overline{2}, \overline{2})$ ☞ $4 = \text{add}(2, 2)$

$\vdots$

6. $\vdash \exists u_1 \, (\varphi_{\text{add}}(\overline{4}, u_1, \overline{2}) \wedge \varphi_{\text{add}}(u_1, 0, \overline{2}))$ ☞ $4 = \text{add}(\text{add}(0, 2), 2)$

$\vdots$

7. $\vdash \exists u_1 \, \exists v_1 \, (\varphi_{\text{add}}(\overline{4}, u_1, \overline{2}) \wedge \varphi_{\text{add}}(u_1, v_1, \overline{2}) \wedge \varphi_{\text{mult}}(v_1, 0, \overline{2}))$
☞ $4 = \text{add}(\text{add}(\text{mult}(0, 2), 2), 2)$

$\vdots$

8. $\vdash \exists u_1 \, (\varphi_{\text{add}}(\overline{4}, u_1, \overline{2}) \wedge \varphi_{\text{mult}}(u_1, \overline{1}, \overline{2}))$ ☞ $4 = \text{add}(\text{mult}(1, 2), 2)$

$\vdots$

9. $\vdash \varphi_{\text{mult}}(\overline{4}, \overline{2}, \overline{2})$ ☞ $4 = \text{mult}(2, 2)$

The instance (5.28) can be derived analogously:

⚙ $\neg \varphi_{\text{mult}}(\overline{5}, \overline{2}, \overline{2})$

1. $\vdash \neg \varphi_s(\overline{5}, \overline{3})$ ☞ $5 \neq s(3)$

$\vdots$

2. $\vdash \neg \exists u_1 \, (\varphi_s(\overline{5}, u_1) \wedge \varphi_s(u_1, \overline{2}))$ ☞ $5 \neq s(s(2))$

$\vdots$

3. $\vdash \neg \exists u_1 \, \exists v_1 \, (\varphi_s(\overline{5}, u_1) \wedge \varphi_s(u_1, v_1) \wedge \varphi_{\text{add}}(v_1, 0, \overline{2}))$ ☞ $5 \neq s(s(\text{add}(0, 2)))$

$\vdots$

4. $\vdash \neg \exists u_1 \, (\varphi_s(\overline{5}, u_1) \wedge \varphi_{\text{add}}(u_1, \overline{1}, \overline{2}))$ ☞ $5 \neq s(\text{add}(1, 2))$

$\vdots$

5. $\vdash \neg \varphi_{\text{add}}(\overline{5}, \overline{2}, \overline{2})$ ☞ $5 \neq \text{add}(2, 2)$

$\vdots$

6. $\vdash \neg \exists u_1 \, (\varphi_{\text{add}}(\overline{5}, u_1, \overline{2}) \wedge \varphi_{\text{add}}(u_1, 0, \overline{2}))$ ☞ $5 \neq \text{add}(\text{add}(0, 2), 2)$

$\vdots$

7. $\vdash \neg \exists u_1 \, \exists v_1 \, (\varphi_{\text{add}}(\overline{5}, u_1, \overline{2}) \wedge \varphi_{\text{add}}(u_1, v_1, \overline{2}) \wedge \varphi_{\text{mult}}(v_1, 0, \overline{2}))$
☞ $5 \neq \text{add}(\text{add}(\text{mult}(0, 2), 2), 2)$

$\vdots$

8. $\vdash \neg \exists u_1 \, (\varphi_{\text{add}}(\overline{5}, u_1, \overline{2}) \wedge \varphi_{\text{mult}}(u_1, \overline{1}, \overline{2}))$ ☞ $5 \neq \text{add}(\text{mult}(1, 2), 2)$

$\vdots$

9. $\vdash \neg \varphi_{\text{mult}}(\overline{5}, \overline{2}, \overline{2})$ ☞ $5 \neq \text{mult}(2, 2)$

In fact, the ability to replicate the calculation within P is not particularly remarkable. Gödel's formal system P is essentially the system of the *Principia Mathematica* and, as such, capable of formalizing all of classical mathematics. However, this does not imply that the derivation is straightforward. If we were to attempt to fill in the remaining proof gaps, we would have a lot of work on our plates. This is why Gödel only sketched the proof, which we take as the opportunity to bid farewell to Theorem V at this juncture.

Finally, Gödel introduces another idiom:

> induktiven Annahme ohne Schwierigkeit beweisen kann. Ein *Relationszeichen r*, welches auf diesem Wege einer rekursiven Relation zugeordnet ist [42]), soll *rekursiv heißen*.
>
> ---
> [42]) Welches also, inhaltlich gedeutet, das Bestehen dieser Relation ausdrückt.

> A *predicate* r which corresponds in this way to a recursive relation [42]) shall be called recursive.
>
> ---
> [42]) Which, therefore, expresses intuitively that this relation holds.

The notation of a relation sign ("Relationszeichen") was introduced earlier on page 148, signifying a formula with free variables, all of which are individual variables. From what has been said so far, the origin of this term becomes evident. It arises from the fact that any formula φ with n free individual variables can be naturally associated with the following relation:

$$R := \{(x_1, \ldots, x_n) \in \mathbb{N}^n \mid \models \varphi(\overline{x_1}, \ldots, \overline{x_n})\}$$

If R is a primitive-recursive relation and the formula φ fulfills the property postulated in Theorem V, that is, if

$$\begin{aligned}(x_1, \ldots, x_n) \in R &\Rightarrow\ \vdash \varphi(\overline{x_1}, \ldots, \overline{x_n}) \\ (x_1, \ldots, x_n) \notin R &\Rightarrow\ \vdash \neg\varphi(\overline{x_1}, \ldots, \overline{x_n}),\end{aligned}$$

holds, φ is called primitive-recursive. In more concise terms, φ is primitive recursive if it syntactically represents a primitive-recursive relation. The concept just introduced also enables us to speak of primitive-recursive *class expressions*, referring to primitive-recursive formulas with a single free individual variable. Those formulas syntactically represent a primitive-recursive set. Remember that Gödel uses the term *recursive* rather than *primitive recursive*, as he does throughout his work.

6 The Limits of Mathematics

"Abandon hope all ye who enter here."

Dante's Inferno

Before setting the stage for the grand finale, let us briefly recapitulate the results we have achieved thus far:

- In Chapter 4, we acquainted ourselves with the formal system P and demonstrated how formulas and proofs can be arithmetized. By assigning Gödel numbers to formulas and sequences of formulas, we could interpret the manipulation of symbol strings, and thus the conduct of a proof, on the arithmetic level.

- In Chapter 5, we introduced the concept of primitive recursion and meticulously derived 45 primitive-recursive functions and relations. Ultimately, we discovered that crucial metamathematical concepts about formal systems are expressible by primitive recursion.

- At the end of Chapter 5, we discussed Theorem V, establishing a link between formulas and primitive-recursive relations. Substantively, this theorem states that any primitive-recursive relation is syntactically representable within P. This means that for any given primitive-recursive relation and number combination, it can either be proved or disproved within P whether or not the relation holds for the given number combination.

Now, Gödel will demonstrate the destructive power unleashed by combining these three partial results in a particular manner. Lean back and enjoy the culmination!

6.1 Gödel's Main Result

Gödel starts by defining a crucial term that will remain prevalent throughout this chapter.

> Wir kommen nun ans Ziel unserer Ausführungen. Sei x eine beliebige Klasse von *Formeln*. Wir bezeichnen mit Flg (x) (Folgerungsmenge von x) die kleinste Menge von *Formeln*, die alle *Formeln*

> aus x und alle *Axiome* enthält und gegen die Relation „*unmittelbare Folge*" abgeschlossen ist. x heißt ω-widerspruchsfrei, wenn es kein

> We now come to the goal of our work. Let x be an arbitrary class of *formulas*. We denote by Flg (x) (consequence set of x) the smallest set of *formulas* which contains all *formulas* of x and all *axioms* and is closed with respect to the relation of "*immediate consequence*".

Pay attention to the italic typeface! It indicates that the set χ and the set $\mathrm{Flg}(\chi)$ are both subsets of the natural numbers. In particular, the set χ contains the Gödel numbers of formulas:

$$\chi = \{\ulcorner \varphi_1 \urcorner, \ulcorner \varphi_2 \urcorner, \ldots\}$$

The set $\mathrm{Flg}(\chi)$ is the superset of χ, which additionally contains the Gödel numbers of all axioms of P, as well as the Gödel numbers of all theorems derivable from the axioms and the formulas encoded in χ. To concisely summarize what has just been said, we adopt the following intuitively obvious notation:

Definition 6.1

- For any set M of formulas, $\mathrm{P} \cup M$ denotes the formal system obtained from P after augmenting the axioms by the formulas from M.

- For any set χ of Gödel numbers, $\mathrm{P} \cup \chi$ denotes the following set:

$$\mathrm{P} \cup \chi := \mathrm{P} \cup \{\varphi \mid \ulcorner \varphi \urcorner \in \chi\}$$

With the simplified notation at hand, the sets χ and $\mathrm{Flg}(\chi)$ can be characterized as follows:

$$\chi \subseteq \{n \in \mathbb{N} \mid n \text{ is the Gödel number of a formula}\}$$
$$\mathrm{Flg}(\chi) = \{n \in \mathbb{N} \mid n \text{ is the Gödel number of a theorem of } \mathrm{P} \cup \chi\} \quad (6.1)$$

The next definition is just as important:

> *Folge*" abgeschlossen ist. x heißt ω-widerspruchsfrei, wenn es kein *Klassenzeichen* a gibt, so daß:
>
> $$(n) \left[Sb\left(a \begin{array}{c} v \\ Z(n) \end{array}\right) \varepsilon \, \mathrm{Flg}\,(x) \right] \& \left[\mathrm{Neg}\,(v \, \mathrm{Gen}\, a) \right] \varepsilon \, \mathrm{Flg}\,(x)$$
>
> wobei v die *freie Variable* des *Klassenzeichens* a ist.

6.1 Gödel's Main Result

> We say that x is ω-consistent if there is no *class expression a* such that
>
> $$(n)\left[Sb\left(a\,{}^{v}_{Z(n)}\right)\,\varepsilon\,\text{Flg}(x)\right]\,\&\,\left[\text{Neg}(v\,\text{Gen}\,a)\right]\,\varepsilon\,\text{Flg}(x)$$
>
> where v is the *free variable* of the *class expression a*.

Let us recall: A class expression is a formula $\varphi(\xi_1)$ with a sole free individual variable, the variable ξ_1. Gödel stipulates that the set χ is ω-consistent if and only if the following two properties do not hold simultaneously for any formula $\varphi(\xi_1)$:

- ■ For every natural number n,   ☞ (n)
  - the formula $\varphi(\overline{n})$ ☞ $Sb\left(a\,{}^{v}_{Z(n)}\right)$
  - is provable in $P \cup \chi$.   ☞ $\varepsilon\,\text{Flg}(x)$
- ■ The formula $\neg \forall \xi_1\,\varphi(\xi_1)$ ☞ $\text{Neg}(v\,\text{Gen}\,a)$
  - is provable in $P \cup \chi$.   ☞ $\varepsilon\,\text{Flg}(x)$

In modern terms:

Definition 6.2   ω-Consistency

The set $P \cup \chi$ is *ω-consistent* if the following holds:

$$\vdash \varphi(\overline{n}) \text{ for all } n \in \mathbb{N} \;\Rightarrow\; \not\vdash \neg \forall x_1\,\varphi(x_1)$$

A set of Gödel numbers χ is ω-consistent if $P \cup \chi$ is ω-consistent.

In the next sentence, Gödel points out that ω-consistency is a stronger property than ordinary consistency. In particular, he states that every ω-consistent system is consistent. But, as will be shown later, the converse does not hold.

> Jedes ω-widerspruchsfreie System ist selbstverständlich auch widerspruchsfrei. Es gilt aber, wie später gezeigt werden wird, nicht das Umgekehrte.

> Every ω-consistent system is obviously also consistent. However, as will be shown later, the converse does not hold.

Every ω-consistent set is consistent because all formulas become theorems in an inconsistent formal system, including those formulas that cause an ω-contradiction, as defined in Definition 6.2. Thus, we have:

Theorem 6.3

> Every ω-consistent set is consistent.

At this point, we are on the verge of reaching the climax of Gödel's work. It comes in the form of Theorem VI, which asserts nothing less than the incompleteness of $P \cup \chi$. This theorem is of such generality that all commonly used variants of the first incompleteness theorem are derivable as corollaries.

6.1.1 Incompleteness of System P

> Das allgemeine Resultat über die Existenz unentscheidbarer Sätze lautet:
> Satz VI: Zu jeder ω-widerspruchsfreien rekursiven Klasse x von *Formeln* gibt es rekursive *Klassenzeichen* r, so daß weder v Gen r noch Neg (v Gen r) zu Flg (x) gehört (wobei v die *freie Variable* aus r ist).

> The general result on the existence of undecidable propositions reads:
> Theorem VI: For every ω-consistent recursive class x of *formulas*, there exists a recursive *class expression* r such that neither v Gen r nor Neg (v Gen r) belongs to Flg (x) (where v is the *free variable* of r).

Let us start by making the theorem more accessible to contemporary readers:

Theorem 6.4　　　　　　　　　　　　　　　　　　　　　　　Gödel's Theorem VI

> Let $\chi = \{\ulcorner \varphi_1 \urcorner, \ulcorner \varphi_2 \urcorner, \ldots\}$ be ω-consistent and primitive recursive. Then, there exists a primitive-recursive formula $\varphi_r(\xi_1)$, for which neither
>
> $$\forall \xi_1 \, \varphi_r(\xi_1) \quad \text{nor} \quad \neg \forall \xi_1 \, \varphi_r(\xi_1)$$
>
> is provable within the system $P \cup \chi$.

The proof commences with the definition of several relations concerning the provability of formulas in $P \cup \chi$:

6.1 Gödel's Main Result

Beweis: Sei x eine beliebige rekursive ω-widerspruchsfreie Klasse von *Formeln*. Wir definieren:

$$Bw_x(x) \equiv (n)\,[n \leq l(x) \rightarrow Ax(n\,Gl\,x) \vee (n\,Gl\,x)\,\varepsilon\,x \vee \quad (5)$$

$$(Ep,q)\,\{0 < p, q < n\,\&\,Fl(n\,Gl\,x, p\,Gl\,x, q\,Gl\,x)\}]\,\&\,l(x) > 0$$

(vgl. den analogen Begriff 44)

$$x\,B_x\,y \equiv Bw_x(x)\,\&\,[l(x)]\,Gl\,x = y \quad (6)$$

$$Bew_x(x) \equiv (Ey)\,y\,B_x\,x \quad (6\cdot 1)$$

(vgl. die analogen Begriffe 45, 46).

Proof. Let x be an arbitrary recursive ω-consistent class of *formulas*. We define:

$$Bw_x(x) \equiv (n)\,[n \leq l(x) \rightarrow Ax(n\,Gl\,x) \vee (n\,Gl\,x)\,\varepsilon\,x \vee \quad (5)$$

$$(Ep,q)\,\{0 < p, q < n\,\&\,Fl(n\,Gl\,x, p\,Gl\,x, q\,Gl\,x)\}]\,\&\,l(x) > 0$$

(cf. the similar concept 44)

$$x\,B_x\,y \equiv Bw_x(x)\,\&\,[l(x)]\,Gl\,x = y \quad (6)$$

$$Bew_x(x) \equiv (Ey)\,y\,B_x\,x \quad (6\cdot 1)$$

(cf. the similar concepts 45, 46).

In modern notation, (5), (6), and (6·1) read as follows:

$$x \in Bw_\chi \;:\Leftrightarrow\; l(x) > 0 \wedge \forall(n \leq l(x)) \begin{pmatrix} Ax(n\,Gl\,x)\ \vee \\ (n\,Gl\,x) \in \chi\ \vee \\ \exists(0 < p, q < n) \\ Fl(n\,Gl\,x, p\,Gl\,x, q\,Gl\,x) \end{pmatrix}$$

$$x\,B_\chi\,y \;:\Leftrightarrow\; Bw_\chi(x) \wedge l(x)\,Gl\,x = y$$

$$x \in Bew_\chi \;:\Leftrightarrow\; \exists y\ y\,B_\chi\,x$$

The definition of Bw_χ lacks the estimate $0 < n$. Adding it leads to:

$$x \in Bw_\chi \;:\Leftrightarrow\; l(x) > 0 \wedge \forall(0 < n \leq l(x)) \begin{pmatrix} Ax(n\,Gl\,x)\ \vee \\ (n\,Gl\,x) \in \chi\ \vee \\ \exists(0 < p, q < n) \\ Fl(n\,Gl\,x, p\,Gl\,x, q\,Gl\,x) \end{pmatrix}$$

The introduced notions are extensions of the primitive-recursive relations 44 and 45 from Section 5.2 and the relation 46 from Section 5.3, respectively. All three follow identical definition patterns, making their substantive meanings almost self-evident:

$$x \in Bw_\chi \Leftrightarrow x \text{ encodes a formal proof chain of the system } P \cup \chi$$
$$x\, B_\chi \ulcorner\varphi\urcorner \Leftrightarrow x \text{ encodes a proof for the formula } \varphi \text{ in } P \cup \chi$$
$$\ulcorner\varphi\urcorner \in \text{Bew}_\chi \Leftrightarrow \varphi \text{ is provable in } P \cup \chi \tag{6.2}$$

> **Es gilt offenbar:**
> $$(x)\,[\text{Bew}_x(x) \sim x\,\varepsilon\,\text{Flg}(x)] \tag{7}$$
> $$(x)\,[\text{Bew}(x) \rightarrow \text{Bew}_x(x)] \tag{8}$$

> **Obviously, we have:**
> $$(x)\,[\text{Bew}_x(x) \sim x\,\varepsilon\,\text{Flg}(x)] \tag{7}$$
> $$(x)\,[\text{Bew}(x) \rightarrow \text{Bew}_x(x)] \tag{8}$$

In modern notation, (7) and (8) take on the following appearance:

$$x \in \text{Bew}_\chi \Leftrightarrow x \in \text{Flg}(\chi) \tag{6.3}$$
$$x \in \text{Bew} \Rightarrow x \in \text{Bew}_\chi \tag{6.4}$$

(6.3) and (6.4) can be expressed even more conciselys as follows:

$$\text{Bew} \subseteq \text{Bew}_\chi = \text{Flg}(\chi)$$

The inclusion $\text{Bew} \subseteq \text{Bew}_\chi$ is immediately apparent because the set of theorems can only increase but never decrease when the set of axioms of a formal system is supplemented with additional formulas. For realizing $\text{Bew}_\chi = \text{Flg}(\chi)$, it is sufficient to have another look at (6.2) and (6.1).

Next, Gödel defines a primitive-recursive relation that will play a central role in the proof of the main result:

$Q(x, y)$ — x is not a proof for y
(Primitive-recursive relation)

188   Kurt Gödel,
Nun definieren wir die Relation:

6.1 Gödel's Main Result

$$Q(x,y) \equiv \overline{x\, B_x \left[Sb\left(y\, \begin{smallmatrix} 19 \\ Z(y) \end{smallmatrix} \right) \right]}. \tag{8.1}$$

Da $x\, B_x\, y$ [nach (6), (5)] und $Sb\left(y\, \begin{smallmatrix} 19 \\ Z(y) \end{smallmatrix} \right)$ (nach Def. 17, 31) rekursiv sind, so auch $Q(x\,y)$. Nach Satz V und (8) gibt es also ein

---

188   Kurt Gödel,

Now we define the relation:

$$Q(x,y) \equiv \overline{x\, B_x \left[Sb\left(y\, \begin{smallmatrix} 19 \\ Z(y) \end{smallmatrix} \right) \right]}. \tag{8.1}$$

Since $x\, B_x\, y$ [according to (6), (5)] and $Sb\left(y\, \begin{smallmatrix} 19 \\ Z(y) \end{smallmatrix} \right)$ (according to Definitions 17, 31) are recursive, so also is $Q(x\,y)$.

The number 19 encodes the symbol y_1. Denoting the formula with the Gödel number y as φ_y allows us to rephrase this definition as follows:

$$(x,y) \in Q :\Leftrightarrow \neg(x\, B_\chi\, \ulcorner \varphi_y[y_1 \leftarrow \overline{y}] \urcorner) \tag{6.5}$$

In colloquial terms, this relationship reads as such:

$$(x,y) \in Q :\Leftrightarrow x \text{ does not encode a proof for the formula } \varphi_y[y_1 \leftarrow \overline{y}] \tag{6.6}$$

In other words, if y is the Gödel number of a formula φ_y with the free variable y_1, then x and y are related if and only if x does *not* encode a formula sequence that derives the diagonal element $\varphi_y(\overline{y})$ within the formal system $P \cup \chi$.

---

rekursiv sind, so auch $Q(x\,y)$. Nach Satz V und (8) gibt es also ein *Relationszeichen q* (mit den *freien Variablen* 17, 19), so daß gilt:

$$\overline{x\, B_x \left[Sb\left(y\, \begin{smallmatrix} 19 \\ Z(y) \end{smallmatrix} \right) \right]} \rightarrow \text{Bew}_x \left[Sb\left(q\, \begin{smallmatrix} 17 & 19 \\ Z(x) & Z(y) \end{smallmatrix} \right) \right] \tag{9}$$

$$x\, B_x \left[Sb\left(y\, \begin{smallmatrix} 19 \\ Z(y) \end{smallmatrix} \right) \right] \rightarrow \text{Bew}_x \left[\text{Neg}\, Sb\left(q\, \begin{smallmatrix} 17 & 19 \\ Z(x) & Z(y) \end{smallmatrix} \right) \right] \tag{10}$$

According to Theorem V and (8), there exists therefore a *predicate q* (with the *free variables* 17 and 19) such that the following hold:

$$\overline{x\, B_x} \left[Sb \left(y\, \overset{19}{Z(y)} \right) \right] \to \text{Bew}_x \left[Sb \left(q\, \overset{17}{Z(x)}\, \overset{19}{Z(y)} \right) \right] \quad (9)$$

$$x\, B_x \left[Sb \left(y\, \overset{19}{Z(y)} \right) \right] \to \text{Bew}_x \left[\text{Neg } Sb \left(q\, \overset{17}{Z(x)}\, \overset{19}{Z(y)} \right) \right] \quad (10)$$

The relation Q is primitive recursive, thus fulfilling the prerequisite of Theorem V. Consequently, there exists a formula with two free variables x_1 and y_1, that syntactically represents Q. Let q be the Gödel number of this formula and $\psi_q(x_1, y_1)$ the formula itself. According to Theorem V, the following applies:

$$(x,y) \in Q \;\Rightarrow\; \vdash \psi_q(\overline{x}, \overline{y})$$
$$(x,y) \notin Q \;\Rightarrow\; \vdash \neg\psi_q(\overline{x}, \overline{y})$$

As per (6.6), this is the same as:

x does not encode a proof for the formula $\varphi_y[y_1 \leftarrow \overline{y}] \;\Rightarrow\; \vdash \psi_q(\overline{x}, \overline{y})$ (6.7)

x encodes a proof for the formula $\varphi_y[y_1 \leftarrow \overline{y}] \;\Rightarrow\; \vdash \neg\psi_q(\overline{x}, \overline{y})$ (6.8)

(6.7) and (6.8) are precisely the statements (9) and (10) in Gödel's article.

> Wir setzen:
> $$p = 17 \text{ Gen } q \quad (11)$$
> (p ist ein *Klassenzeichen* mit der *freien Variablen* 19) und

> We set:
> $$p = 17 \text{ Gen } q \quad (11)$$
> (p is a *class expression* with the *free variable* 19) and

The number 17 encodes the symbol x_1, thus making p the Gödel number of the following formula:
$$\varphi_p(y_1) := \forall x_1\; \psi_q(x_1, y_1)$$
Having a single free variable, $\varphi_p(y_1)$ is a class expression in Gödel's terminology.

> $$r = Sb \left(q\, \overset{19}{Z(p)} \right) \quad (12)$$
> (r ist ein rekursives *Klassenzeichen* mit der *freien Variablen* 17 [43]).

6.1 Gödel's Main Result

> [43] r entsteht ja aus dem rekursiven *Relationszeichen* q durch Ersetzen einer *Variablen* durch eine bestimmte Zahl (p).

$$r = Sb\left(q \, \substack{19 \\ Z(p)}\right) \tag{12}$$

(r is a recursive *class expression* with the *free variable* 17 [43]).

---

[43] r arises from the *recursive predicate* q by replacing one *variable* by a definite numeral (p).

The number 19 encodes the symbol y_1, thus making r the Gödel number of the following formula:

$$\varphi_r(x_1) := \psi_q(x_1, \bar{p}) \tag{6.9}$$
$$= \psi_q(x_1, \ulcorner \forall x_1 \, \psi_q(x_1, y_1) \urcorner)$$

Dann gilt:

$$Sb\left(p \, \substack{19 \\ Z(p)}\right) = Sb\left([17 \text{ Gen } q] \, \substack{19 \\ Z(p)}\right) = 17 \text{ Gen } Sb\left(q \, \substack{19 \\ Z(p)}\right) \tag{13}$$
$$= 17 \text{ Gen } r \; ^{44})$$

[wegen (11) und (12)] ferner:

---

[44] Die Operationen Gen, Sb sind natürlich immer vertauschbar, falls sie sich auf verschiedene *Variable* beziehen.

Then we have:

$$Sb\left(p \, \substack{19 \\ Z(p)}\right) = Sb\left([17 \text{ Gen } q] \, \substack{19 \\ Z(p)}\right) = 17 \text{ Gen } Sb\left(q \, \substack{19 \\ Z(p)}\right) \tag{13}$$
$$= 17 \text{ Gen } r \; ^{44})$$

[by virtue of (11) and (12)]

---

[44] The operations Gen and Sb naturally always commute with each other, in case they refer to different variables.

In modern notation, Gödel states the following:

$$\varphi_p[y_1 \leftarrow \bar{p}] \qquad\qquad \mathrel{\text{☞}} Sb\left(p \, \substack{19 \\ Z(p)}\right)$$

$$= \left(\forall x_1 \, \psi_q(x_1, y_1)\right)[y_1 \leftarrow \bar{p}] \qquad \mathrel{\text{☞}} Sb\left([17 \text{ Gen } q] \, \substack{19 \\ Z(p)}\right)$$

$$= \forall x_1\, \psi_q(x_1, \bar{p}) \qquad\qquad \text{☞ 17 Gen } Sb\left(q\,\begin{smallmatrix}19\\Z(p)\end{smallmatrix}\right)$$

$$= \forall x_1\, \varphi_r(x_1) \qquad\qquad \text{☞ 17 Gen } r$$

By choosing the number p for y, (6.7) and (6.8) can be rewritten as follows:

x does not encode a proof for the formula $\varphi_p[y_1 \leftarrow \bar{p}] \;\Rightarrow\; \vdash\; \psi_q(\bar{x}, \bar{p})$

x encodes a proof for the formula $\varphi_p[y_1 \leftarrow \bar{p}] \;\Rightarrow\; \vdash\; \neg\psi_q(\bar{x}, \bar{p})$

According to what has just been said, this is the same as:

x does not encode a proof for the formula $\forall x_1\, \varphi_r(x_1) \;\Rightarrow\; \vdash\; \psi_q(\bar{x}, \bar{p})$ (6.10)

x encodes a proof for the formula $\forall x_1\, \varphi_r(x_1) \;\Rightarrow\; \vdash\; \neg\psi_q(\bar{x}, \bar{p})$ (6.11)

[wegen (11) und (12)] **ferner:**

$$Sb\left(q\,\begin{smallmatrix}17\\Z(x)\end{smallmatrix}\,\begin{smallmatrix}19\\Z(p)\end{smallmatrix}\right) = Sb\left(r\,\begin{smallmatrix}17\\Z(x)\end{smallmatrix}\right) \qquad (14)$$

[nach (12)]. Setzt man nun in (9) und (10) p für y ein, so entsteht unter Berücksichtigung von (13) und (14):

$$\overline{x\, B_x\, (17\text{ Gen } r)} \;\to\; \text{Bew}_x\left[Sb\left(r\,\begin{smallmatrix}17\\Z(x)\end{smallmatrix}\right)\right] \qquad (15)$$

$$x\, B_x\, (17\text{ Gen } r) \;\to\; \text{Bew}_x\left[\text{Neg } Sb\left(r\,\begin{smallmatrix}17\\Z(x)\end{smallmatrix}\right)\right] \qquad (16)$$

and, furthermore:

$$Sb\left(q\,\begin{smallmatrix}17\\Z(x)\end{smallmatrix}\,\begin{smallmatrix}19\\Z(p)\end{smallmatrix}\right) = Sb\left(r\,\begin{smallmatrix}17\\Z(x)\end{smallmatrix}\right) \qquad (14)$$

[from (12)]. If one now substitutes p for y in (9) and (10), then, taking into account (13) and (14), we have the result:

$$\overline{x\, B_x\, (17\text{ Gen } r)} \;\to\; \text{Bew}_x\left[Sb\left(r\,\begin{smallmatrix}17\\Z(x)\end{smallmatrix}\right)\right] \qquad (15)$$

$$x\, B_x\, (17\text{ Gen } r) \;\to\; \text{Bew}_x\left[\text{Neg } Sb\left(r\,\begin{smallmatrix}17\\Z(x)\end{smallmatrix}\right)\right] \qquad (16)$$

Gödel's equation (14) corresponds to

$$\psi_q(\bar{x}, \bar{p}) \;=\; \varphi_r(\bar{x}),$$

6.1 Gödel's Main Result

which can be utilized to further rephrase (6.10) and (6.11):

x does not encode a proof for the formula $\forall x_1\, \varphi_r(x_1) \;\Rightarrow\; \vdash \varphi_r(\overline{x})$ (6.12)

x encodes a proof for the formula $\forall x_1\, \varphi_r(x_1) \;\Rightarrow\; \vdash \neg\varphi_r(\overline{x})$ (6.13)

(6.12) and (6.13) correspond to Gödel's equations (15) and (16), respectively.

$\forall x_1\, \varphi_r(x_1)$ is the undecidable formula we have been looking for: Neither itself nor its negation is provable within $P \cup \chi$. To see why, let us distinguish the two possible cases:

■ Case 1: $\vdash \forall x_1\, \varphi_r(x_1)$

If $\forall x_1\, \varphi_r(x_1)$ were provable, some Gödel number, say n, would encode the proof of this formula. Then, according to (6.13):

$$\vdash \neg\varphi_r(\overline{n}) \qquad (6.14)$$

On the other hand, the assumption $\vdash \forall x_1\, \varphi_r(x_1)$ would allow us to derive the following theorem:

$$\vdash \varphi_r(\overline{n})$$

Because of (6.14), this is only possible if χ is inconsistent. Then, a fortiori χ would be ω-inconsistent, contrary to our assumption.

Gödels phrases the presented arguments as follows:

Über formal unentscheidbare Sätze der Principia Mathematica etc.   189

Daraus ergibt sich:

    1. 17 Gen r ist nicht x-*beweisbar*[45]). Denn wäre dies der Fall, so gäbe es (nach 6·1) ein n, so daß $n\, B_x\, (17\ \text{Gen}\ r)$. Nach (16) gälte also: $\text{Bew}_x \left[\text{Neg}\, S b \left(r\, \dfrac{17}{Z(n)} \right) \right]$, während andererseits aus der x-*Beweisbarkeit* von 17 Gen r auch die von $S b \left(r\, \dfrac{17}{Z(n)} \right)$ folgt. x wäre also widerspruchsvoll (umsomehr ω-widerspruchsvoll).

---

[45]) x ist x-*beweisbar*, soll bedeuten: $x \varepsilon\, \text{Flg}\, (\chi)$, was nach (7) dasselbe besagt wie: $\text{Bew}_x\, (x)$.

On formally undecidable propositions of Principia Mathematica etc.   189

From this follows:

> 1. 17 Gen r is not x-*provable*[45]). For, were this the case, then (according to 6·1) there would exist an n such that $n\, B_x\, (17\,\text{Gen}\, r)$. Hence, according to (16), $\text{Bew}_x \left[\text{Neg}\, Sb\left(r\, {}^{17}_{Z(n)} \right) \right]$ would hold, while, on the other hand, from the x-*provability* of 17 Gen r that of $Sb\left(r\, {}^{17}_{Z(n)} \right)$ would also follow. Therefore, x would be inconsistent (a fortiori, ω-inconsistent)
>
> ---
>
> [45]) x is x-*provable* shall mean: $x\,\varepsilon\,\text{Flg}\,(\chi)$, which, according to (7), has the same meaning as $\text{Bew}_x\,(x)$.

■ Case 2: $\vdash \neg \forall x_1\, \varphi_r(x_1)$

We have just shown that the formula $\forall x_1\, \varphi_r(x_1)$ cannot be proven, implying that no natural number encodes a proof for this formula. Thus, according to (6.12), the following holds:

$$\vdash \varphi_r(\overline{0}),\ \vdash \varphi_r(\overline{1}),\ \vdash \varphi_r(\overline{2}),\ \vdash \varphi_r(\overline{3}),\ \vdash \varphi_r(\overline{4}),\ \ldots$$

At this point, we have crossed the finish line. On the one hand, $\neg \forall x_1\, \varphi_r(x_1)$ is provable by assumption. On the other hand, all instances $\varphi_r(\overline{n})$ are also provable. χ would then be ω-inconsistent, contrary to our assumption.

Now, it is easy to follow Gödel's original words:

> 2. Neg (17 Gen r) ist nicht x-*beweisbar*. Beweis: Wie eben bewiesen wurde, ist 17 Gen r nicht x-*beweisbar*, d. h. (nach 6·1) es gilt $\overline{(n)\, n\, B_x\, (17\, \text{Gen}\, r)}$. Daraus folgt nach (15) $(n)\, \text{Bew}_x \left[Sb\left(r\, {}^{17}_{Z(n)} \right) \right]$, was zusammen mit $\text{Bew}_x\, [\text{Neg}\, (17\, \text{Gen}\, r)]$ gegen die ω-Widerspruchsfreiheit von x verstoßen würde.
>
> 17 Gen r ist also aus x unentscheidbar, womit Satz VI bewiesen ist.

> 2. Neg (17 Gen r) is not x-*provable*. Proof: As was just proved, 17 Gen r is not x-*provable*, i.e. (according to 6·1), $\overline{(n)\, n\, B_x\, (17\, \text{Gen}\, r)}$ holds. From this, we deduce, according to (15), $(n)\, \text{Bew}_x \left[Sb\left(r\, {}^{17}_{Z(n)} \right) \right]$, which, together with $\text{Bew}_x\, [\text{Neg}\, (17\, \text{Gen}\, r)]$, would contradict the ω-consistency of x.
>
> Hence, 17 Gen r is undecidable from x, which proves Theorem VI.

This concludes the proof of the most paramount theorem in Gödel's work.

6.1 Gödel's Main Result

However, it is not yet the time to rest as the main result allows us to draw two fundamental conclusions:

> **Corollary 6.5**
>
> The formal system P is incomplete.

This assertion is a consequence of Theorem VI when χ is taken as the empty set. Note that the corollary implicitly assumes that the axioms of system P are ω-consistent. The reason why this assumption is not explicitly mentioned is simple: P is essentially a formalized variant of classical mathematics. Thus, if we could derive a contradiction in P, we could also make this contradiction visible in ordinary mathematics. Conversely, this means that we can safely assume the consistency of P as long as we trust ordinary mathematics.

The close relationship between P and ordinary mathematics makes it possible to replicate the notions of P and the utilized ways of reasoning in any formal system expressive enough to formalize ordinary mathematics. Thus, Corollary 6.5 also applies to these systems:

> **Corollary 6.6**
>
> Every consistent formal system expressive enough to formalize ordinary mathematics is incomplete.

This is precisely the formulation of Theorem 1.6. Substantively, it corresponds to the variant of the first incompleteness theorem that Gödel presented at the 2nd Conference for Epistemology of the Exact Sciences in Königsberg.

In the following paragraphs, Gödel further generalizes his main result. Section 6.2 will become particularly exciting as Gödel demonstrates that the prerequisite of his main result, namely that a formal system is at least as expressive as P, can be considerably weakened. Eventually, this brings us to the renowned variant of the main result, now known as the first incompleteness theorem.

6.1.2 Consequences of the Main Result

The forthcoming section primarily holds historical significance. Gödel highlights that the proof of the main result relies solely on arguments recognized as legitimate by intuitionists.

> Man kann sich leicht überzeugen, daß der eben geführte Beweis konstruktiv ist[45a]), d. h. es ist intuitionistisch einwandfrei

> folgendes bewiesen: Sei eine beliebige rekursiv definierte Klasse x
>
> ⁴⁵ᵃ⁾ Denn alle im Beweise vorkommenden Existentialbehauptungen beruhen auf Satz V, der, wie leicht zu sehen, intuitionistisch einwandfrei ist.

> One can easily convince oneself that the proof we have just given is constructive ⁴⁵ᵃ⁾, i.e. the following has been proved in an intuitionistically unobjectionable way:
>
> ⁴⁵ᵃ⁾ For, all the existential assertions occurring in the proof rest upon Theorem V, which, as is easy to see, is intuitionistically unobjectionable.

In fact, it is easy to convince oneself that the just conducted proof is constructive. First, recall that all formulas used in the proof of Theorem V can be written down explicitly, which means that Theorem V can be strengthened:

Theorem 6.7 — Theorem V (constructive)

For every primitive-recursive relation R, a formula can be constructed that syntactically represents R in P.

This theorem allows us to rewrite (6.12) and (6.13) as follows:

$$x \text{ does not encode a proof for the formula } \forall x_1\, \varphi_r(x_1)$$
$$\Rightarrow \text{ a proof for } \varphi_r(\overline{x}) \text{ can be constructed} \qquad (6.15)$$
$$x \text{ encodes a proof for the formula } \forall x_1\, \varphi_r(x_1)$$
$$\Rightarrow \text{ a proof for } \neg\varphi_r(\overline{x}) \text{ can be constructed} \qquad (6.16)$$

Now, the two cases in the proof of Theorem VI read like this:

■ **Case 1:** $\vdash \forall x_1\, \varphi_r(x_1)$

If $\forall x_1\, \varphi_r(x_1)$ were provable, some Gödel number, say n, would encode the proof of this formula. Then, according to (6.16),

$$\textit{a proof for } \neg\varphi_r(\overline{n}) \textit{ can be constructed.} \qquad (6.17)$$

On the other hand, the assumption $\vdash \forall x_1\, \varphi_r(x_1)$ would allow us to

$$\textit{construct a proof for } \varphi_r(\overline{n}).$$

Because of (6.17), this is only possible if χ is inconsistent. Then, a fortiori χ would be ω-inconsistent, contrary to our assumption.

6.1 Gödel's Main Result

■ Case 2: $\vdash \neg \forall x_1 \, \varphi_r(x_1)$

We have just shown that the formula $\forall x_1 \, \varphi_r(x_1)$ cannot be proven, implying that no natural number encodes a proof for this formula. Thus, according to (6.16), the following holds:

A proof for the formula $\varphi_r(\overline{0})$ can be constructed.
A proof for the formula $\varphi_r(\overline{1})$ can be constructed.
A proof for the formula $\varphi_r(\overline{2})$ can be constructed.
...

Then, χ would be ω-inconsistent, contrary to our assumption.

In Gödel's words, the same argument reads as follows:

> folgendes bewiesen: Sei eine beliebige rekursiv definierte Klasse x von *Formeln* vorgelegt. Wenn dann eine formale Entscheidung (aus x) für die (effektiv aufweisbare) *Satzformel* 17 Gen r vorgelegt ist, so

> Assume given an arbitraty recursively defined class x of *formulas*. Then, if we are presented with a formal decision (from x) of the (effectively presentable) *sentence* 17 Gen r,

17 Gen r is the Gödel number of $\forall x_1 \, \varphi_r(x_1)$. When Gödel says a formal decision of the sentential formula 17 Gen r is presented, he means that either a proof for $\forall x_1 \, \varphi_r(x_1)$ or a proof for $\neg \forall x_1 \, \varphi_r(x_1)$ is given.

The discussion above has shown the following: If, on the one hand, there is a proof for $\forall x_1 \, \varphi_r(x_1)$, then a proof can be constructed for any given formula. If, on the other hand, there is a proof for $\neg \forall x_1 \, \varphi_r(x_1)$, then proofs can be constructed for all formulas of the form $\varphi_r(\overline{n})$. Therefore, in both cases, the following holds:

> für die (effektiv aufweisbare) *Satzformel* 17 Gen r vorgelegt ist, so kann man effektiv angeben:
> 1. Einen *Beweis* für Neg (17 Gen r).

> we can effectively give:
> 1. A *proof* of Neg (17 Gen r).

☞ Neg (17 Gen r) is the Gödel number of $\neg \forall x_1 \, \varphi_r(x_1)$

2. Für jedes beliebige n einen *Beweis* für $Sb\left(r\,{17\atop Z(n)}\right)$ d. h. eine

2. For any arbitrary n, a *proof* of $Sb\left(r\,{17\atop Z(n)}\right)$

☞ $Sb\left(r\,{17\atop Z(n)}\right)$ is the Gödel number of $\varphi_r(\overline{n})$

2. Für jedes beliebige n einen *Beweis* für $Sb\left(r\,{17\atop Z(n)}\right)$ d. h. eine formale Entscheidung von 17 Gen r würde die effektive Aufweisbarkeit eines ω-Widerspruchs zur Folge haben.

i.e. a formal decision for 17 Gen r would have as a consequence the effective exhibition of an ω-inconsistency.

Next, Gödel discusses several generalizations of the main result. In the original formulation of Theorem VI, he had assumed that χ is a primitive-recursive set, and from our current knowledge, it is clear why this assumption is needed. In the proof of Theorem VI, we relied on the relation Q being syntactically representable by a formula ψ_q, which Theorem V guarantees for primitive-recursive relations. However, since we do not need the property of primitive recursivity at any other point in the proof, we can weaken the premise of Theorem VI.

Gödel follows the same line of reasoning but with a different terminology. He refers to syntactically representable functions and relations as "entscheidungsdefinit" or simply as "decidable" in the English translation.

Wir wollen eine Relation (Klasse) zwischen natürlichen Zahlen $R(x_1\ldots x_n)$ entscheidungsdefinit nennen, wenn es ein n-stelliges *Relationszeichen* r gibt, so daß (3) und (4) (vgl. Satz V) gilt. Insbesondere ist also nach Satz V jede rekursive Relation entscheidungsdefinit. Analog soll ein *Relationszeichen* entscheidungsdefinit heißen, wenn es auf diese Weise einer entscheidungsdefiniten Relation zugeordnet ist. Es genügt nun für die Existenz von aus χ unentscheidbarer Sätze, von der Klasse χ vorauszusetzen, daß sie ω-

6.1 Gödel's Main Result

> widerspruchsfrei und entscheidungsdefinit ist. Denn die Entscheidungs-
> definitheit überträgt sich von x auf $x\,B_x\,y$ (vgl. (5), (6)) und auf $Q(x,y)$
>
> 190   Kurt Gödel,
>
> (vgl. (8·1)) und nur dies wurde in obigem Beweise verwendet. Der un-
> entscheidbare Satz hat in diesem Fall die Gestalt $v\,\text{Gen}\,r$, wo r ein
> entscheidungsdefinites *Klassenzeichen* ist (es genügt übrigens sogar,
> daß x in dem durch x erweiterten System entscheidungsdefinit ist).

> We shall call a relation (class) among natural numbers
> $R(x_1\ldots x_n)$ decidable if there exists an n-place *predicate* r
> such that (3) and (4) (cf. Theorem V) hold. In particular, according to
> Theorem V, every recursive relation is decidable. Similarly a *predicate*
> will be called decidable when it corresponds in this way to a decidable
> relation. Now it suffices for the existence of [from x] undecidable sen-
> tences to assume that the class x is ω-consistent and decidable. For the
> decidability carries over from x to $x\,B_x\,y$ (cf. (5), (6)) and to $Q(x,y)$
>
> 190   Kurt Gödel,
>
> (cf. (8·1)), and only this was used in the proof above. The undecidable
> proposition has, in this case, the form $v\,\text{Gen}\,r$, where r is a decidable
> *class expression* (moreover, it even suffices that x be decidable in the
> system extended by x).

As a theorem, the result reads as follows:

Theorem 6.8   Theorem VI, stronger variant

Let $\chi = \{\ulcorner\varphi_1\urcorner, \ulcorner\varphi_2\urcorner, \ldots\}$ be ω-consistent and syntactically representable.
Then, there exists a formula $\varphi_r(\xi_1)$, for which neither

$$\forall \xi_1\, \varphi_r(\xi_1) \quad \text{nor} \quad \neg\forall \xi_1\, \varphi_r(\xi_1)$$

is provable within $\mathsf{P} \cup \chi$.

Next, Gödel derives a variant of the main result, which replaces the requirement
of ω-consistency with the weaker requirement of consistency. Before presenting the theorem, let us examine the consequences by revisiting the two cases
distinguished on page 303:

■ Case 1: $\vdash \forall x_1\, \varphi_r(x_1)$

In the argument made, ω-consistency plays no role at all; it hinges solely on the consistency of the formal system. Therefore, we can adopt the proof steps one by one.

■ Case 2: $\nvdash \forall x_1\, \varphi_r(x_1)$

The line of reasoning follows up to the point where the formula instances $\varphi_r(\overline{n})$ are shown to be provable:

$$\vdash \varphi_r(\overline{0}),\ \vdash \varphi_r(\overline{1}),\ \vdash \varphi_r(\overline{2}),\ \vdash \varphi_r(\overline{3}),\ \vdash \varphi_r(\overline{4}),\ \ldots \tag{6.18}$$

The consistency assumption doesn't suffice to deduce $\nvdash \neg \forall x_1\, \varphi_r(x_1)$. However, it does lead to the conclusion that none of the formulas referenced in (6.18) is derivable in negated form.

$$\nvdash \neg\varphi_r(\overline{0}),\ \nvdash \neg\varphi_r(\overline{1}),\ \nvdash \neg\varphi_r(\overline{2}),\ \nvdash \neg\varphi_r(\overline{3}),\ \nvdash \neg\varphi_r(\overline{4}),\ \ldots \tag{6.19}$$

Now, we are prepared to articulate the anticipated variant of the main result:

Theorem 6.9 — Theorem VI for consistent sets

Let $\chi = \{\ulcorner\varphi_1\urcorner, \ulcorner\varphi_2\urcorner, \ldots\}$ be consistent and syntactically representable. Then, there exists a formula $\varphi_r(\xi_1)$ with the following properties:

- $\forall \xi_1\, \varphi_r(\xi_1)$ is unprovable. ☞ $\nvdash \forall \xi_1\, \varphi_r(\xi_1)$

- No counterexample can be given. ☞ $\nvdash \neg\varphi_r(\overline{n})$ for all $n \in \mathbb{N}$

In Gödel's article, this theorem is hidden in the following passage:

> Setzt man von χ statt ω-Widerspruchsfreiheit, bloß Widerspruchsfreiheit voraus, so folgt zwar nicht die Existenz eines unentscheidbaren Satzes, wohl aber die Existenz einer Eigenschaft (r), für die weder ein Gegenbeispiel angebbar, noch beweisbar ist, daß sie allen Zahlen zukommt. Denn zum Beweise, daß 17 Gen r nicht

> If one assumes merely the consistency of χ, instead of its ω-consistency, then, to be sure, the existence of an undecidable proposition does not follow; however, we do obtain the existence of a property (r) for which neither a counterexample can be given nor can it be proved that it holds for all numbers.

Given the previous discussion, Gödel's proof is easy to understand, as it follows the very same line of reasoning:

6.1 Gödel's Main Result

sie allen Zahlen zukommt. Denn zum Beweise, daß 17 Gen r nicht $\varkappa$-*beweisbar* ist, wurde nur die Widerspruchsfreiheit von $\varkappa$ verwendet (vgl. S. 189) und aus $\overline{\text{Bew}_\varkappa}(17 \text{ Gen } r)$ folgt nach (15), daß für jede Zahl x $Sb\left(r\,\dfrac{17}{Z(x)}\right)$, folglich für keine Zahl Neg $Sb\left(r\,\dfrac{17}{Z(x)}\right)$ $\varkappa$-*beweisbar* ist.

For in the proof that 17 Gen r is not $\varkappa$-*provable*, only the consistency of $\varkappa$ was used (cf. p. 189), and from $\overline{\text{Bew}_\varkappa}(17 \text{ Gen } r)$ it follows, according to (15), that $Sb\left(r\,\dfrac{17}{Z(x)}\right)$ holds for all numbers x; consequently, for no number x is Neg $Sb\left(r\,\dfrac{17}{Z(x)}\right)$ $\varkappa$-*provable*.

Next, Gödel delivers on a promise made on page 295 by proving the existence of formal systems that are consistent but not ω-consistent. We can readily obtain such a system by adjoining the formula $\neg \forall x_1\, \varphi_r(x_1)$ to the axioms, or, which is the same, by expanding the set χ with the Gödel number $\ulcorner \neg \forall x_1\, \varphi_r(x_1) \urcorner$:

$$\chi' := \chi \cup \{\ulcorner \neg \forall x_1\, \varphi_r(x_1) \urcorner\}$$

Adjungiert man Neg (17 Gen r) zu $\varkappa$, so erhält man eine widerspruchsfreie aber nicht ω-widerspruchsfreie *Formelklasse* $\varkappa'$. $\varkappa'$ ist widerspruchsfrei, denn sonst wäre 17 Gen r $\varkappa$-*beweisbar*. $\varkappa'$ ist aber nicht ω-widerspruchsfrei, denn wegen $\overline{\text{Bew}_\varkappa}$ (17 Gen r) und (15) gilt: $(x)\,\text{Bew}_\varkappa\, Sb\left(r\,\dfrac{17}{Z(x)}\right)$, umsomehr also: $(x)\,\text{Bew}_{\varkappa'}\, Sb\left(r\,\dfrac{17}{Z(x)}\right)$ und andererseits gilt natürlich: $\text{Bew}_{\varkappa'}\,[\text{Neg }(17 \text{ Gen } r)])\,{}^{46)}$.

---

[46] Die Existenz widerspruchsfreier und nicht ω-widerspruchsfreier $\varkappa$ ist damit natürlich nur unter der Voraussetzung bewiesen, daß es überhaupt widerspruchsfreie $\varkappa$ gibt (d. h. daß P widerspruchsfrei ist).

If one adjoins Neg (17 Gen r) to $\varkappa$, then one obtains a consistent, but not an ω-consistent, class of *formulas* $\varkappa'$. $\varkappa'$ is consistent, for, otherwise, 17 Gen r would be $\varkappa$-*provable*. $\varkappa'$ is however not ω-consistent, for, by virtue of $\overline{\text{Bew}_\varkappa}$ (17 Gen r) and (15), we have $(x)\,\text{Bew}_\varkappa\, Sb\left(r\,\dfrac{17}{Z(x)}\right)$, and, a fortiori, $(x)\,\text{Bew}_{\varkappa'}\, Sb\left(r\,\dfrac{17}{Z(x)}\right)$. On the other hand, of course, $\text{Bew}_{\varkappa'}\,[\text{Neg }(17 \text{ Gen } r)])$ holds. [46]

---

[46] The existence of consistent and non-ω-consistent x is, of course, only proved under the assumption that there exists any consistent x at all (i.e. that P is consistent).

Let us examine the line of reasoning more closely. Based on the above discussion, it is apparent that within $P \cup \chi$, neither the formula $\forall x_1\ \varphi_r(x_1)$ nor the formula $\neg \forall x_1\ \varphi_r(x_1)$ is provable. According to theorem 4.8, this means that the addition of $\neg \forall x_1\ \varphi_r(x_1)$ does not generate any contradictions. Thus, χ' is a consistent set.

However, it is easy to recognize that χ' is not ω-consistent. The unprovability of $\forall x_1\ \varphi_r(x_1)$ implies that no natural number x encodes the Gödel number of a proof. Consequently, because of (6.12), the following holds for the formal system $P \cup \chi$:

$$\vdash \varphi_r(\overline{0}),\ \vdash \varphi_r(\overline{1}),\ \vdash \varphi_r(\overline{2}),\ \vdash \varphi_r(\overline{3}),\ \vdash \varphi_r(\overline{4}),\ \ldots \tag{6.20}$$

These formulas are thus provable in $P \cup \chi'$. But there is more to say about $P \cup \chi'$. In $P \cup \chi'$, the formula $\neg \forall x_1\ \varphi_r(x_1)$ is an axiom and thus a theorem, too:

$$\vdash \neg \forall x_1\ \varphi_r(x_1) \tag{6.21}$$

Together, (6.20) and (6.21) show that the formal system is not ω-consistent. Hence, the substantive meaning of Theorem 6.3 cannot be reversed:

Theorem 6.10

Not every consistent set is ω-consistent.

Did you wonder why Gödel bothered to include the set χ in his theorems? This set isn't necessary to demonstrate the incompleteness of the formal system P, as the main result yields this conclusion by substituting χ with the empty set. At first glance, the case $\chi = \emptyset$ appears to be the most intriguing.

In fact, including the set χ gives the main result an unexpected punch. Since χ can be chosen almost freely, the gaps in the formal system P cannot be closed by introducing additional axioms.

This applies as long as the axioms constitute a syntactically representable set. In particular, the axioms of all modern formal systems meet this conceivably weak criterion, and we would need a high degree of destructive ingenuity to come up with a set of formulas that does not. Thus, as long as we stay within the realm of serious mathematics, any extension of P must remain as inherently incomplete as P itself:

6.1 Gödel's Main Result

Corollary 6.11

The formal system P cannot be completed.

Due to the boldness of his discoveries, Gödel seemed uncertain whether his reading audience would truly grasp Theorem VI's weak premise. Probably to be safe, he decided to list several specific examples.

> Ein Spezialfall von Satz VI ist der, daß die Klasse x aus endlich vielen *Formeln* (und ev. den daraus durch *Typenerhöhung* entstehenden) besteht. Jede endliche Klasse α ist natürlich rekursiv. Sei a die größte in α enthaltene Zahl. Dann gilt in diesem Fall für x:
>
> $$x \, \varepsilon \, \varkappa \sim (E m, n) \, [m \leq x \, \& \, n \leq a \, \& \, n \, \varepsilon \, \alpha \, \& \, x = m \, Th \, n]$$
>
> x ist also rekursiv. Das erlaubt z. B. zu schließen, daß auch mit Hilfe des Auswahlaxioms (für alle Typen) oder der verallgemeinerten Kontinuumshypothese nicht alle Sätze entscheidbar sind, vorausgesetzt, daß diese Hypothesen ω-widerspruchsfrei sind.

> A special case of Theorem VI occurs when the class x consists of finitely many *formulas* (and possibly also those arising therefrom by *type elevation*). Of course every finite class α is recursive. Let a be the largest number in α. Then, in this case, we have for x:
>
> $$x \, \varepsilon \, \varkappa \sim (E m, n) \, [m \leq x \, \& \, n \leq a \, \& \, n \, \varepsilon \, \alpha \, \& \, x = m \, Th \, n]$$
>
> Hence, x is recursive. This allows us to deduce that even with the aid of the axiom of choice (for all types) or of the generalized continuum hypothesis not all sentences are decidable, assuming that these hypotheses are ω-consistent.

In the quoted paragraph, Gödel explicitly mentions the axiom of choice, addressed in detail in Section 2.4.3.1, and the generalized continuum hypothesis, elaborated upon in Section 2.4.1.

> Beim Beweise von Satz VI wurden keine anderen Eigenschaften des Systems P verwendet als die folgenden:
> 1. Die Klasse der Axiome und die Schlußregeln (d. h. die Relation „unmittelbare Folge") sind rekursiv definierbar (sobald man die Grundzeichen in irgend einer Weise durch natürliche Zahlen ersetzt).
> 2. Jede rekursive Relation ist innerhalb des Systems P definierbar (im Sinn von Satz V).
>
> Daher gibt es in jedem formalen System, das den Voraussetzungen 1, 2 genügt und ω-widerspruchsfrei ist, unentscheidbare Sätze der Form $(x) \, F(x)$, wo F eine rekursiv definierte Eigenschaft

natürlicher Zahlen ist, und ebenso in jeder Erweiterung eines solchen

---

Über formal unentscheidbare Sätze der Principia Mathematica etc. 191

Systems durch eine rekursiv definierbare ω-widerspruchsfreie Klasse von Axiomen. Zu den Systemen, welche die Voraussetzungen 1, 2 erfüllen,
gehören, wie man leicht bestätigen kann, das Zermelo-Fraenkelsche und das v. Neumannsche Axiomensystem der Mengenlehre[47]), ferner

---

[47]) Der Beweis von Voraussetzung 1. gestaltet sich hier sogar einfacher als im Falle des Systems P, da es nur eine Art von Grundvariablen gibt (bzw. zwei bei J. v. Neumann).

In the proof of Theorem VI no properties of the system P were used other than the following:
1. The class of axioms and the rules of inference (i.e. the relation "immediate consequence") are recursively definable (when the primitive symbols are replaced in some manner by natural numbers).
2. Every recursive relation is definable within the system P (in the sense of Theorem V).

Hence, in every formal system which satisfies assumptions 1, 2 and is ω-consistent, there exist undecidable propositions of the form $(x) F(x)$, where F is a recursively defined property of natural numbers, and likewise in every extension of such a

---

On formally undecidable propositions of Principia Mathematica etc. 191

system by recursively definable ω-consistent class of axioms. To the systems which satisfy assumptions 1, 2 belong, as one can easily confirm, the Zermelo-Fraenkel and the v. Neumann axiom systems for set theory,[47])

---

[47]) The proof of assumption 1 turns out to be even simpler here than in the case of the system P, since there is only one kind of primitive variable (resp. two in J. v. Neumann's system).

Gödel explicitly mentioned the Zermelo-Fraenkel set theory, discussed in Section 2.4, and the von Neumann axiom system, which serves as a precursor to the NBG set theory, also referenced in Section 2.4. The NBG set theory did not exist until around 1940 and is therefore not cited in Gödel's paper.

---

und das v. Neumannsche Axiomensystem der Mengenlehre[47]), ferner das Axiomensystem der Zahlentheorie, welches aus den Peanoschen Axiomen, der rekursiven Definition [nach Schema (2)] und den logischen

6.2 The First Incompleteness Theorem

> Regeln besteht⁴⁸). Die Voraussetzung 1. erfüllt überhaupt jedes System, dessen Schlußregeln die gewöhnlichen sind und dessen Axiome (analog wie in P) durch Einsetzung aus endlich vielen Schemata entstehen⁴⁸ᵃ).
>
> ---
>
> ⁴⁸) Vgl. Problem III in D. Hilberts Vortrag: Probleme der Grundlegung der Mathematik. Math. Ann. 102.
>
> ⁴⁸ᵃ) Der wahre Grund für die Unvollständigkeit, welche allen formalen Systemen der Mathematik anhaftet, liegt, wie im II. Teil dieser Abhandlung gezeigt werden wird, darin, daß die Bildung immer höherer Typen sich ins Transfinite fortsetzen läßt. (Vgl. D. Hilbert, Über das Unendliche, Math. Ann. 95, S. 184), während in jedem formalen System höchstens abzählbar viele vorhanden sind. Man kann nämlich zeigen, daß die hier aufgestellten unentscheidbaren Sätze durch Adjunktion passender höherer Typen (z. B. des Typus ω zum System P) immer entscheidbar werden. Analoges gilt auch für das Axiomensystem der Mengenlehre.

> and, in addition, the axiom system for number theory which consists of Peano's axioms, recursive definitions [according to schema (2)] and the logical rules.⁴⁸). Assumption 1 is fulfilled in general by every system whose rules of inference are the usual ones and whose axioms (as in P) result from substitution in finitely many schemata.⁴⁸ᵃ)
>
> ---
>
> ⁴⁸) Cf. Problem III in D. Hilbert's address: "Probleme der Grundlegung der Mathematik", Math. Ann. 102.
>
> ⁴⁸ᵃ) The true reason for the incompleteness which attaches to all formal systems of mathematics lies, as will be shown in Part II of this paper, in the fact that the formation of higher and higher types can be continued into the transfinite (cf. D. Hilbert, "Über das Unendliche", Math. Ann. 95, S. 184), while, in every formal system, only countably many are available. Namely, one can show that the undecidable sentences which have been constructed here always become decidable through adjunction of sufficiently high types (e. g. of the type ω to the system P). A similar result holds for the axiom systems of set theory.

6.2 The First Incompleteness Theorem

6.2.1 Incompleteness of Arithmetic

Up to this point, Gödel has based his investigations on the formal system P, which has the expressive power of ordinary mathematics. He proceeds by demonstrating that even formal systems with much less expressive power are afflicted with incompleteness. Soon, we will realize that undecidable propositions exist within arithmetic. In particular, a formal system already suffers from incompleteness if it is expressive enough to talk about the additive and multiplicative properties of the natural numbers.

First, Gödel introduces the notion of arithmetic relations and arithmetic sets:

3.

Wir ziehen nun aus Satz VI weitere Folgerungen und geben zu diesem Zweck folgende Definition:

Eine Relation (Klasse) heißt **arithmetisch**, wenn sie sich allein mittels der Begriffe $+$, $\cdot$ [Addition und Multiplikation, bezogen auf natürliche Zahlen [49])] und den logischen Konstanten $\lor$, $\overline{}$, (x), $=$ definieren läßt, wobei (x) und $=$ sich nur auf natürliche Zahlen beziehen dürfen [50]). Entsprechend wird der Begriff „arithmetischer Satz" definiert. Insbesondere sind z. B. die Relationen „größer" und „kongruent nach einem Modul" arithmetisch, denn es gilt:

$$x > y \sim \overline{(E z)} \, [y = x + z]$$
$$x \equiv y \,(\text{mod } n) \sim (E z) \, [x = y + z \cdot n \lor y = x + z \cdot n]$$

---

[49]) Die Null wird hier und im folgenden immer mit zu den natürlichen Zahlen gerechnet.

[50]) Das Definiens eines solchen Begriffes muß sich also allein mittels der angeführten Zeichen, Variablen für natürliche Zahlen $x, y, \ldots$ und den Zeichen 0, 1 aufbauen (Funktions- und Mengenvariable dürfen nicht vorkommen). (In den Präfixen darf statt x natürlich auch jede andere Zahlvariable stehen.)

3.

We shall now derive further consequences from Theorem VI, and, to this end, we give the following definition:

A relation (class) is called **arithmetical**, if it can be defined by means of only the concepts $+$, $\cdot$ [addition and multiplication of natural numbers [49])] and the logical constants $\lor$, $\overline{}$, (x), $=$, where (x) and $=$ are to refer to natural numbers [50]). The concept "arithmetical proposition" is defined in a corresponding manner. In particular, the relations "greater" and "congruent with respect to a modulus", for example, are arithmetical; for we have:

$$x > y \sim \overline{(E z)} \, [y = x + z]$$
$$x \equiv y \,(\text{mod } n) \sim (E z) \, [x = y + z \cdot n \lor y = x + z \cdot n]$$

---

[49]) Zero is, here and in the sequal, always counted among the natural numbers.

[50]) The definiens of such a concept must therefore be constructed only by means of the indicated symbols, variables for natural numbers $x, y, \ldots$ and the symbols 0, 1 (function variables and set variables must not occur). (Of course, any other number variable may occur in the prefixes instead of x.)

Gödel speaks of an arithmetic relation if it can be characterized by a mathematical expression that uses no constructs other than those listed above. Examples are the order relation '$>$' and the congruence relation '$\equiv$' (modulo n). In modern notation, Gödel's definitions read as follows:

$$x > y \; :\Leftrightarrow \; \neg \exists z \, (y = x + z) \tag{6.22}$$

6.2 The First Incompleteness Theorem

$$x \equiv y \pmod{n} :\Leftrightarrow \exists z \, (x = y + z \cdot n \lor y = x + z \cdot n) \tag{6.23}$$

Note that the right-hand sides of these definitions contain ordinary mathematical expressions rather than formulas of a particular formal system. On the next pages, we will follow Gödel's words as usual while adopting a more formal approach. Specifically, we will treat Gödel's expressions as formulas of a formal system equipped with the appropriate language constructs. The new formal system is called PA, which is short for *Peano Arithmetic*. From a distance, PA resembles Gödel's system P. Upon closer examination, however, notable differences become apparent:

- PA is a so-called *first-order logic*, as it contains only individual variables. In contrast to P, the new system assigns no special semantic meaning to variables marked with an index, that is, x_1, x_2, x_3, etc., are all individual variables. The sole purpose of indices is to enlarge the pool of available symbols.

- In PA, the numerical operators '+' and '×' are native language constructs. We have already seen such a system in Section 4.4.5, where we have added these operators to Gödel's system P to simplify the derivation of numerical theorems.

- The equality operator '=' is also native to PA because first-order logic is too weak to capture the equality relation. Leibniz's identity principle, thoroughly discussed on page 155, sheds light on why. To formally define equality, it is necessary to quantify a variable of the second type, which is not available in PA.

Apart from the language extensions discussed, PA is a subset of P, making the formal definition of syntax and semantics straightforward. Figure 6.1 summarizes the result.

In PA, we employ the same syntactic simplifications as in the formal system P, allowing for the omission of specific pairs of parentheses and the use of the symbols '∃', '∧', '→', and '↔' as syntactic abbreviations in the usual way.

For closed formulas, PA and P share a crucial property. A closed PA formula is either true for every arithmetic interpretation or false for every arithmetic interpretation. Hence, rather than $I \models \varphi$ or $I \not\models \varphi$, we can revert to the simpler notation $\models \varphi$ or $\not\models \varphi$.

With the groundwork laid, we can translate the right-hand sides of (6.22) and (6.23) into PA formulas in a one-to-one manner:

$$\varphi_{\text{gr}}(x, y) := \neg \exists z \, (y = x + z) \tag{6.24}$$
$$\varphi_{\text{mod}}(x, y, n) := \exists z \, (x = y + z \times n \lor y = x + z \times n) \tag{6.25}$$

Syntax of PA

The set of *arithmetic terms* is defined inductively:

- $0, x, y, z, \ldots$ are arithmetic terms.
- If σ and τ are arithmetic terms, so are $f\,\sigma$, $(\sigma + \tau)$, and $(\sigma \times \tau)$.

The set of *arithmetic formulas* is defined inductively:

- If σ, τ are arithmetic terms, then $(\sigma = \tau)$ is an arithmetic formula.
- If φ and ψ are arithmetic formulas, so are

$$\neg(\varphi),\ (\varphi) \vee (\psi),\ \text{and}\ \forall \xi\, (\varphi)\ \text{with}\ \xi \in \{x, y, z, \ldots\}$$

Semantics of PA

An *arithmetic interpretation* is a mapping I with:

$$\begin{aligned}
I(\xi) &\in \mathbb{N} \quad \text{for each variable } \xi \\
I(0) &= 0 \\
I(f\,\sigma) &= I(\sigma) + 1 \\
I(\sigma + \tau) &= I(\sigma) + I(\tau) \\
I(\sigma \times \tau) &= I(\sigma) \cdot I(\tau)
\end{aligned}$$

The semantics of PA is given by the *model relation* '$\models$':

$$\begin{aligned}
I &\models (\sigma_1 = \sigma_2) &:\Leftrightarrow\ & I(\sigma_1) = I(\sigma_2) \\
I &\models \neg(\varphi) &:\Leftrightarrow\ & \not\models \varphi \\
I &\models (\varphi) \vee (\psi) &:\Leftrightarrow\ & \models \varphi\ \text{or}\ \models \psi \\
I &\models \forall \xi\, (\varphi) &:\Leftrightarrow\ & \models \varphi[\xi \leftarrow \overline{n}]\ \text{for all}\ n \in \mathbb{N}
\end{aligned}$$

Figure 6.1: Syntax and Semantics of Peano Arithmetic

The formula φ_{gr} is true precisely when the free variables x and y are interpreted as two numbers x and y with $x > y$. Something similar applies to the formula φ_{mod}. It is true precisely when the free variables x, y, and n are interpreted as three numbers x, y, and n, such that x and y differ by a multiple of n. This is meant by saying the relations "greater" and "congruent modulo" are arithmetically represented by the formulas φ_{gr} and φ_{mod}, respectively.

As a more complex example, let us try to arithmetically represent the factorial function, defined by the following primitive-recursive scheme:

$$\text{factorial}(0) = s(0)$$

6.2 The First Incompleteness Theorem

$$\text{factorial}(k+1) = \text{mult}(s(k), \text{factorial}(k))$$

For the sake of simplicity, temporarily assume that PA provides unary function symbols and the necessary means to bind those function symbols with a quantifier. If f is such a symbol, $f(0)$, $f(1)$, etc., denote the individual function values, and the expression $\exists f$ takes on the following substantive meaning:

$$\exists f \ldots \mathrel{\hat=} \text{"There exists a function } f : \mathbb{N} \to \mathbb{N} \text{ with } \ldots\text{"}$$

Now, it is easy to translate the primitive-recursive definition of the factorial function directly into a formula:

$$\exists f \ (f(0) = \mathsf{f}\ 0 \wedge \forall \mathsf{k}\ (f(\mathsf{k}+1) = (\mathsf{f}\ \mathsf{k}) \times f(\mathsf{k})) \wedge \mathsf{x_0} = f(\mathsf{x_1})) \qquad (6.26)$$

The formula holds true precisely when its two free variables $\mathsf{x_0}$ and $\mathsf{x_1}$ are interpreted as two numbers x_0 and x_1 with $x_0 = x_1!$.

Let us focus on the universally quantified variable k in (6.26). On closer inspection, it becomes apparent that it suffices to quantify over all natural numbers smaller than x_1 rather than all natural numbers. Consequently, the following formula serves the same purpose:

$$\exists f \ (f(0) = \mathsf{f}\ 0 \wedge \forall \mathsf{k}\ (\mathsf{k} < \mathsf{x_1} \to f(\mathsf{k}+1) = (\mathsf{f}\ \mathsf{k}) \times f(\mathsf{k})) \wedge \mathsf{x_0} = f(\mathsf{x_1}))$$

Writing f_x instead of $f(x)$, as Gödel is about to do, this formula changes into:

$$\exists f \ (f_0 = \mathsf{f}\ 0 \wedge \forall \mathsf{k}\ (\mathsf{k} < \mathsf{x_1} \to f_{\mathsf{k}+1} = (\mathsf{f}\ \mathsf{k}) \times f_{\mathsf{k}}) \wedge \mathsf{x_0} = f_{\mathsf{x_1}}) \qquad (6.27)$$

This formula can be generalized by substituting the base case $\mathsf{f}\ 0$ and the recursion case $(\mathsf{f}\ \mathsf{k}) \times f_{\mathsf{k}}$ with the two placeholders ψ and μ, respectively. We then obtain:

$$\exists f \ (f_0 = \psi \wedge \forall \mathsf{k}\ (\mathsf{k} < \mathsf{x_1} \to f_{\mathsf{k}+1} = \mu(\mathsf{k}, f_{\mathsf{k}})) \wedge \mathsf{x_0} = f_{\mathsf{x_1}}) \qquad (6.28)$$

If ψ and μ are themselves primitive recursive, we can translate them into two formulas $S(\xi)$ and $T(\xi, \zeta, \nu)$ in a similar way. In particular, we can rewrite formula (6.28) as such:

$$\exists f \ (S(f_0) \wedge \forall \mathsf{k}\ (\mathsf{k} < \mathsf{x_1} \to T(f_{\mathsf{k}+1}, \mathsf{k}, f_{\mathsf{k}})) \wedge \mathsf{x_0} = f_{\mathsf{x_1}}) \qquad (6.29)$$

By turning our considerations into a formal inductive proof, it can be shown that all primitive-recursive relations are arithmetically representable in the way described. In fact, this is what Gödel does:

> **Es gilt der**
> **Satz VII:** Jede rekursive Relation ist arithmetisch.

Wir beweisen den Satz in der Gestalt: Jede Relation der Form $x_0 = \varphi(x_1\ldots x_n)$, wo φ rekursiv ist, ist arithmetisch, und wenden vollständige Induktion nach der Stufe von φ an. φ habe die s-te Stufe $(s > 1)$. Dann gilt entweder:

1. $\varphi(x_1\ldots x_n) = \rho\,[\chi_1(x_1\ldots x_n), \chi_2(x_1\ldots x_n)\ldots \chi_m(x_1\ldots x_n)]$ [51]

(wo ρ und sämtliche χ_i kleinere Stufe haben als s) oder:

2. $\varphi(0, x_2\ldots x_n) = \psi(x_2\ldots x_n)$
$\varphi(k+1, x_2\ldots x_n) = \mu\,[k, \varphi(k, x_2\ldots x_n), x_2\ldots x_n]$

(wo ψ, μ niedrigere Stufe als s haben).

Im ersten Falle gilt:

$$x_0 = \varphi(x_1\ldots x_n) \sim (Ey_1\ldots y_m)\,[R(x_0\,y_1\ldots y_m)\,\&$$
$$\&\,S_1(y_1, x_1\ldots x_n)\,\&\ldots\&\,S_m(y_m, x_1\ldots x_n)],$$

wo R bzw. S_i die nach induktiver Annahme existierenden mit $x_0 = \rho(y_1\ldots y_m)$ bzw. $y = \chi_i(x_1\ldots x_n)$ äquivalenten arithmetischen Relationen sind. Daher ist $x_0 = \varphi(x_1\ldots x_n)$ in diesem Fall arithmetisch.

Im zweiten Fall wenden wir folgendes Verfahren an: Man kann die Relation $x_0 = \varphi(x_1\ldots x_n)$ mit Hilfe des Begriffes „Folge von Zahlen" (f) [52] folgendermaßen ausdrücken:

$$x_0 = \varphi(x_1\ldots x_n) \sim (Ef)\,\{f_0 = \psi(x_2\ldots x_n)\,\&\,(k)\,[k < x_1 \rightarrow$$
$$f_{k+1} = \mu(k, f_k, x_2\ldots x_n)]\,\&\,x_0 = f_{x_1}\}$$

Wenn $S(y, x_2\ldots x_n)$ bzw. $T(z, x_1\ldots x_{n+1})$ die nach induktiver Annahme existierenden mit $y = \psi(x_2\ldots x_n)$ bzw. $z = \mu(x_1\ldots x_{n+1})$ äquivalenten arithmetische Relationen sind, gilt daher:

$$x_0 = \varphi(x_1\ldots x_n) \sim (Ef)\,\{S(f_0, x_2\ldots x_n)\,\&\,(k)\,[k < x_1 \rightarrow$$
$$T(f_{k+1}, k, f_k, x_2\ldots x_n)]\,\&\,x_0 = f_{x_1}\} \qquad (17)$$

---

[51] Es brauchen natürlich nicht alle $x_1\ldots x_n$ in den χ_i tatsächlich vorzukommen [vgl. das Beispiel in Fußnote [27]].

[52] f bedeutet hier eine Variable, deren Wertebereich die Folgen natürl. Zahlen sind. Mit f_k wird das $k+1$-te Glied einer Folge f bezeichnet (mit f_0 das erste).

The following proposition is true:

Theorem VII: Every recursive relation is arithmetical.

We prove the theorem in the form: Every relation of the form $x_0 = \varphi(x_1\ldots x_n)$, where φ is recursive, is arithmetical and we apply complete induction on the rank of φ. Let φ have rank s $(s > 1)$. Then we have either:

6.2 The First Incompleteness Theorem

> 1. $\varphi(x_1 \ldots x_n) = \rho\,[\chi_1(x_1 \ldots x_n), \chi_2(x_1 \ldots x_n) \ldots \chi_m(x_1 \ldots x_n)]$ [51]
>
> (where ρ and all the χ_i have lower rank than s) or:
>
> 2. $\varphi\,(0, x_2 \ldots x_n) = \psi\,(x_2 \ldots x_n)$
> $\varphi\,(k+1, x_2 \ldots x_n) = \mu\,[k, \varphi\,(k, x_2 \ldots x_n), x_2 \ldots x_n]$
>
> (where ψ, μ have lower rank than s).
>
> In the first case we have:
>
> $x_0 = \varphi\,(x_1 \ldots x_n) \sim (E\,y_1 \ldots y_m)\,[R\,(x_0\,y_1 \ldots y_m)\,\&$
> $\&\,S_1\,(y_1, x_1 \ldots x_n)\,\&\,\ldots\,\&\,S_m\,(y_m, x_1 \ldots x_n)],$
>
> where R, and the S_i are the arithmetical relations which, according to inductive hypothesis, are equivalent to $x_0 = \rho\,(y_1 \ldots y_m)$, and $y = \chi_i\,(x_1 \ldots x_n)$, respectively. Hence, in this case, $x_0 = \varphi\,(x_1 \ldots x_n)$ is arithmetical.
>
> In the second case we apply the following procedure: one can express the relation $x_0 = \varphi\,(x_1 \ldots x_n)$ with the help of the concept "sequence of numbers" (f) [52] in the following manner:
>
> $x_0 = \varphi\,(x_1 \ldots x_n) \sim (Ef)\,\{f_0 = \psi\,(x_2 \ldots x_n)\,\&\,(k)\,[k < x_1 \to$
> $f_{k+1} = \mu\,(k, f_k, x_2 \ldots x_n)]\,\&\,x_0 = f_{x_1}\}$
>
> If $S\,(y, x_2 \ldots x_n)$, $T\,(z, x_1 \ldots x_{n+1})$ are the arithmetical relations which, according to the inductive hypothesis, are equivalent to $y = \psi\,(x_2 \ldots x_n)$, and $z = \mu\,(x_1 \ldots x_{n+1})$ respectively, then we have:
>
> $x_0 = \varphi\,(x_1 \ldots x_n) \sim (Ef)\,\{S(f_0, x_2 \ldots x_n)\,\&\,(k)\,[k < x_1 \to$
> $T\,(f_{k+1}, k, f_k, x_2 \ldots x_n)]\,\&\,x_0 = f_{x_1}\}$ (17)
>
> ---
>
> [51]) Naturally, not all the variables $x_1 \ldots x_n$ need actually occur in the χ_i [cf. the example in footnote [27])].
>
> [52]) f denotes here a variable whose domain is the sequence of natural numbers. The $(k+1)$-st term of a sequence f is designated f_k (and the first, f_0).

The right-hand side of Gödel's formula (17) is the generalization of the previously derived formula (6.29).

Yet Theorem VII is only partially proved, as the construction relies on function symbols not native to PA. Completing the proof requires the function symbol f to be eliminated from the constructed formulas, which Gödel is about to do next.

> Nun ersetzen wir den Begriff „Folge von Zahlen" durch „Paar von Zahlen", indem wir dem Zahlenpaar n, d die Zahlenfolge $f^{(n,\,d)}$ $(f_k^{(n,\,d)} = [n]_{1+(k+1)\,d})$ zuordnen, wobei $[n]_p$ den kleinsten nicht negativen Rest von n modulo p bedeutet.

> Now we replace the concept "sequence of numbers" by "pairs of numbers" by correlating with the number pair n, d the sequence of numbers $f^{(n,\,d)}$ ($f_k^{(n,\,d)} = [n]_{1+\,(k+1)\,d}$) where $[n]_p$ denotes the smallest non-negative remainder of n modulo p.

For two given numbers n and d, Gödel constructs the number sequence

$$f_1^{(n,d)}, f_2^{(n,d)}, f_3^{(n,d)}, f_4^{(n,d)}, \ldots$$

with

$$f_i^{(n,d)} := n \bmod (1 + (i+1) \cdot d) \tag{6.30}$$

Today, this function is now known as *Gödel's β-function*. The function is so important that we formally introduce it in a separate definition:

Definition 6.12 — Gödel's β-function

Gödel's β-function $\beta : \mathbb{N} \times \mathbb{N} \times \mathbb{N} \to \mathbb{N}$ is defined as the following function:

$$\beta(n, d, i) := n \bmod (1 + (i+1) \cdot d)$$

Be cautious not to mistake Gödel's β-function with the β-function discussed on page 220. Despite sharing the same name, these functions are entirely unrelated. It is important to note that in Gödel's work, only the function from page 220 is denoted with β, whereas in contemporary texts about Gödel's work, the term β-function consistently refers to the function defined in Definition 6.12.

The next theorem reveals why Gödel's β-function is so important: It allows the generation of any initial segment of any given number sequence:

Theorem 6.13

Let k be a natural number. For every sequence of numbers $f_0, f_1, \ldots, f_{k-1}$ of length k there exist two natural numbers n and d with

$$f_i = \beta(n, d, i).$$

To justify this far-reaching assertion, two auxiliary theorems are needed and will be proven first:

Theorem 6.14

For any natural number l, the numbers

$$1 + 1 \cdot l!, \ 1 + 2 \cdot l!, \ 1 + 3 \cdot l!, \ \ldots, \ 1 + l \cdot l!$$

6.2 The First Incompleteness Theorem

are pairwise coprime.

Proof: If a prime number p divided both

$$(1 + i \cdot l!) \text{ and } (1 + j \cdot l!), \qquad (1 \leq i < j \leq l)$$

p would also be a divisor of the difference

$$(1 + j \cdot l!) - (1 + i \cdot l!) = (j - i) \cdot l!.$$

Then, p would divide at least one of the numbers $(j - i)$ or $l!$. We now show that both assumptions lead to a contradiction:

- Suppose $p \mid l!$. Then p is also a divisor of $i \cdot l!$, contradicting the assumption that p divides $1 + i \cdot l!$.

- Suppose $p \mid (j - i)$. Because $j - i$ is smaller than l, $j - i$ is a divisor of $l!$. This implies that p is also a divisor of $l!$, contradicting the first case. □

The next theorem makes a statement about the solvability of a particular class of simultaneous congruences:

Theorem 6.15 — Chinese Remainder Theorem

Let $m_0, \ldots, m_n$ be pairwise coprime natural numbers and $a_0, \ldots, a_n$ be natural numbers with $a_i < m_i$. Then, the system of simultaneous congruences

$$x \equiv a_0 \bmod m_0 \qquad x \equiv a_1 \bmod m_1 \quad \ldots \quad x \equiv a_n \bmod m_n$$

has a solution, and the solution is unique modulo $m_0 \cdot \ldots \cdot m_n$.

Proof: We consider the mapping

$$\pi : \{0, \ldots, (m_0 \cdot \ldots \cdot m_n) - 1\} \to \{0, \ldots, m_0 - 1\} \times \ldots \times \{0, \ldots, m_n - 1\}$$

with:

$$\pi(x) := (x \bmod m_0, \ldots, x \bmod m_n)$$

The theorem is proved once it is shown that π is bijective.

- **Injectivity**

  Let $x, y \in \{0, \ldots, (m_0 \cdot \ldots \cdot m_n) - 1\}$ with $y > x$ and $\pi(x) = \pi(y)$. Then,

  $$(x \bmod m_0, \ldots, x \bmod m_n) = (y \bmod m_0, \ldots, y \bmod m_n)$$

which implies $m_0 \mid (y - x), \ldots, m_n \mid (y - x)$. Since the moduli $m_0, \ldots, m_n$ are pairwise coprime, we can conclude

$$m_0 \cdot \ldots \cdot m_n \mid (y - x). \tag{6.31}$$

y and x are both smaller than the product $(m_0 \cdot \ldots \cdot m_1)$, and so is the difference $y - x$. Thus, equation (6.31) only has a solution for $y - x = 0$, contradicting the assumption that x and y are different numbers. Consequently, π must be injective.

■ **Surjectivity**

The domain of π is finite and contains as many elements as the range. Any injective function with this property is necessarily surjective. □

Now, the assertion of Theorem 6.13 almost follows by itself. For the sequence

$$f_0, f_1, \ldots, f_{k-1}$$

let us define the number l as

$$l := \max\{k, f_0, f_1, \ldots, f_{k-1}\} \tag{6.32}$$

and consider the following system of simultaneous congruences:

$$\begin{aligned} n &\equiv f_0 \quad \mod \ (1 + 1 \cdot l!) \\ n &\equiv f_1 \quad \mod \ (1 + 2 \cdot l!) \\ & \ldots \\ n &\equiv f_{k-1} \quad \mod \ (1 + k \cdot l!) \end{aligned}$$

We know from Theorem 6.14 that the moduli are pairwise coprime. Then, according to the Chinese remainder theorem, the simultaneous congruence is solvable with a number n. Setting d to $l!$ implies

$$f_i = n \bmod (1 + (i+1) \cdot d) = \beta(n, d, i),$$

which was to be proven. □

In Gödel's words, the proof sounds a bit more succinct:

> Es gilt dann der
>
> **Hilfssatz 1:** Ist f eine beliebige Folge natürlicher Zahlen und k eine beliebige natürliche Zahl, so gibt es ein Paar von natürlichen Zahlen n, d, so daß $f^{(n,\,d)}$ und f in den ersten k Gliedern übereinstimmen.
>
> **Beweis:** Sei l die größte der Zahlen $k, f_0, f_1 \ldots f_{k-1}$. Man bestimme n so, daß:
>
> $$n \equiv f_i \ [\mathrm{mod} \ (1 + (i+1) \ l!)] \ \text{für} \ i = 0, 1 \ldots k-1$$

6.2 The First Incompleteness Theorem

> Über formal unentscheidbare Sätze der Principia Mathematica etc. 193
>
> was möglich ist, da je zwei der Zahlen $1+(i+1)\, l!\ (i = 0, 1 \ldots k-1)$ relativ prim sind. Denn eine in zwei von diesen Zahlen enthaltene Primzahl müßte auch in der Differenz $(i_1 - i_2)\, l!$ und daher wegen $|i_1 - i_2| < l$ in $l!$ enthalten sein, was unmöglich ist. Das Zahlenpaar $n, l!$ leistet dann das Verlangte.

> Then:
> Lemma 1: If f is an arbitrary sequence of natural numbers and k is an arbitrary natural number, then there exists a pair of natural numbers n, d such that $f^{(n,\, d)}$ and f coincide in their first k terms.
> Proof: Let l be the greatest of the numbers $k, f_0, f_1 \ldots f_{k-1}$. Determine n so that
> $$n \equiv f_i\, [\mathrm{mod}\, (1+(i+1)\, l!)] \text{ für } i = 0, 1 \ldots k-1,$$

> On formally undecidable propositions of Principia Mathematica etc. 193
>
> which is possible, since any two of the numbers $1+(i+1)\, l!$ $(i = 0, 1 \ldots k-1)$ are relatively prime. For, a prime dividing two of these numbers must also divide the difference $(i_1 - i_2)\, l!$ and therefore, since $|i_1 - i_2| < l$, must also divide $l!$, which is impossible. The number pair $n, l!$ fulfills our requirement.

At this juncture, let us revisit assignment (6.32), which sets l to the value from Gödel's article. While being on the safe side with the chosen magnitude, significantly smaller numbers do suffice in most cases. What has been said above remains valid as long as l meets the following two conditions:

$$l \geq k$$
$$1 + (i+1) \cdot l! > f_i$$

In particular, this is the case when l satisfies the following:

$$l \geq k$$
$$l! \geq \max\{f_0, f_1, \ldots, f_{k-1}\}$$

Let's attempt to apply Theorem 6.13 to generate some initial segments of the factorial sequence:

$$0!, 1!, 2!, 3!, 4!, \ldots = 1, 1, 2, 6, 24, \ldots$$

Figure 6.2 summarizes the system of simultaneous congruences that needs to be solved. Each congruence yields a natural number n, allowing us to reconstruct the initial segments by evaluating Gödel's β-function with n and the previously determined value of d. The calculations in Figure 6.3 demonstrate that the original sequences are indeed reconstructable with this approach.

To employ Gödel's β-function for our purposes, we need to find a way to define it within PA. The following formula demonstrates that this is indeed possible with little effort:

$$\varphi_\beta(\mathsf{y},\mathsf{n},\mathsf{d},\mathsf{i}) := \underbrace{\varphi_{\mathrm{mod}}(\mathsf{y},\mathsf{n},\mathsf{f}\,((\mathsf{f}\,\mathsf{i}) \times \mathsf{d}))}_{\mathsf{y} \equiv n \bmod 1+(i+1)d} \wedge \underbrace{\varphi_{\mathrm{gr}}(\mathsf{f}\,((\mathsf{f}\,\mathsf{i}) \times \mathsf{d}),\mathsf{y})}_{\mathsf{y} < 1+(i+1)d} \qquad (6.33)$$

φ_{gr} and φ_{mod} are the two formulas (6.24) and (6.25), defined on page 317.

If the variables n, d, and i are interpreted as the three natural numbers n, d, and i, respectively, then $\varphi_\beta(\mathsf{y},\mathsf{n},\mathsf{d},\mathsf{i})$ is substantively true precisely when the fourth variable y is interpreted as the number $\beta(n,d,i)$. In short: The function φ_β defines Gödel's β-function inside PA.

At this point, we have successfully attained our objective, as we are now able to turn formula (6.27) into a genuine formula of PA:

$$\exists \mathsf{n}\, \exists \mathsf{d}\, (\varphi_\beta(\mathsf{f}\,0,\mathsf{n},\mathsf{d},0) \wedge$$
$$\forall \mathsf{k}\, (\varphi_{\mathrm{gr}}(\mathsf{x}_1,\mathsf{k}) \to \forall \mathsf{w}\, (\varphi_\beta(\mathsf{w},\mathsf{n},\mathsf{d},\mathsf{k}) \to \varphi_\beta((\mathsf{f}\,\mathsf{k}) \times \mathsf{w},\mathsf{n},\mathsf{d},\mathsf{f}\,\mathsf{k}))) \wedge$$
$$\varphi_\beta(\mathsf{x}_0,\mathsf{n},\mathsf{d},\mathsf{x}_1))$$

This formula holds true if and only if its free variables x_0 and x_1 are interpreted as two numbers x_0 and x_1, satisfying $x_0 = x_1!$. In other words, the formula defines the factorial function within Peano arithmetic.

Being able to generate any initial segment of a given number sequence with Gödel's β-function puts us in the position to define every primitive-recursive function arithmetically and, consequently, every primitive-recursive relation.

Gödel expresses this final step of reasoning as follows:

> Da die Relation $x = [n]_p$ durch:
>
> $$x \equiv n\, (\bmod\, p)\, \&\, x < p$$
>
> definiert und daher arithmetisch ist, so ist auch die folgendermaßen definierte Relation $P(x_0, x_1 \ldots x_n)$:
>
> $$P(x_0 \ldots x_n) \equiv (E\,n,d)\,\{S\,([n]_{d+1}, x_2 \ldots x_n)\, \&\, (k)\,[k < x_1 \to$$
> $$T\,([n]_{1+d\,(k+2)}, k, [n]_{1+d\,(k+1)}, x_2 \ldots x_n)]\, \&\, x_0 = [n]_{1+d\,(x_1+1)}\}$$
>
> arithmetisch, welche nach (17) und Hilfssatz 1 mit: $x_0 = \varphi(x_1 \ldots x_n)$ äquivalent ist (es kommt bei der Folge f in (17) nur auf ihren Verlauf bis zum x_1+1-ten Glied an). Damit ist Satz VII bewiesen.

6.2 The First Incompleteness Theorem

Initial segment $1, 1$	Initial segment $1, 1, 2$
☞ $d = 2!$	☞ $d = 3!$
$n \equiv 1 \mod (1 \cdot 2! + 1)$ $n \equiv 1 \mod (2 \cdot 2! + 1)$	$n \equiv 1 \mod (1 \cdot 3! + 1)$ $n \equiv 2 \mod (2 \cdot 3! + 1)$ $n \equiv 6 \mod (3 \cdot 3! + 1)$
☞ Result: $n = 1$	☞ Result: $n = 1275$

Initial segment $1, 1, 2, 6$	Initial segment $1, 1, 2, 6, 24$
☞ $d = 4!$	☞ $d = 5!$
$n \equiv 1 \mod (1 \cdot 4! + 1)$ $n \equiv 1 \mod (2 \cdot 4! + 1)$ $n \equiv 2 \mod (3 \cdot 4! + 1)$ $n \equiv 6 \mod (4 \cdot 4! + 1)$	$n \equiv 1 \mod (1 \cdot 5! + 1)$ $n \equiv 1 \mod (2 \cdot 5! + 1)$ $n \equiv 2 \mod (3 \cdot 5! + 1)$ $n \equiv 6 \mod (4 \cdot 5! + 1)$ $n \equiv 24 \mod (5 \cdot 5! + 1)$
☞ Result: $n = 4610901$	☞ Result: $n = 2234239447342$

Figure 6.2: Encoding of initial segments of the factorial number series

Initial segment $1, 1$	Initial segment $1, 1, 2$
$\beta(1, 2!, 0) = 1$ $\beta(1, 2!, 1) = 1$ $\beta(1, 2!, 2) = 1$ $\beta(1, 2!, 3) = 1$ $\beta(1, 2!, 4) = 1$ $\beta(1, 2!, 5) = 1$	$\beta(1275, 3!, 0) = 1$ $\beta(1275, 3!, 1) = 1$ $\beta(1275, 3!, 2) = 2$ $\beta(1275, 3!, 3) = 0$ $\beta(1275, 3!, 4) = 4$ $\beta(1275, 3!, 5) = 17$

Initial segment $1, 1, 2, 6$	Initial segment $1, 1, 2, 6, 24$
$\beta(4610901, 4!, 0) = 1$ $\beta(4610901, 4!, 1) = 1$ $\beta(4610901, 4!, 2) = 2$ $\beta(4610901, 4!, 3) = 6$ $\beta(4610901, 4!, 4) = 75$ $\beta(4610901, 4!, 5) = 46$	$\beta(2234239447342, 5!, 0) = 1$ $\beta(2234239447342, 5!, 1) = 1$ $\beta(2234239447342, 5!, 2) = 2$ $\beta(2234239447342, 5!, 3) = 6$ $\beta(2234239447342, 5!, 4) = 24$ $\beta(2234239447342, 5!, 5) = 497$

Figure 6.3: Reconstruction of the initial segments via Gödel's β-function

> Since the relation $x = [n]_p$ is defined by
>
> $$x \equiv n \,(\text{mod } p) \,\&\, x < p$$
>
> and is therefore arithmetical, then so also is the relation $P(x_0, x_1 \ldots x_n)$ defined as follows:
>
> $$P(x_0 \ldots x_n) \equiv (En, d)\, \{S([n]_{d+1}, x_2 \ldots x_n) \,\&\, (k)[k < x_1 \rightarrow T([n]_{1+d\,(k+2)}, k, [n]_{1+d\,(k+1)}, x_2 \ldots x_n)] \,\&\, x_0 = [n]_{1+d\,(x_1+1)}\}$$
>
> which, according to (17) and Lemma 1, is equivalent to $x_0 = \varphi(x_1 \ldots x_n)$ (in the sequence f in (17) only its values up to the (x_1+1)th term matter). Thus, Theorem VII is proved.

Viewing Theorem VII in the light of the main result leads to a stunning conclusion. It implies that undecidable formulas aren't secluded entities confined to a remote corner of specialized mathematics; instead, they permeate the very essence of mathematics, residing at the heart of elementary number theory.

> Gemäß Satz VII gibt es zu jedem Problem der Form $(x)\,F(x)$ (F rekursiv) ein äquivalentes arithmetisches Problem und da der ganze Beweis von Satz VII sich (für jedes spezielle F) innerhalb des Systems P formalisieren läßt, ist diese Äquivalenz in P beweisbar. Daher gilt:

> According to Theorem VII, for every problem of the form $(x)\,F(x)$ (F recursive), there is an equivalent arithmetical problem, and since the whole proof of Theorem VII can be formulated (for each particular F) within the system P, this equivalence is provable in P. Therefore, we have:

Let us carefully review what Gödel is saying here. With Theorem VII, he has proven that all primitive-recursive relations are arithmetic. For example, suppose $F(x)$ is a unary primitive-recursive relation. In that case, Theorem VII implies the existence of an arithmetic formula $\psi_F(\mathsf{x}_1)$, which is true precisely when x_1 is interpreted as a number x with $x \in F$:

$$x \in F \Leftrightarrow \models \psi_F(\overline{x})$$

From this, it follows:

$$x \in F \text{ for all } x \in \mathbb{N} \Leftrightarrow \models \forall \mathsf{x}_1\, \psi_F(\mathsf{x}_1) \qquad (6.34)$$

Since the formal system P can formalize Peano arithmetic, the arithmetic for-

6.2 The First Incompleteness Theorem

mula ψ_F can be translated into a substantially equivalent formula φ_F of P:

$$\models \forall x_1\, \psi_F(x_1) \;\Leftrightarrow\; \models \forall x_1\, \varphi_F(x_1)$$

Note that the symbol '$\models$' refers to the model relation of the formal system PA on the left-hand side of the equivalence and to the model relation of the formal system P on the right side.

Now, according to (6.34), the following relationship holds:

$$x \in F \text{ for all } x \in \mathbb{N} \;\Leftrightarrow\; \models \forall x_1\, \varphi_F(x_1)$$

For example, F may be chosen to be the primitive-recursive relation described on page 301 by the formula with the Gödel number r. To replicate the proof of Theorem VII for this relation within P means to deduce the following theorem from the axioms of P:

$$\vdash \forall x_1\, \varphi_r(x_1) \leftrightarrow \forall x_1\, \varphi_F(x_1)$$

The formula $\forall x_1\, \varphi_r(x_1)$ on the left-hand side is the formula we previously identified as undecidable; neither itself nor its negation is provable within the formal system P. Then, the formulas $\forall x_1\, \varphi_F(x_1)$ and $\neg \forall x_1\, \varphi_F(x_1)$ can't be theorems either, implying the existence of an arithmetic statement that is neither provable nor disprovable within P. Overall, we have:

Satz VIII: In jedem der in Satz VI genannten formalen Systeme [53] gibt es unentscheidbare arithmetische Sätze.

Dasselbe gilt (nach den Bemerkungen auf Seite 190) für das Axiomensystem der Mengenlehre und dessen Erweiterungen durch ω-widerspruchsfreie rekursive Klassen von Axiomen.

---

[53] Das sind diejenigen ω-widerspruchsfreien Systeme, welche aus P durch Hinzufügung einer rekursiv definierbaren Klasse von Axiomen entstehen.

Theorem VIII: There exist undecidable arithmetical propositions in each of the formal systems [53] mentioned in Theorem VI. The same holds also (according to the remark on page 190) for the axiom system of set theory and its extensions by ω-consistent recursive classes of axioms.

---

[53] They are those ω-consistent systems which result from P by addition of a recursively definable class of axioms.

In a slightly different formulation, the theorem appears like this:

Theorem 6.16 — First Incompleteness Theorem, Gödel 1931

Every ω-consistent formal system expressive enough to formalize Peano arithmetic is negation-incomplete.

At the time of writing, Gödel did not succeed in replacing the requirement of ω-consistency with the weaker requirement of consistency. The fact that Theorem VIII remains valid under the weaker assumption was only established in 1936 by John Barkley Rosser, about five years after the publication of the incompleteness theorems. The American mathematician modified Gödel's formula $\varphi_r(x_1)$ in a way that allowed him to weaken the premise without departing from Gödel's general line of reasoning [89, 96]. In contemporary literature, this technique of substituting Gödel's formula with another, more suitable one is called *Rosser's trick*.

Theorem 6.17 — First Incompleteness Theorem, Rosser 1936

Every consistent formal system expressive enough to formalize Peano arithmetic is negation-incomplete.

This is a typical formulation of Gödel's first incompleteness theorem, as found in many textbooks. To appreciate Rosser's contribution, some authors refer to Theorem 6.17 as the *Gödel-Rosser-Theorem*.

6.2.2 Implications for the Restricted Function Calculus

Next, Gödel will prove a theorem about a system known as *engerer Funktionenkalkül* in the German original. This term was coined by the Hilbert school and utilized for over 30 years in the renowned textbook *Grundzüge der theoretischen Logik* authored by Hilbert and Ackermann [52]. In the 1959 edition, Ackermann replaced the term with *Prädikatenkalkül* (predicate calculus). Essentially, this calculus aligns with what is today recognized as *first-order predicate logic*, commonly abbreviated as PL1. We have mentioned this term several times in passing, but a formal definition is yet to be given. At this point, we want to make up for what we have missed so far.

6.2.2.1 Syntax of First-Order Predicate Logic

We define the syntax of first-order predicated logic in three steps. In Step 1, we introduce the notion of a *logic signature*. Based on this concept, we define *first-order terms* in Step 2. Finally, in Step 3, we demonstrate how terms can be combined to form *first-order formulas*.

6.2 The First Incompleteness Theorem

Definition 6.18 — Signature (First-order predicate logic)

A signature Σ is a triple $(V_\Sigma, F_\Sigma, P_\Sigma)$ with

- a set V_Σ of *variables*, e. g. $\{x, y, z, \ldots\}$,
- a set F_Σ of *function symbols*, e. g. $\{f, g, h, \ldots\}$,
- a set P_Σ of *predicates*, e. g. $\{P, Q, R, \ldots\}$.

Each function and each predicate has a fixed arity ≥ 0.

Simply put, a signature provides a stock of symbols for constructing terms. In first-order predicate logic, terms must obey the following rules:

Definition 6.19 — Term (First-order predicate logic)

Let $\Sigma = (V_\Sigma, F_\Sigma, P_\Sigma)$ be a signature of first-order predicate logic. The set of *terms* is inductively defined:

- Every variable $\xi \in V_\Sigma$ is a term.
- Every 0-ary function symbol $f \in F_\Sigma$ is a term.
- If $\sigma_1, \ldots, \sigma_n$ are terms and $f \in F_\Sigma$ is an n-ary function symbol, then $f(\sigma_1, \ldots, \sigma_n)$ is a term.

For example, with two variables, x and y, and a binary function symbol f, we can build the following terms:

$$x, y,$$
$$f(x, x), f(x, y),$$
$$f(f(x, y), x),$$
$$f(x, f(x, y)),$$
$$f(f(x, x), f(x, y)), \ldots$$

Terms are primitive building blocks that can be combined into *formulas* according to the following rules:

Definition 6.20 — Syntax of first-order predicate logic

Let ξ be a variable and $\sigma_1, \ldots, \sigma_n$ be terms. The set of *atomic formulas* is defined as follows:

- If P is an n-ary predicate, then $P(\sigma_1, \ldots, \sigma_n)$ is an atomic formula.

In *first-order logic with equality*, the following additionally applies:

- $(\sigma_1 \doteq \sigma_2)$ is an atomic formula.

The *formulas of first-order predicate logic* are defined inductively:

- ■ 0, 1, and every atomic formula are formulas.
- ■ If φ and ψ are formulas, so are $\neg(\varphi)$, $(\varphi) \vee (\psi)$, $\forall \xi\, (\varphi)$.

In first-order predicate logic, the operators '$\exists$', '$\wedge$', '$\rightarrow$', and '$\leftrightarrow$' play the same role as in P or PA. They act as syntactic sugar, letting us write down formulas in a more concise and easier-to-understand form.

As in all the other formal systems we've seen, there is no need to scope variables by a quantifier. For instance, in the formula P(x), variable x is *free* or *unbound*, whereas in the formula ∀x P(x), variable x is *bound*. Formulas with no free variables are called *closed*; all others are called *open*.

6.2.2.2 Semantics of First-Order Predicate Logic

As usual, we define the semantics of first-order predicate logic through the *model relation* '$\models$'. However, to coin this term accurately, we need to adjust the notion of interpretation to suit the requirements of predicate logic.

Definition 6.21 — Interpretation (first-order logic)

Let $\Sigma = (V_\Sigma, F_\Sigma, P_\Sigma)$ be a signature. An *interpretation* over Σ is a tuple (U, I) with the following properties:

- ■ U is a non-empty set.
- ■ I is a mapping that assigns
  - to each variable symbol $\xi \in V_\Sigma$ an element $I(\xi) \in U$,
  - to each function symbol $f \in F_\Sigma$ a function $I(f) : U^n \to U$, and
  - to each predicate symbol $P \in P_\Sigma$ a relation $I(P) \subseteq U^n$.

Herein, n is the arity of f or P as given by Σ.

In the literature, the set U is referred to by various synonyms. Some authors term it the *set of individuals*, while others call it the *domain* or *universe*.

Note that the actual mapping of variables to elements of U is irrelevant for closed formulas. For open formulas, however, it ensures that all free variables

6.2 The First Incompleteness Theorem

are assigned individual elements. Further, note that the definition encompasses function and predicate symbols of arity 0. The function symbols of arity 0 formally represent functions of the form $U^0 \to U$, thus acting as *constants*. The predicate symbols of arity 0 represent relations over the set U^0. They are atomic statements that are interpreted either as true or false, thus playing the role of *propositional variables*.

The mapping I, which assigns a function $I(f)$ to each function symbol f, extends naturally to complex terms according to the following inductive scheme:

$$I(f(\sigma_1, \ldots, \sigma_n)) := I(f)(I(\sigma_1), \ldots, I(\sigma_n))$$

Now, after having the groundwork laid, the semantics of first-order predicate logic flows right from the pen:

Definition 6.22 — Semantics of first-order predicate logic

Let φ and ψ be formulas of first-order predicate logic and (U, I) be an interpretation. The semantics of first-order predicate logic is given by the *model relation* '$\models$', which is inductively defined over the formula structure:

$$(U, I) \models 1$$
$$(U, I) \not\models 0$$
$$(U, I) \models P(\sigma_1, \ldots, \sigma_n) :\Leftrightarrow (I(\sigma_1), \ldots, I(\sigma_n)) \in I(P)$$
$$(U, I) \models (\sigma_1 \doteq \sigma_2) :\Leftrightarrow I(\sigma_1) = I(\sigma_2)$$
$$(U, I) \models \neg(\varphi) :\Leftrightarrow (U, I) \not\models \varphi$$
$$(U, I) \models (\varphi) \vee (\psi) :\Leftrightarrow (U, I) \models \varphi \text{ or } (U, I) \models \psi$$
$$(U, I) \models \forall \xi \, (\varphi) :\Leftrightarrow (U, I_{[\xi/u]}) \models \varphi \text{ for all } u \in U$$

An interpretation (U, I) with $(U, I) \models \varphi$ is called a *model* for φ.

We have already used the notation $I_{[\xi/u]}$ on page 151 in Definition 4.6. If (U, I) is an interpretation of first-order predicate logic, then $(U, I_{[\xi/u]})$ is the interpretation that assigns the individual element u to the variable ξ and is identical to (U, I) otherwise.

A significant characteristic of predicate logic is that the domain of an interpretation is not confined to the natural numbers as it is in P or PA. The range of possible interpretations is significantly broader, with far-reaching consequences. Unlike in the formal systems P or PA, it can no longer be claimed that every closed formula is either substantively true or substantively false.

As an example, let us consider the closed formula

$$\forall x \, \exists y \, P(f(x, y))$$

$$\forall x \, \exists y \, P(f(x,y))$$
$$(V_\Sigma = \{x,y\}, F_\Sigma = \{f\}, P_\Sigma = \{P\})$$

First Interpretation (U, I)	Second Interpretation (U', I')
$U := \mathbb{Z}$ $I(f) := (x,y) \mapsto x+y$ $I(P) := \{0\}$ $\mathbb{Z}$ ⋯┼┼┼┼┼┼┼┼┼┼┼┼┼┼┼⋯ $y \quad\; 0 \quad\; x$ "For all x, there exists a y such that $x+y=0$" is a true statement in $\mathbb{Z}$. $(U, I) \models \forall x \, \exists y \, P(f(x,y))$	$U' := \mathbb{N}$ $I'(f) := (x,y) \mapsto x+y$ $I'(P) := \{0\}$ $\mathbb{N}$ ⋯┼┼┼┼┼┼┼┼┼┼┼┼┼┼┼⋯ $y \notin \mathbb{N} \quad 0 \quad\; x$ "For all x, there exists a y such that $x+y=0$" is a false statement in $\mathbb{N}$. $(U', I') \not\models \forall x \, \exists y \, P(f(x,y))$

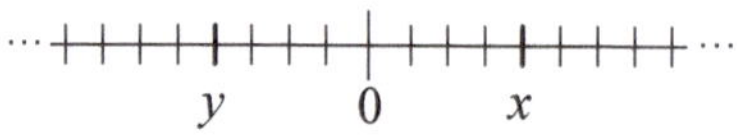

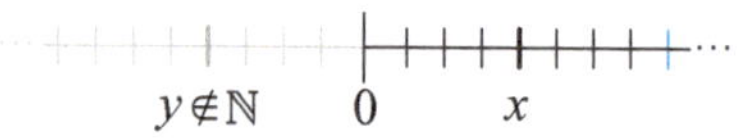

Figure 6.4: Two interpretations for the formula $\forall x \, \exists y \, P(f(x,y))$

and the two interpretations depicted in Figure 6.4. Both interpret the function symbol f as ordinary addition and the predicate symbol P as the set $\{0\}$, that is, P(x) is true exactly when x is interpreted as the number 0. Thus, the formula has the following substantive meaning:

"For all x there exists a y such that $x+y=0$."

Because both interpretations map into different domains, the formula is true under the first and false under the second.

The following definition takes this circumstance into account:

Definition 6.23 — Satisfiable, unsatisfiable, universally valid

A formula φ is called

- **satisfiable**, if at least one interpretation is a model of φ,

  (☞ some (U, I) satisfies $(U, I) \models \varphi$)

- **unsatisfiable**, if it is not satisfiable,

  (☞ no (U, I) satisfies $(U, I) \models \varphi$)

- **universally valid**, if every interpretation is a model of φ.

  (☞ all (U, I) satisfy $(U, I) \models \varphi$)

6.2 The First Incompleteness Theorem

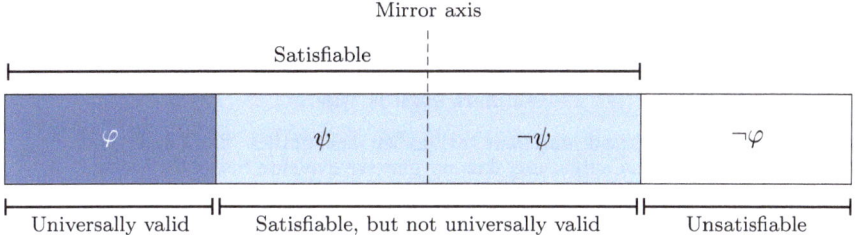

Figure 6.5: Satisfiable, unsatisfiable and universally valid formulas.

Figure 6.5 illustrates the interplay between satisfiable, unsatisfiable, and universally valid formulas.

First-order predicate logic exhibits the intriguing property that it allows for the definition of correct and, at the same time, complete formal systems when these terms are understood to refer to universally valid formulas. In fact, we have already encountered such a formal system on page 84 in Figure 2.1. The observation that all universally valid formulas are derivable in this system is the content of *Gödel's completeness theorem*, which Gödel proved in his 1929 dissertation. A revised version was later published in the *Monatshefte für Mathematik* [32] (Figure 6.6).

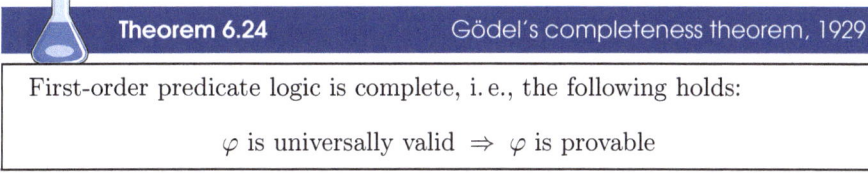

| Theorem 6.24 | Gödel's completeness theorem, 1929 |

First-order predicate logic is complete, i.e., the following holds:

$$\varphi \text{ is universally valid} \Rightarrow \varphi \text{ is provable}$$

With this knowledge in our pocket, we are well prepared to follow Gödel's footprints again:

> Wir leiten schließlich noch folgendes Resultat her:
> Satz IX: In allen in Satz VI genannten formalen Systemen[53]) gibt es unentscheidbare Probleme des engeren Funktionenkalküls[54]) (d. h. Formeln des engeren Funktionenkalküls, für die weder Allgemeingültigkeit noch Existenz eines Gegenbeispiels beweisbar ist)[55]).
>
> ---
>
> [53]) Das sind diejenigen ω-widerspruchsfreien Systeme, welche aus P durch Hinzufügung einer rekursiv definierbaren Klasse von Axiomen entstehen.
> [54]) Vgl. Hilbert-Ackermann, Grundzüge der theoretischen Logik. Im System P sind unter Formeln des engeren Funktionenkalküls diejenigen zu verstehen, welche aus den Formeln des engeren Funktionenkalküls der PM durch die auf S. 176 angedeutete Ersetzung der Relationen durch Klassen höheren Typs entstehen.
> [55]) In meiner Arbeit: Die Vollständigkeit der Axiome des logischen Funktionenkalküls, Monatsh. f. Math. u. Phys. XXXVII, 2, habe ich gezeigt, daß jede Formel des engeren Funktionenkalküls entweder als allgemeingültig nach-

> # Die Vollständigkeit der Axiome des logischen Funktionenkalküls[1]).
>
> Von **Kurt Gödel** in Wien.
>
> Whitehead und Russell haben bekanntlich die Logik und Mathematik so aufgebaut, daß sie gewisse evidente Sätze als Axiome an die Spitze stellten und aus diesen nach einigen genau formulierten Schlußprinzipien auf rein formalem Wege (d. h. ohne weiter von der Bedeutung der Symbole Gebrauch zu machen) die Sätze der Logik und Mathematik deduzierten. Bei einem solchen Vorgehen erhebt sich natürlich sofort die Frage, ob das an die Spitze gestellte System von Axiomen und Schlußprinzipien vollständig ist, d. h. wirklich dazu ausreicht, jeden logisch-mathematischen Satz zu deduzieren, oder ob vielleicht wahre (und nach anderen Prinzipien ev. auch beweisbare) Sätze denkbar sind, welche in dem betreffenden System nicht abgeleitet werden können. Für den Bereich der logischen Aussageformeln ist diese Frage in positivem Sinn entschieden, d. h. man hat gezeigt[2]), daß tatsächlich jede richtige Aussageformel aus den in den Principia Mathematica angegebenen Axiomen folgt. Hier soll dasselbe für einen weiteren Bereich von Formeln, nämlich für die des „engeren Funktionenkalküls"[,3]), geschehen, d. h. es soll gezeigt werden:
>
> **Satz I:** Jede allgemeingültige[4]) Formel des engeren Funktionenkalküls ist beweisbar.
>
> Dabei legen wir folgendes Axiomensystem[5]) zugrunde:
>
> Undefinierte Grundbegriffe: v, $\overline{}$, (x). [Daraus lassen sich in bekannter Weise $\&$, $\rightarrow$, $\sim$, (Ex) definieren.]

Figure 6.6: Gödel's completeness theorem states that the axiom system in Table 2.1 on page 84 is complete. All universally valid formulas are derivable from the axioms.

weisbar ist oder ein Gegenbeispiel existiert; die Existenz dieses Gegenbeispiels ist aber nach Satz IX n i c h t immer nachweisbar (in den angeführten formalen Systemen).

Finally, we derive the following result:

Theorem IX: In all of the formal systems[53]) mentioned in Theorem VI there exist undecidable problems of the restricted functional calculus[54]) (i. e. formulas of the restricted functional calculus for which neither the universal validity nor the existence of a counter-example is provable).[55])

---

[53]) They are those ω-consistent systems which result from P by addition of a recursively definable class of axioms.

[54]) Cf. Hilbert-Ackermann, Grundzüge der theoretischen Logik. In the system P by formulas of the restricted function calculus we are to understand those which arise from formulas of the restricted function calculus of PM by the substitution indicated on p. 176 of classes of higher type for relations.

[55]) In my paper: "Die Vollständigkeit der Axiome des logischen Funktionenkalküls", Monatsh. f. Math. u. Phys. XXXVII, 2, I have shown that every formula of the restricted functional calculus either can be proved to be universally

6.2 The First Incompleteness Theorem

> valid or has a counter-example; the existence of this counter-example is, however, according to Theorem IX, not always provable (in the given formal systems).

Let us analyze step by step what Gödel is saying here. He commences with a statement about the formal systems to which Theorem VI applies. For simplicity's sake, we assume Gödel refers to P. He continues to say that deciding the universality of PL1 formulas is impossible within P.

Deciding universal validity within P requires the capability to talk about this concept within P. Thus, let us assume that for each PL-1-formula φ, there exists a P-formula Val_φ, which is universally valid precisely when φ is. In this context, deciding the universal validity within P means that for any given PL1- formula φ, either Val_φ or $\neg \text{Val}_\varphi$ is derivable, contingent upon whether φ is universally valid or not. Briefly put:

$$\varphi \text{ is a universally valid PL1-formula} \Rightarrow \vdash \text{Val}_\varphi \qquad (6.35)$$
$$\varphi \text{ is not a universally valid PL1-formula} \Rightarrow \vdash \neg \text{Val}_\varphi \qquad (6.36)$$

Within P, we could thus prove every PL1-formula to be universally valid or not universally valid. Gödel's Theorem IX, however, states that this is impossible: There is at least one formula φ for which neither (6.35) nor (6.36) applies.

Hold on! Isn't this contradictory to Gödel's completeness theorem? This theorem affirms that every universally valid PL1-formula φ is derivable from the axioms of PL1. Replicating this proof within P would yield the formula Val_φ. Indeed, (6.35) does follow from the completeness theorem, and there is a simple reason why this is not a contradiction: The completeness theorem does not guarantee that formulas that are not universally valid are provable as such. Consequently, (6.36) remains untouched by this theorem.

Before moving on, let us slightly rephrase (6.35) and (6.36). According to Definition 6.23, a PL1 formula φ is universally valid precisely when $\neg \varphi$ is unsatisfiable. Using Sat_φ as a shorthand for $\neg \text{Val}_{\neg \varphi}$, (6.35) and (6.36) can be rewritten as follows:

$$\varphi \text{ is a universally valid PL1 formula} \Rightarrow \vdash \neg \text{Sat}_{\neg \varphi}$$
$$\varphi \text{ is not a universally valid PL1 formula} \Rightarrow \vdash \text{Sat}_{\neg \varphi}$$

Or, which is equivalent:

$$\neg \varphi \text{ is a universally valid PL1 formula} \Rightarrow \vdash \neg \text{Sat}_\varphi \qquad (6.37)$$
$$\neg \varphi \text{ is not a universally valid PL1 formula} \Rightarrow \vdash \text{Sat}_\varphi \qquad (6.38)$$

To prove Theorem IX, Gödel commences by reducing whether a primitive-recursive relation $F(x)$ applies to all natural numbers x to a first-order satisfiability problem. This is the content of Theorem X:

> 194 Kurt Gödel,
>
> Dies beruht auf:
> Satz X: Jedes Problem der Form $(x)\, F(x)$ (F rekursiv) läßt sich zurückführen auf die Frage nach der Erfüllbarkeit einer Formel des engeren Funktionenkalküls (d. h. zu jedem rekursiven F kann man eine Formel des engeren Funktionenkalküls angeben, deren Erfüllbarkeit mit der Richtigkeit von $(x)\, F(x)$ äquivalent ist).

> 194 Kurt Gödel,
>
> This is based upon:
> Theorem X: Every problem of the form $(x)\, F(x)$ (F recursive) can be reduced to the question of the satisfiability of a formula of the restricted functional calculus (i.e. for each recursive F one can produce a formula of the restricted functional calculus whose satisfiability is equivalent to the truth of $(x)\, F(x)$).

$(x)\, F(x)$ serves as the shorthand for the mathematical statement

$$x \in F \text{ for all } x \in \mathbb{N}$$

with $F \subseteq \mathbb{N}$ representing any unary primitive-recursive relation over the natural numbers. Thus, we can formulate Theorem X as follows:

Theorem 6.25 — Gödel's Theorem X

For every primitive-recursive relation $F \subseteq \mathbb{N}$, there exists a PL1-formula φ with the following property:

$$x \in F \text{ for all } x \in \mathbb{N} \Leftrightarrow \varphi \text{ is satisfiable}$$

Gödel precedes the proof of Theorem X with a passage in which he briefly defines the formal system under consideration:

> Zum engeren Funktionenkalkül (e. F.) rechnen wir diejenigen Formeln, welche sich aus den Grundzeichen: $\overline{}$, $\vee$, (x), $=$; x, y ... (Individuenvariable) $F(x)$, $G(x\, y)$, $H(x, y, z)$ ... (Eigenschafts- und

6.2 The First Incompleteness Theorem

> Relationsvariable) aufbauen [56]), wobei (x) und $=$ sich nur auf Individuen beziehen dürfen. Wir fügen zu diesen Zeichen noch eine
>
> ---
>
> [56]) D. Hilbert und W. Ackermann rechnen in dem eben zitierten Buch das Zeichen $=$ nicht zum engeren Funktionenkalkül. Es gibt aber zu jeder Formel, in der das Zeichen $=$ vorkommt, eine solche ohne dieses Zeichen, die mit der ursprünglichen gleichzeitig erfüllbar ist (vgl. die in Fußnote [55]) zitierte Arbeit).

> We consider as formulas of the restricted functional calculus (r. f.) those formulas which are build up from the primitive symbols: $\overline{}$, $\vee$, (x), $=$; x, y ... (individual variables); $F(x)$, $G(x\,y)$, $H(x,y,z)$ ... (variables for properties and relations), where (x) and $=$ refer only to individuals. [56])
>
> ---
>
> [56]) D. Hilbert und W. Ackermann, in the book cited above, do not consider the symbol $=$ as belonging to the restricted functional calculus. However, for every formula in which the symbol $=$ occurs, there exists a formula without this symbol which is satisfiable if and only if the original one is (cf. the paper cited in footnote [55])).

The true nature of Gödel's *restricted functional calculus* becomes evident. It is a variant of first-order predicate logic with equality that prohibits the usage of function symbols.

In Footnote 56, Gödel points out that whether the equal sign is permitted or prohibited is irrelevant, as his concern lies solely in the satisfiability of the constructed formulas. In this case, we can rely on a well-known result from first-order predicate logic, which asserts that for every PL1 formula φ containing the equality sign, an *equisatisfiable* formula ψ without the equality sign exists. Here, equisatisfiability refers to the following property:

$$\varphi \text{ is satisfiable} \Leftrightarrow \psi \text{ is satisfiable} \tag{6.39}$$

It is important to note that φ and ψ need not necessarily be equivalent to fulfill (6.39). Equisatisfiability merely states that the existence of a model of φ implies the existence of a model of ψ and vice versa. The models of the two formulas may well be different, though.

First-order predicate logic with equality, as defined by us, is called the *restricted functional calculus in the extended sense* in Gödel's work:

> viduen beziehen dürfen. Wir fügen zu diesen Zeichen noch eine dritte Art von Variablen $\varphi(x)$, $\psi(x\,y)$, $\chi(x\,y\,z)$ etc. hinzu, die Gegenstandsfunktionen vertreten (d. h. $\varphi(x)$, $\psi(x\,y)$ etc.) bezeichnen eindeutige Funktionen, deren Argumente und Werte Individuen sind [57]). Eine Formel, die außer den zuerst angeführten Zeichen des e. F. noch Variable dritter Art ($\varphi(x)$, $\psi(x\,y)$... etc.) enthält, soll eine

340   6 The Limits of Mathematics

> Formel im weiteren Sinne (i. w. S.) heißen⁵⁸). Die Begriffe „erfüllbar", „allgemeingültig" übertragen sich ohneweiters auf Formeln i. w. S. und es gilt der Satz, daß man zu jeder Formel i. w. S. A eine
>
> ⁵⁷) Und zwar soll der Definitionsbereich immer der g a n z e Individuenbereich sein.
> ⁵⁸) Variable dritter Art dürfen dabei an allen Leerstellen für Individuenvariable stehen, z. B.: $y = \varphi(x)$, $F(x, \varphi(y))$, $G[\psi(x, \varphi(y)), x]$ usw.

> We add to these symbols a third kind of variable $\varphi(x)$, $\psi(x\,y)$, $\chi(x\,y\,z)$, etc., which represent objective functions (i.e. $\varphi(x)$, $\psi(x\,y)$, etc. denote single valued functions whose arguments and values are individuals⁵⁷). A formula which, in addition to the symbols of the r.f. initially mentioned above also contains variables of the third kind ($\varphi(x)$, $\psi(x\,y)$... etc.), shall be called a formula in the wider sense (i. w. s.).⁵⁸) The concepts "satisfiable", "universally valid" carry over without any further ado to formulas i. w. s.,
>
> ⁵⁷) And, in addition, the domain of definition shall always be the entire domain of individuals.
> ⁵⁸) Variables of the third kind are permitted to replace individual variables at all argument places, e. g.: $y = \varphi(x)$, $F(x, \varphi(y))$, $G[\psi(x, \varphi(y)), x]$ etc.

Gödel will now demonstrate how any PL1-formula containing function symbols or the equality sign translates into an equisatisfiable formula lacking these symbols.

> i. w. S. und es gilt der Satz, daß man zu jeder Formel i. w. S. A eine gewöhnliche Formel des e. F. B angeben kann, so daß die Erfüllbarkeit von A mit der von B äquivalent ist. B erhält man aus A, indem man

> and we have the theorem that, for every formula i. w. s. A, one can give an ordinary formula B of the r.f. such that the satisfiability of A is equivalent with that of B.

To eliminate function symbols, it suffices to interpret every n-ary function f as a relation F of arity $n + 1$, with the left-most argument representing the function value. However, for the relation F to be considered a function, it must meet two requirements:

- For every $x_1, \ldots, x_n$, there exists a y satisfying $F(y, x_1, \ldots, x_n)$.

6.2 The First Incompleteness Theorem

- For every $x_1, \ldots, x_n$, the element y is unique.

It is easy to formalize these properties within first-order predicate logic with equality. For instance, the following formula serves our needs for unary functions:

$$\forall x \exists y\, (F(y,x) \wedge \forall z\, (F(z,x) \rightarrow y = z))$$

We are now well prepared to translate a formula such as

$$\exists x\, (P(f(x)) \vee P(g(g(x)))) \tag{6.40}$$

into a formula without function symbols. This formula reads as follows:

$$
\begin{aligned}
&\forall x \exists y\, (F(y,x) \wedge \forall z\, (F(z,x) \rightarrow y = z)) &&\text{☜ } F \text{ is a function}\\
\wedge\ &\forall x \exists y\, (G(y,x) \wedge \forall z\, (G(z,x) \rightarrow y = z)) &&\text{☜ } G \text{ is a function}\\
\wedge\ &F(u,x) &&\text{☜ } u = f(x)\\
\wedge\ &G(v,x) &&\text{☜ } v = g(x)\\
\wedge\ &G(w,v) &&\text{☜ } w = g(g(x))\\
\wedge\ &\exists x\, (P(u) \vee P(w)) &&\text{☜ Formula (6.40)}
\end{aligned}
$$

This formula is equisatisfiable to (6.40), as every model of one formula directly leads to a model of the other. However, both formulas are not logically equivalent. Since they are composed of different symbols, their models differ for this reason alone.

Gödel describes the transformation in less detail. He mainly refers to §14 of the first volume of the *Principia Mathematica*, which discusses a similar transformation. The symbol 'ι' also originates from there; Russel employed it for defining *"descriptive functions"*. From today's perspective, neither the term nor the symbol holds relevance.

> von A mit der von B äquivalent ist. B erhält man aus A, indem man die in A vorkommenden Variablen dritter Art $\varphi(x)$, $\psi(x\,y)$ .. durch Ausdrücke der Form: $(\iota z)\, F(z\,x)$, $(\iota z)\, G(z, x\,y)\ldots$ ersetzt, die „beschreibenden" Funktionen im Sinne der PM. I * 14 auflöst und die so erhaltene Formel mit einem Ausdruck logisch multipliziert [59]), der besagt, daß sämtliche an Stelle der φ, ψ.. gesetzte F, G.. hinsichtlich der ersten Leerstelle genau eindeutig sind.
>
> ---
> [59]) D. h. die Konjunktion bildet.

> One obtains B from A by replacing the variables of the third kind $\varphi(x)$, $\psi(x\,y)$ .. occurring in A by expressions of the form $(\iota z)\, F(z\,x)$, $(\iota z)\, G(z, x\,y)\ldots$, then by eliminating the "descriptive" functions in the sense of PM. I * 14, and by logically

> multiplying [59]) the formula thus obtained by an expression which says that the F, G.. replacing φ, ψ.. are single valued with respect to the first argument.
>
> ---
>
> [59]) I. e. forming the conjunction.

Overall, our considerations have led to the following result: If we are only interested in the satisfiability of formulas, it is irrelevant whether we allow the usage of the equal sign or function symbols. For each formula of first-order predicate logic, we can construct an equisatisfiable formula in which neither the equality sign nor any function symbol appears.

At this point, it gets exciting: We are approaching the proof of Theorem X:

> Wir zeigen nun, daß es zu jedem Problem der Form $(x)\, F(x)$ (F rekursiv) ein äquivalentes betreffend die Erfüllbarkeit einer Formel i. W. S. gibt, woraus nach der eben gemachten Bemerkung Satz X folgt.

> We shall now show that, for every problem of the form $(x)\, F(x)$ (F recursive), there is an equivalent problem concerning the satisfiability of a formula i. w. s., from which, according to the remark just made, Theorem X follows.

Gödel announces to demonstrate that for every primitive-recursive relation F, there exists a corresponding PL1 formula φ_F, characterized by the following property:

$$x \in F \text{ for all } x \in \mathbb{N} \iff \varphi_F \text{ is satisfiable} \quad (6.41)$$

To prove this statement, we briefly turn back to page 217. Definition 5.3 stipulated that the relation F is primitive-recursive precisely when a primitive-recursive function ϕ with the following property exists:

$$x \in F \iff \phi(x) = 0 \quad (6.42)$$

Thus, (6.41) can be reformulated as follows:

$$\phi(x) = 0 \text{ for all } x \in \mathbb{N} \iff \varphi_F \text{ is satisfiable} \quad (6.43)$$

The function ϕ is primitive recursive if and only if a series of formulas

$$\phi_1, \phi_2, \ldots, \underbrace{\phi_n}_{\phi_n = \phi}$$

6.2 The First Incompleteness Theorem

exists, which is constructed according to the schemata (PR1) to (PR5) from Definition 5.2, with the last formula being identical to ϕ. Gödel starts to argue the same way:

> Da F rekursiv ist, gibt es eine rekursive Funktion ϕ (x), so daß $F(x) \sim [\phi(x) = 0]$, und für ϕ gibt es eine Reihe von Funktionen ϕ$_1$, ϕ$_2$ ... ϕ$_n$, so daß: ϕ$_n$ = ϕ, ϕ$_1(x) = x+1$ und für jedes

> Since F is recursive, there is a recursive function ϕ (x) such that $F(x) \sim [\phi(x) = 0]$, and for ϕ there is a sequence of functions ϕ$_1$, ϕ$_2$ ... ϕ$_n$ such that ϕ$_n$ = ϕ, ϕ$_1(x) = x+1$

Gödel assumes that every series begins with the successor function

$$\phi_1(x) = x+1 \qquad (6.44)$$

This requirement poses no problems for us, as (6.44) can be added at the beginning of any formula series without altering the final function. In our example on page 215, the successor function was already the first function, so this series meets Gödel's requirement right from the start. Let us recall the beginning of this series:

$$f_1(x) := s(x) \qquad \text{(PR2)}$$
$$\qquad \qquad \qquad \qquad \text{☞} \ f_1(x) = x+1$$
$$f_2(x_1, x_2, x_3) := \pi_2^3(x_1, x_2, x_3) \qquad \text{(PR3)}$$
$$\qquad \qquad \qquad \qquad \text{☞} \ f_2(x_1, x_2, x_3) = x_2$$
$$f_3(x_1, x_2, x_3) := f_1(f_2(x_1, x_2, x_3)) \qquad \text{(PR4)}$$
$$\qquad \qquad \qquad \qquad \text{☞} \ f_3(x_1, x_2, x_3) = x_2 + 1$$
$$f_4(x) := \pi_1^1(x) \qquad \text{(PR3)}$$
$$\qquad \qquad \qquad \qquad \text{☞} \ f_4(x) = x$$
$$f_5(0, x) := f_4(x) \qquad \text{(PR5)}$$
$$f_5(k+1, x) := f_3(k, f_5(k, x), x) \qquad \text{☞} \ f_5(k, x) = x + k$$

Next, Gödel revisits the various schemata available for generating primitive-recursive functions.

> ionen ϕ$_1$, ϕ$_2$ ... ϕ$_n$, so daß: ϕ$_n$ = ϕ, ϕ$_1(x) = x+1$ und für jedes ϕ$_k$ $(1 < k \leq n)$ entweder:
>
> 1. $(x_2 \ldots x_m)\ [\phi_k(0, x_2 \ldots x_m) = \phi_p(x_2 \ldots x_m)]$ \quad (18)

$$(x, x_2 \ldots x_m) \{\phi_k[\phi_1(x), x_2 \ldots x_m] = \phi_q[x, \phi_k(x, x_2 \ldots x_m), x_2 \ldots x_m]\}$$
$$p, q < k$$

and, for every ϕ_k $(1 < k \leq n)$, either:

1. $(x_2 \ldots x_m) [\phi_k(0, x_2 \ldots x_m) = \phi_p(x_2 \ldots x_m)]$ (18)
$$(x, x_2 \ldots x_m) \{\phi_k[\phi_1(x), x_2 \ldots x_m] = \phi_q[x, \phi_k(x, x_2 \ldots x_m), x_2 \ldots x_m]\}$$
$$p, q < k$$

Schema 1 can be rewritten as follows:

$$\phi_k(0, x_2, \ldots, x_m) = \phi_p(x_2, \ldots, x_m) \qquad (p < k)$$
$$\phi_k(\phi_1(x), x_2, \ldots, x_m) = \phi_q(x, \phi_k(x, x_2, \ldots, x_m), x_2, \ldots, x_m) \qquad (q < k)$$

After replacing $\phi_1(x)$ with $x + 1$, the definition takes on a familiar face:

$$\phi_k(0, x_2, \ldots, x_m) = \phi_p(x_2, \ldots, x_m) \qquad (p < k)$$
$$\phi_k(x+1, x_2, \ldots, x_m) = \phi_q(x, \phi_k(x, x_2, \ldots, x_m), x_2, \ldots, x_m) \qquad (q < k)$$

This is (PR5), the schema of primitive recursion from Definition 5.2.

Über formal unentscheidbare Sätze der Principia Mathematica etc.   195

oder:

2. $(x_1 \ldots x_m) [\phi_k(x_1 \ldots x_m) = \phi_r(\phi_{i_1}(\mathfrak{x}_1) \ldots \phi_{i_s}(\mathfrak{x}_s))]$ [60]) (19)
$r < k$, $i_v < k$ (für $v = 1, 2 \ldots s$)

---

[60]) $\mathfrak{x}_i$ ($i = 1 \ldots s$) vertreten irgend welche Komplexe der Variablen $x_1, x_2 \ldots x_m$, z. B. $x_1\ x_3\ x_2$.

On formally undecidable propositions of Principia Mathematica etc.   195

or:

2. $(x_1 \ldots x_m) [\phi_k(x_1 \ldots x_m) = \phi_r(\phi_{i_1}(\mathfrak{x}_1) \ldots \phi_{i_s}(\mathfrak{x}_s))]$ [60]) (19)
$r < k$, $i_v < k$ (for $v = 1, 2 \ldots s$)

---

[60]) $\mathfrak{x}_i$ ($i = 1 \ldots s$) represent any complexes made up of the variables $x_1, x_2 \ldots x_m$; e. g. $x_1\ x_3\ x_2$.

Here, Gödel refers to (PR4), the substitution schema from Definition 5.2.

6.2 The First Incompleteness Theorem

> oder:
>
> 3. $(x_1 \ldots x_m) \, [\phi_k (x_1 \ldots x_m) = \phi_1 (\phi_1 \ldots \phi_1 (0))]$ \hfill (20)

> or:
>
> 3. $(x_1 \ldots x_m) \, [\phi_k (x_1 \ldots x_m) = \phi_1 (\phi_1 \ldots \phi_1 (0))]$ \hfill (20)

This schema generates the constant functions $f(x_1, \ldots, x_m) = n$ for any natural number n. It is absent in Definition 5.2 as it is dispensable; the constant functions can be obtained from the zero function through repeated application of the substitution scheme.

Rather, Definition 5.2 included the projection functions π for properly handling parameter lists of different lengths.

> Ferner bilden wir die Sätze:
>
> $(x) \, \overline{\phi_1 (x) = 0} \; \& \; (x \, y) \, [\phi_1 (x) = \phi_1 (y) \rightarrow x = y]$ \hfill (21)
> $(x) \, [\phi_n (x) = 0]$ \hfill (22)

> Furthermore, we form the sentences:
>
> $(x) \, \overline{\phi_1 (x) = 0} \; \& \; (x \, y) \, [\phi_1 (x) = \phi_1 (y) \rightarrow x = y]$ \hfill (21)
> $(x) \, [\phi_n (x) = 0]$ \hfill (22)

Let's have a closer look at both propositions:

■ **Proposition (21)**

$$\phi_1(x) \neq 0$$
$$\phi_1(x) = \phi_1(y) \Rightarrow x = y$$

Given that ϕ_1 is the successor function, both statements represent trivial truth. The reason why Gödel mentioned them explicitly will only become apparent later. He will integrate them into a formula of first-order predicate logic to ensure that the function symbol for ϕ_1 represents an injective function that never takes on the value 0.

■ **Proposition (22)**

$$\phi_n(x) = 0 \text{ for all } x \in \mathbb{N}$$

This proposition demands that ϕ_n is the zero function. Because of (6.42), this is equivalent to the statement

$$x \in F \text{ for all } x \in \mathbb{N}$$

Next, Gödel translates the series $\phi_1, \phi_2, \ldots, \phi_n$ into a formula of first-order predicate logic:

> Wir ersetzen nun in allen Formeln (18), (19), (20) (für $k = 2$, $3\ldots n$) und in (21) (22) die Funktionen ϕ_i durch Funktionsvariable φ_i, die Zahl 0 durch eine sonst nicht vorkommende Individuenvariable x_0 und bilden die Konjunktion C sämtlicher so erhaltener Formeln.

> Now, in all the formulas (18), (19), (20) (for $k = 2$, $3\ldots n$) and in (21), (22), we replace the functions ϕ_i by function variables φ_i, the number 0 by an individual variable x_0 which does not occur elsewhere, and we form the conjunction C of all the formulas so obtained.

Gödel advises us to replace the functions ϕ_i with function variables φ_i and the number 0 with an otherwise not occurring individual variable x_0. For our example, the outcome reads as follows:

$$C_2 := \forall x_1 \, \forall x_2 \, \forall x_3 \, (\varphi_2(x_1, x_2, x_3) = x_2)$$
$$C_3 := \forall x_1 \, \forall x_2 \, \forall x_3 \, (\varphi_3(x_1, x_2, x_3) = \varphi_1(\varphi_2(x_1, x_2, x_3)))$$
$$C_4 := \forall x \, (\varphi_4(x) = x)$$
$$C_5 := \forall x_2 \, (\varphi_5(x_0, x_2) = \varphi_4(x_2)) \wedge$$
$$\quad\quad \forall x \, \forall x_2 \, (\varphi_5(\varphi_1(x), x_2) = \varphi_3(x, \varphi_5(x, x_2), x_2))$$

$$C_{(21)} := \forall x \, \neg(\varphi_1(x) = x_0) \wedge \forall x \, \forall y \, (\varphi_1(x) = \varphi_1(y) \to x = y)$$
$$C_{(22)} := \forall x \, (\varphi_5(x) = x_0)$$

Now, we form the conjunction C of all formulas thus obtained:

$$C = C_2 \wedge C_3 \wedge C_4 \wedge C_5 \wedge C_{(21)} \wedge C_{(22)}$$

Next, we will show that the formula

$$\varphi_F := \exists x_0 \, C$$

satisfies property (6.43), namely:

$$\phi(x) = 0 \text{ for all } x \in \mathbb{N} \Leftrightarrow \varphi_F \text{ is satisfiable}$$

6.2 The First Incompleteness Theorem

The direction from left to right is straightforward. If $\phi(x) = 0$ for all $x \in \mathbb{N}$, we obtain a model for C by interpreting the function symbols φ_i as the functions ϕ_i and the individual variable x_i as the number 0. In Gödel's words:

> Die Formel $(E x_0)$ C hat dann die verlangte Eigenschaft, d. h.
> 1. Wenn $(x) [\phi (x) = 0]$ gilt, ist $(E x_0)$ C erfüllbar, denn die Funktionen $\phi_1, \phi_2 \ldots \phi_n$ ergeben dann offenbar in $(E x_0)$ C für $\varphi_1, \varphi_2 \ldots \varphi_n$ eingesetzt einen richtigen Satz.

> Then the formula $(E x_0)$ C has the required property, i.e.
> 1. If $(x) [\phi (x) = 0]$ holds, then $(E x_0)$ C is satisfiable, for, when the functions $\phi_1, \phi_2 \ldots \phi_n$ are substituted for $\varphi_1, \varphi_2 \ldots \varphi_n$ in $(E x_0)$ C, then a true sentence obviously results.

The direction from right to left is more elaborate. The satisfiability of φ_F implies the existence of a model for the formula $\exists x_0\, C$. Consequently, the following remains to be shown:

$$\exists x_0\, C \text{ has a model} \Rightarrow \phi(x) = 0 \text{ for all } x \in \mathbb{N}$$

Gödel commences the proof by saying:

> 2. Wenn $(E x_0)$ C erfüllbar ist, gilt $(x) [\phi (x) = 0]$.
> Beweis: Seien $\psi_1, \psi_2 \ldots \psi_n$ die nach Voraussetzung existierenden Funktionen, welche in $(E x_0) C$ für $\varphi_1, \varphi_2 \ldots \varphi_n$ eingesetzt einen richtigen Satz liefern. Ihr Individuenbereich sei $\mathfrak{J}$. Wegen der Richtigkeit

> 2. If $(E x_0)$ C is satisfiable, then $(x) [\phi (x) = 0]$ holds.
> Proof: Let $\psi_1, \psi_2 \ldots \psi_n$ be the functions which, according to our hypothesis, yield a true sentence when substituted for $\varphi_1, \varphi_2 \ldots \varphi_n$ in $(E x_0) C$. Let their domain of individuals be $\mathfrak{J}$.

Gödel assumes the existence of a model $(\mathfrak{J}, I)$ for the formula $\exists x_0\, C$. The set $\mathfrak{J}$ is the domain of the model, and I is the mapping that assigns each individual variable an element from $\mathfrak{J}$, and each function symbol φ_i a function $\psi_i : \mathfrak{J}^n \to \mathfrak{J}$, where n is the arity of φ_i. The property of an interpretation being a model means that the formula becomes substantively true for this interpretation. In symbolic form, this can be succinctly expressed as follows:

$$(\mathfrak{J}, I) \models \exists x_0\, C$$

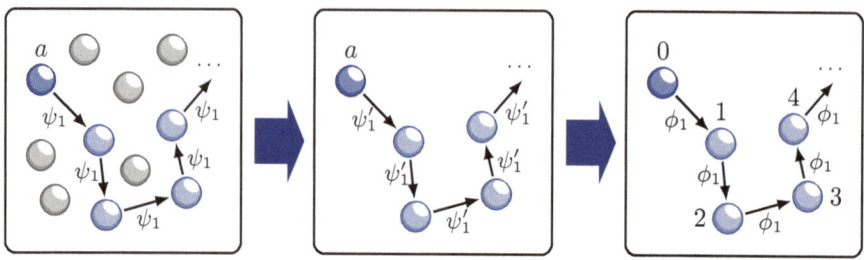

Figure 6.7: The construction in detail. Starting with a model of $\exists x_0 \, C$, Gödel constructs a model with the natural numbers as domain, interpreting the function symbols $\varphi_1, \varphi_2, \ldots$ as the functions $\phi_1, \phi_2, \ldots$.

Consequently, an individual element a with the following property must exist:

$$(\mathfrak{J}, I_{[x_0/a]}) \models C_2 \wedge C_3 \wedge \ldots \wedge C_{(21)} \wedge C_{(22)}$$

Recall that $(\mathfrak{J}, I_{[x_0/a]})$ denotes the interpretation that maps the variable x_0 to the individual element $a \in \mathfrak{J}$ and is otherwise identical to $(\mathfrak{J}, I)$. Gödel argues the very same way:

> richtigen Satz liefern. Ihr Individuenbereich sei $\mathfrak{J}$. Wegen der Richtigkeit von $(E x_0) \, C$ für die Funktionen ψ_i gibt es ein Individuum a (aus $\mathfrak{J}$), so daß sämtliche Formeln (18) bis (22) bei Ersetzung der ϕ_i durch ψ_i und von 0 durch a in richtige Sätze (18') bis (22') übergehen.

> Because $(E x_0) \, C$ holds for the functions ψ_i, there exists an individual a (in $\mathfrak{J}$) such that all the formulas (18) – (22) become true sentences (18') – (22') when the ϕ_i are replaced by the ψ_i and 0 by a.

Let us try to visualize the model $(\mathfrak{J}, I)$. Our starting point is the left part of Figure 6.7, where each sphere symbolizes an individual element of the set $\mathfrak{J}$. The element a resides somewhere in this set. Progressing from a to $\psi_1(a)$, then to $\psi_1(\psi_1(a))$, and so forth, unveils a uniquely determined sequence of elements. The subformulas $C_{(20)}$ and $C_{(21)}$ ensure the traversed elements form a sequence that starts with a and extends to infinity. Gödel argues that $(\mathfrak{J}, I)$ remains a model if the set of individuals $\mathfrak{J}$ is replaced with $\mathfrak{J}'$ and the domain of the functions ψ_i is restricted accordingly. This way, each function ψ_i transforms into a function ψ_i', as depicted in the middle of Figure 6.7.

> Wir bilden nun die kleinste Teilklasse von $\mathfrak{J}$, welche a enthält und gegen die Operation $\psi_1(x)$ abgeschlossen ist. Diese Teilklasse ($\mathfrak{J}'$)

6.2 The First Incompleteness Theorem

> hat die Eigenschaft, daß jede der Funktionen ψ_i auf Elemente aus $\mathfrak{J}'$ angewendet wieder Elemente aus $\mathfrak{J}'$ ergibt. Denn für ψ_1 gilt dies nach Definition von $\mathfrak{J}'$ und wegen (18'), (19'), (20') überträgt sich diese Eigenschaft von ψ_i mit niedrigerem Index auf solche mit höherem. Die Funktionen, welche aus ψ_i durch Beschränkung auf den Individuenbereich $\mathfrak{J}'$ entstehen, nennen wir ψ_i'. Auch für diese Funktion gelten sämtliche Formeln (18) bis (22) (bei der Ersetzung von 0 durch a und ϕ_i durch ψ_i').

> Now we form the smallest subclass of $\mathfrak{J}$ which contains a and is closed with respect to the operation $\psi_1(x)$. This subclass ($\mathfrak{J}'$) has the property that each of the functions ψ_i, when applied to elements of $\mathfrak{J}'$, yields an element of $\mathfrak{J}'$. For, this holds for ψ_1 by definition of $\mathfrak{J}'$, and, by virtue of (18'), (19'), (20'), this property is transmitted from those ψ_i with lower subscript to those with higher subscript. The functions which arise from the ψ_i by restriction to the domain of individuals $\mathfrak{J}'$ are called ψ_i'. These functions also satisfy all the formulas (18) – (22) (after substitution of a for 0 and ψ_i' for ϕ_i).

Next, Gödel utilizes the model $(\mathfrak{J}', I')$ to construct an isomorphic image within the natural numbers (Figure 6.7 right). The construction is straightforward, as it suffices to identify a with 0, $\psi_1'(a)$ with 1, $\psi_1'(\psi_1'(a))$ with 2, and so forth. Proceeding this way lets the function ψ_1' become the successor function ϕ_1. The same applies to the more complex functions in the new model $(\mathbb{N}, I'')$, that is, the functions ψ_i' become the primitive recursive functions ϕ_i. It then holds:

$$I''(\mathsf{x}_0) = 0 \qquad (6.45)$$
$$I''(\varphi_i) = \phi_i \qquad (6.46)$$

Like in any model, all subformulas of C are true statements in $(\mathbb{N}, I'')$. Consequently, the subformula $C_{(22)}$ is true a fortiori:

$$(\mathbb{N}, I'') \models \forall \mathsf{x}\, (\varphi_n(\mathsf{x}) = \mathsf{x}_0)$$

Due to (6.45) and (6.46), this is equivalent to

$$\phi_n(x) = 0 \text{ for all } x \in \mathbb{N},$$

and because of $\phi_n = \phi$, this is the same as

$$\phi(x) = 0 \text{ for all } x \in \mathbb{N},$$

which had to be proven. □

Gödel articulates the final proof step in the following manner:

> Wegen der Richtigkeit von (21) für ψ_1' und a kann man die Individuen aus $\mathfrak{I}'$ eineindeutig auf die natürlichen Zahlen abbilden u. zw. so, daß a in 0 und die Funktion ψ_1' in die Nachfolgerfunktion ϕ_1 übergeht. Durch diese Abbildung gehen aber sämtliche Funktionen ψ_i' in die Funktionen ϕ_i über und wegen der Richtigkeit von (22)

> 196   Kurt Gödel,
>
> für ψ_n' und a gilt $(x)\,[\phi_n(x) = 0]$ oder $(x)\,[\phi(x) = 0]$, was zu beweisen war[61]).
>
> ---
>
> [61]) Aus Satz X folgt z. B., daß das Fermatsche und das Goldbachsche Problem lösbar wären, wenn man das Entscheidungsproblem des e. F. gelöst hätte.

> Because of the truth of (21) for ψ_1' and a, one can map the individuals of $\mathfrak{I}'$ in a one-to-one manner onto the natural numbers, and, moreover, in such a way that a goes over into 0 and the function ψ_1' into the successor function ϕ_1. Under this mapping, however, all the functions ψ_i' go over into the functions ϕ_i, and by the truth of (22)

> 196   Kurt Gödel,
>
> for ψ_n' and a, we have $(x)\,[\phi_n(x) = 0]$ or $(x)\,[\phi(x) = 0]$, which was to be proved[61]).
>
> ---
>
> [61]) From Theorem X it follows, for example, that the Fermat and Goldbach problems would have been solvable, if one had solved the decision problem of the r. f.

In footnote 61, Gödel hints at why we must acknowledge the contents of Theorem X with regret. The ability to determine the satisfiability of every formula of first-order predicate logic could potentially resolve a vast amount of number-theoretical problems. Gödel mentions Fermat's Last Theorem and Goldbach's conjecture as prominent examples.

- Goldbach's conjecture asserts that every even number $n > 2$ can be written as the sum of two prime numbers. It holds true if and only if the following primitive-recursive relation encompasses all natural numbers:

6.2 The First Incompleteness Theorem

| Gb(x) | Goldbach's Conjecture |

(Primitive-recursive relation)

$$x \in \text{Gb} :\Leftrightarrow x \leq 2 \vee \text{odd}(x) \vee \exists\,(y,z < x)\,(\text{Prim}(y) \wedge \text{Prim}(z) \wedge x = y + z)$$

- Fermat's Last Theorem states that the equation $a^n + b^n = c^n$ has no solutions in positive integers for $n > 2$. To express this theorem in relational form, we draw upon a well-established principle from recursion theory. It states the existence of primitive-recursive functions $\pi_1, \ldots, \pi_4$ such that

$$x \mapsto (\pi_1(x), \pi_2(x), \pi_3(x), \pi_4(x))$$

bijectively maps the set $\mathbb{N}$ into the set $\mathbb{N}^+ \times \mathbb{N}^+ \times \mathbb{N}^+ \times \mathbb{N}^+$. With these functions at hand, we can easily craft a suitable primitive-recursive relation:

| Fermat(x) | Fermat's Theorem |

(Primitive-recursive relation)

$$x \in \text{Fermat} :\Leftrightarrow \pi_1(x) \leq 2 \vee \pi_2(x)^{\pi_1(x)} + \pi_3(x)^{\pi_1(x)} \neq \pi_4(x)^{\pi_1(x)}$$

Keep in mind that the absence of a decision procedure for first-order predicate logic does not necessarily imply the falsity or unprovability of the mentioned problems. For instance, Fermat's theorem was proven in 1995 by other means. A proof for Goldbach's conjecture, however, still stands out to this day.

At this juncture, we are ready to finalize the proof of Theorem IX. Gödel's line of reasoning resembles the passage quoted on page 329, where he elucidates that the proof can also be carried out within the system P:

> Da man die Überlegungen, welche zu Satz X führen, (für jedes spezielle F) auch innerhalb des Systems P durchführen kann, so ist die Äquivalenz zwischen einem Satz der Form $(x)\,F(x)$ (F rekursiv) und der Erfüllbarkeit der entsprechenden Formel des e. F. in P beweisbar und daher folgt aus der Unentscheidbarkeit des einen die des anderen, womit Satz IX bewiesen ist.[62])
>
> ---
> [62]) Satz IX gilt natürlich auch für das Axiomensystem der Mengenlehre und dessen Erweiterungen durch rekursiv definierbare ω-widerspruchsfreie Klassen von Axiomen, da es ja auch in diesen Systemen unentscheidbare Sätze der Form $(x)\,F(x)$ (F rekursiv) gibt.

> Since one can carry out within the system P the argument which led to Theorem X (for every special F), then the equivalence

> between a sentence of the form $(x)\,F(x)$ (F recursive) and the satisfiability of the corresponding formula of the r. f. is provable in P, and therefore the undecidability of the former implies that of the latter, which proves Theorem IX.[62]
>
> ---
>
> [62] Theorem IX holds naturally also for the axiom system of set theory and its extensions by recursively definable ω-consistent classes of axioms, since there also exist undecidable sentences of the form $(x)\,F(x)$ (F recursive) in these systems.

For F, we can once more select the primitive-recursive relation described on page 301 by the formula with the Gödel number r. To replicate the proof for the selected relation in P means to derive a theorem of the following form:

$$\vdash \forall x_1\, \varphi_r(x_1) \leftrightarrow \mathrm{Sat}_{\varphi_r}$$

On the left-hand side is the formula $\forall x_1\, \varphi_r(x_1)$, whose undecidability has been established earlier; neither the formula itself nor its negation is provable within P. Accordingly, Sat_φ and $\neg\mathrm{Sat}_\varphi$ cannot be theorems either:

$$\nvdash \mathrm{Sat}_\varphi$$
$$\nvdash \neg\mathrm{Sat}_\varphi$$

The undecidability of Sat_φ implies that proving or disproving the universal validity of $\neg\forall x_1\, \varphi_r(x_1)$ within P would contradict (6.37) or (6.38). Thus, we have established Theorem IX: PL1 formulas exist whose universal validity can neither be proved nor disproved within P.

6.3 The Second Incompleteness Theorem

We are on the verge of reaching the next and final climax of Gödel's work. The subject of our interest is Theorem XI, today referred to as *Gödel's second incompleteness theorem*.

> 4.
>
> Aus den Ergebnissen von Abschnitt 2 folgt ein merkwürdiges Resultat, bezüglich eines Widerspruchslosigkeitsbeweises des Systems P (und seiner Erweiterungen), das durch folgenden Satz ausgesprochen wird:
>
> Satz XI: Sei x eine beliebige rekursive widerspruchsfreie[63] Klasse von *Formeln*, dann gilt: Die *Satzformel*, welche besagt, daß x widerspruchsfrei ist, ist nicht x-*beweisbar*; insbesondere ist die Widerspruchsfreiheit von P in P unbeweisbar[64],

6.3 The Second Incompleteness Theorem

> vorausgesetzt, daß P widerspruchsfrei ist (im entgegengesetzten Fall ist natürlich jede Aussage beweisbar).
>
> ---
>
> [63]) x ist widerspruchsfrei (abgekürzt als Wid (x)) wird folgendermaßen definiert: Wid (x) $\equiv$ $(E\,x)$ [Form (x) & $\overline{\mathrm{Bew}}_x\,(x)$].
> [64]) Dies folgt, wenn man für x die leere Klasse von *Formeln* einsetzt.

> 4.
>
> From the results of Section 2 there follows a remarkable result concerning a consistency proof for the system P (and its extensions), which is expressed by the following theorem:
>
> Theorem XI: Let x be an arbitrary recursive consistent class [63]) of *formulas*. Then the *sentence* which asserts that x is consistent is not x-*provable*; in particular, the consistency of P is unprovable in P,[64]) assuming that P is consistent (in the contrary case, of course, every statement is provable).
>
> ---
>
> [63]) x is consistent (abbreviated Wid (x)) is defined as follows: Wid (x) $\equiv$ $(E\,x)$ [Form (x) & $\overline{\mathrm{Bew}}_x\,(x)$].
> [64]) This follows when one substitutes for x the empty class of *formulas*.

Slightly more informal yet easier to grasp, Theorem XI can be phrased as follows:

Theorem 6.26   Second Incompleteness Theorem

No formal system with at least the expressive power of P can prove its own consistency.

First, let us elucidate the true significance of Theorem XI. Our starting point is footnote 63, where Gödel explains how the consistency of a set of formulas χ can be expressed within the formal system P. In modern notation, his definition reads as follows:

$$\mathrm{Wid}(\chi) \;:\Leftrightarrow\; \exists x\,(\mathrm{Form}(x) \wedge \neg\,\mathrm{Bew}_\chi(x)) \tag{6.47}$$

Gödel capitalizes on the fact that every inconsistent formal system, as long as it includes the ordinary propositional logic apparatus, can prove all formulas without exception. Conversely, $P \cup \chi$ is consistent if

- a formula exists   ☞ $\exists x\,\mathrm{Form}(x)$

- which is unprovable in $P \cup \chi$.   ☞ $\neg\,\mathrm{Bew}_\chi(x)$

> Der Beweis ist (in Umrissen skizziert) der folgende: Sei κ eine beliebige für die folgenden Betrachtungen ein für allemal gewählte rekursive Klasse von *Formeln* (im einfachsten Falle die leere Klasse). Zum Beweise der Tatsache, daß 17 Gen r nicht κ-*beweisbar* ist [65]), wurde, wie aus 1. Seite 189 hervorgeht, nur die Widerspruchsfreiheit von κ benutzt, d. h. es gilt:
>
> $$\text{Wid}(\kappa) \rightarrow \overline{\text{Bew}_\kappa}\,(17\ \text{Gen}\ r) \qquad (23)$$
>
> d. h. nach (6·1):
>
> $$\text{Wid}(\kappa) \rightarrow \overline{(x)\ x\ B_\kappa\,(17\ \text{Gen}\ r)}$$
>
> ---
> [65]) r hängt natürlich (ebenso wie p) von κ ab.

> The proof is (in outline) the following: Let κ be an arbitrary recursive class of *formulas* (in the simplest case, the empty class) which, for the following considerations, is chosen once and for all. In the proof of the fact that 17 Gen r is not κ-*provable*, [65]) only the consistency of κ is used, as can be seen from 1. on page 189; that is, we have:
>
> $$\text{Wid}(\kappa) \rightarrow \overline{\text{Bew}_\kappa}\,(17\ \text{Gen}\ r) \qquad (23)$$
>
> i.e., by virtue of (6·1):
>
> $$\text{Wid}(\kappa) \rightarrow \overline{(x)\ x\ B_\kappa\,(17\ \text{Gen}\ r)}$$
>
> ---
> [65]) Of course, r (as well as p) depends upon κ.

Let us recall: On page 303, we have shown that neither the formula $\forall x_1\, \varphi_r(x_1)$ nor its negation $\neg \forall x_1\, \varphi_r(x_1)$ is provable. In the first case, we only relied on the consistency of P, allowing us to draw the following conclusion:

$$\text{Wid}(\chi) \Rightarrow \underbrace{\ulcorner \forall x_1\, \varphi_r(x_1) \urcorner}_{17\ \text{Gen}\ r} \notin \text{Bew}_\chi \qquad (6.48)$$

This conclusion is the statement (23) in Gödel's work.

The relation Bew_χ was defined on page 297 as follows:

$$x \in \text{Bew}_\chi :\Leftrightarrow \exists y\ y\ B_\chi\ x$$

Thus, (6.48) can be rewritten as such:

$$\text{Wid}(\chi) \Rightarrow \neg \exists y\ y\ B_\chi\ \ulcorner \forall x_1\, \varphi_r(x_1) \urcorner$$

6.3 The Second Incompleteness Theorem

A minor reformulation yields the statement in Gödel's article:

$$\text{Wid}(\chi) \Rightarrow \forall x \, \neg(x \, B_\chi \, \ulcorner \forall \mathsf{x}_1 \, \varphi_r(\mathsf{x}_1) \urcorner) \tag{6.49}$$

> Nach (13) ist $17 \, \text{Gen} \, r = S b \left(p \, \overset{19}{Z(p)} \right)$ und daher:
>
> Über formal unentscheidbare Sätze der Principia Mathematica etc. 197
>
> $$\overline{\text{Wid}(x) \rightarrow \overline{(x) \, x \, B_\text{x} \, S b \left(p \, \overset{19}{Z(p)} \right)}}$$
>
> d. h. nach (8·1):
>
> $$\text{Wid}(x) \rightarrow (x) \, Q(x, p) \tag{24}$$

> By (13), $17 \, \text{Gen} \, r = S b \left(p \, \overset{19}{Z(p)} \right)$ and therefore:
>
> On formally undecidable propositions of Principia Mathematica etc. 197
>
> $$\overline{\text{Wid}(x) \rightarrow \overline{(x) \, x \, B_\text{x} \, S b \left(p \, \overset{19}{Z(p)} \right)}}$$
>
> i.e., by (8·1):
>
> $$\text{Wid}(x) \rightarrow (x) \, Q(x, p) \tag{24}$$

In Gödel's work, (13) references the equation

$$\forall \mathsf{x}_1 \, \varphi_r(\mathsf{x}_1) = \varphi_p(\bar{p}),$$

which enables us to transform (6.49) into

$$\text{Wid}(\chi) \Rightarrow \forall x \, \neg(x \, B_\chi \, \ulcorner \varphi_p(\bar{p}) \urcorner).$$

Because of (6.5), this is the same as:

$$\text{Wid}(\chi) \Rightarrow (x, p) \in Q \text{ for all } x \tag{6.50}$$

> Wir stellen nun folgendes fest: Sämtliche in Abschnitt 2 [66]) und Abschnitt 4 bisher definierte Begriffe (bzw. bewiesene Behauptungen) sind auch in P ausdrückbar (bzw. beweisbar). Denn es wurden überall

> nur die gewöhnlichen Definitions- und Beweismethoden der klassischen Mathematik verwendet, wie sie im System P formalisiert sind. Insbesondere ist x (wie jede rekursive Klasse) in P definierbar. Sei
>
> ---
> [66]) Von der Definition für „rekursiv" auf Seite 179 bis zum Beweis von Satz VI inkl.

> Now we establish the following: All the defined concepts (proved assertions) of Section 2[66]) and Section 4 are expressible (provable) in P. For, we have used throughout only the ordinary methods of definition and proof of classical mathematics, as they are formalized in the system P. In particular, x (like every recursive class) is definable in P.
>
> ---
> [66]) From the definition of "recursive" on p. 179 until the proof of Theorem VI, inclusive.

This passage contains the key idea: Gödel again takes advantage of the fact that all previously defined notions (resp. proven assertions) are also expressible (resp. provable) in P. In particular, this implies the existence of a closed formula Wid(χ), which formalizes the relation Wid(χ) within P. Gödel denotes the Gödel number of this formula with w:

> besondere ist x (wie jede rekursive Klasse) in P definierbar. Sei w die *Satzformel*, durch welche in P Wid (x) ausgedrückt wird. Die Relation $Q(x,y)$ wird gemäß (8·1), (9), (10) durch das *Relationszeichen* q ausgedrückt, folglich $Q(x,p)$ durch r [da nach (12) $r =$ $= Sb\left(q \begin{smallmatrix}19\\Z(p)\end{smallmatrix}\right)$] und der Satz $(x) Q(x,p)$ durch 17 Gen r.

> Let w be the *sentence* by which Wid (x) is expressed in P. The relation $Q(x,y)$ is, according to (8·1), (9), (10), expressed by the *predicate* q, and, consequently, $Q(x,p)$ by r [since, by (12), $r =$ $= Sb\left(q \begin{smallmatrix}19\\Z(p)\end{smallmatrix}\right)$], and the sentence $(x) Q(x,p)$ by 17 Gen r.

Gödel explains that all other statements, previously formulated at the metalevel, can also be formally expressed within P. The ones listed in Table 6.1 are particularly significant for us.

6.3 The Second Incompleteness Theorem

Mathematical statement	Formalization in P	
$\mathrm{Wid}(\chi)$	$\mathrm{Wid}(\chi)$	
$(x,y) \in Q$	$\psi_q(\overline{x}, \overline{y})$	
$(x,p) \in Q$	$\varphi_r(\overline{x})$	because of (6.9)
$(x,p) \in Q$ for all x	$\forall x_1\, \varphi_r(x_1)$	

Table 6.1: Mathematical statements and their formal representation in P

By formalizing the conclusion leading to (6.50), the following theorem is derivable within P:

$$\vdash \mathrm{Wid}(\chi) \to \forall x_1\, \varphi_r(x_1) \tag{6.51}$$

At this juncture, we have crossed the finish line. If we could establish the consistency of P within P, specifically, if we had

$$\vdash \mathrm{Wid}(\chi),$$

this would, in conjunction with (6.51), lead to a proof for $\forall x_1\, \varphi_r(x_1)$:

$$\vdash \forall x_1\, \varphi_r(x_1) \tag{6.52}$$

The first incompleteness theorem states that (6.52) cannot be proved within P, provided that P is consistent. As a result, $\mathrm{Wid}(\chi)$ also remains unprovable within P. Gödel articulates this conclusion as follows:

> Wegen (24) ist also w Imp (17 Gen r) in P beweisbar [67]) (um so mehr $\varkappa$-*beweisbar*). Wäre nun w $\varkappa$-*beweisbar*, so wäre auch 17 Gen r $\varkappa$-*beweisbar* und daraus würde nach (23) folgen, daß $\varkappa$ nicht widerspruchsfrei ist.
>
> ---
> [67]) Daß aus (23) auf die Richtigkeit von w Imp (17 Gen r) geschlossen werden kann, beruht einfach darauf, daß der unentscheidbare Satz 17 Gen r, wie gleich zu Anfang bemerkt, seine eigene Unbeweisbarkeit behauptet.

> Hence w Imp (17 Gen r) is, by virtue of (24), provable in P^{67}) (a fortiori, $\varkappa$-*provable*). Now, were w to be $\varkappa$-*provable*, then 17 Gen r would also be $\varkappa$-*provable*, whence, by (23), it would follow that $\varkappa$ is not consistent.
>
> ---
> [67]) That the truth of w Imp (17 Gen r) can be deduced from (23) rests simply on the fact that the undecidable proposition 17 Gen r, as was remarked at the very beginning, asserts its own unprovability.

This concludes the proof of Theorem XI.

Be sure to draw the proper conclusions from the second incompleteness theorem. This theorem is often interpreted to suggest that the provability of Wid(χ) implies the consistency of the underlying formal system. However, this is by no means true. In any contradictory formal system that includes the standard propositional logic apparatus, every formula can be derived from the axioms, including Wid(χ). Consequently, the opposite conclusion is correct: If it is possible to prove consistency within a formal system that meets the requirements of the second incompleteness theorem, it must necessarily be contradictory. Therefore, the second incompleteness theorem can only serve to demonstrate the inconsistency of a formal system but never its consistency.

Gödel starts the next section by highlighting that the provided proof is constructive. After that, he points out its applicability to all formal systems possessing at least the expressive power of Gödel's system P and explicitly mentions axiomatic set theory and classical mathematics as examples. The distinction made by Gödel between these two terms is rooted in history. It wasn't until the latter half of the twentieth century that axiomatic set theory gained widespread acceptance, which eventually led to both concepts being identified.

> Es sei bemerkt, daß auch dieser Beweis konstruktiv ist, d. h. er gestattet, falls ein *Beweis* aus x für w vorgelegt ist, einen Widerspruch aus x effektiv herzuleiten. Der ganze Beweis für Satz XI läßt sich wörtlich auch auf das Axiomensystem der Mengenlehre M und der klassischen Mathematik [68]) A übertragen und liefert auch hier das Resultat: Es gibt keinen Widerspruchslosigkeitsbeweis für M bzw. A, der innerhalb von M bzw. A formalisiert werden könnte, vorausgesetzt daß M bzw. A widerspruchsfrei ist. Es sei ausdrück-
>
> ---
> [68]) Vgl. J. v. Neumann, Zur Hilbertschen Beweistheorie, Math. Zeitschr. 26, 1927.

> Notice that this proof is also constructive, i. e. it permits us to effectively derive a contradiction from x, if we are presented with a *proof* of w from x. The whole proof of Theorem XI can be carried over, word for word, to the axiom systems of set theory M and of classical mathematics [68]) A, and yields here also the result: There exists no consistency proof for M which can be formalized within M assuming that M is consistent, and similarly for A.
>
> ---
> [68]) Cf. J. v. Neumann, "Zur Hilbertschen Beweistheorie", Math. Zeitschr. 26, 1927.

Today, we know that even even more limited systems are subject to the second incompleteness theorem. Back in 1939, David Hilbert and Paul Bernays proved the theorem for two variants of Peano arithmetic [56], thereby establishing precise criteria that were later simplified by the German mathematician Martin

6.3 The Second Incompleteness Theorem

Löb [66]. When a formal system meets these criteria, it can formalize and prove the first incompleteness theorem. This, in turn, is sufficient to derive the disastrous formula (6.51), rendering the formal system susceptible to the second incompleteness theorem. The systems meeting these criteria include Peano arithmetic, as introduced in Section 6.2.1, implying that the second incompleteness theorem affects all formal systems expressive enough to talk about the additive and multiplicative properties of the natural numbers.

Does this result suggest that we should distrust Peano arithmetic? The short answer is No. Even though the second incompleteness theorem dashes the hope of safeguarding PA with inference methods that are more basic and thus potentially more trustworthy than Peano arithmetic itself, there is no compelling reason for distrust. Overall, the axioms are too simple, and the natural numbers are overly familiar to us.

But what about set theory? Are the precautions taken here really sufficient to exclude all antinomies? Once again, the prevailing view is that mathematics can be built on ZF or ZFC without contradictions, yet formal proof is lacking. The second incompleteness theorem underscores that such a proof is only attainable in formal systems more intricate than ZF or ZFC, ultimately transferring the question to another system. Indeed, the second incompleteness theorem destroys any hope of ever answering the question of consistency in a trustworthy way.

The second incompleteness theorem dramatically affected Hilbert's program by unequivocally demonstrating that a formal system as expressive as P cannot prove its own consistency. As a result, the consistency of mathematics is not provable with standard mathematical methods. But that was precisely the plan Hilbert had been pursuing so vehemently for years: to prove the consistency of classical mathematics with finite means. But did Gödel's second incompleteness theorem really extinguish all possibilities of fulfilling Hilbert's objectives?

Gödel comments on this matter with a surprising statement:

> vorausgesetzt daß M bzw. A widerspruchsfrei ist. Es sei ausdrücklich bemerkt, daß Satz XI (und die entsprechenden Resultate über M, A) in keinem Widerspruch zum Hilbertschen formalistischen Standpunkt stehen. Denn dieser setzt nur die Existenz eines mit finiten Mitteln geführten Widerspruchsfreiheitsbeweises voraus und es wäre denkbar, daß es finite Beweise gibt, die sich in P (bzw. M, A) nicht darstellen lassen.

> It should be expressly noted that Theorem XI (and the corresponding results about M and A) in no way contradicts Hilbert's formalistic standpoint. For the latter presupposes only the existence of a consistency proof carried out

> by finitary methods, and it is conceivable that there might be finitary proofs which cannot be represented in P (or in M or A).

Apparently, unlike other renowned mathematicians such as John von Neumann, Gödel did not see Hilbert's program as a failure. From a formal standpoint, we cannot object to Gödel's presented argument. Even if the consistency of ordinary mathematics is unprovable with methods of ordinary mathematics itself, this does not entirely rule out the existence of a less complex system in which a consistency proof could be carried out.

But what form would such a system take? Primarily, it would need to incorporate novel proof methods not present in ordinary mathematics. Furthermore, these new proof methods would have to be simple enough to qualify as *finite means*; that is, they would have to be valid from obvious considerations. Even though Gödel's incompleteness theorems do not rule out its existence, no one has yet discovered such a system, let alone formulated an idea of its potential structure. In essence, only a few experts still believe in the existence of such a system.

The subsequent paragraph in Gödel's work holds marginal significance. Gödel utilizes the second incompleteness theorem to weaken the prerequisite of his main result slightly.

> Da für jede widerspruchsfreie Klasse $\varkappa$ w nicht $\varkappa$-*beweisbar* ist, so gibt es schon immer dann (aus $\varkappa$) unentscheidbare Sätze (nämlich w), wenn Neg (w) nicht $\varkappa$-*beweisbar* ist; m. a. W. man kann in Satz VI

198   Kurt Gödel, Über formal unentscheidbare Sätze etc.

> die Voraussetzung der ω-Widerspruchsfreiheit ersetzen durch die folgende: Die Aussage „$\varkappa$ ist widerspruchsvoll" ist nicht $\varkappa$-beweisbar. (Man beachte, daß es widerspruchsfreie $\varkappa$ gibt, für die diese Aussage $\varkappa$-beweisbar ist.)

> Since w is not $\varkappa$-*provable* for any consistent class $\varkappa$, then there already will exist sentences (namely, w) undecidable from $\varkappa$, if Neg (w) is not $\varkappa$-*provable*; in other words, in Theorem VI one can replace

198   Kurt Gödel, Über formal unentscheidbare Sätze etc.

> the assumption of ω-consistency by the following: The proposition "$\varkappa$ is inconsistent" is not $\varkappa$-provable. (One should observe that there exist consistent $\varkappa$ for which this proposition is $\varkappa$-provable.)

6.3 The Second Incompleteness Theorem

Gödels adopts the following line of argument: We have just demonstrated the unprovability of Wid(χ) in P $\cup$ χ, given that χ is consistent. If ¬Wid(χ) is also unprovable, then Wid(χ) and ¬Wid(χ) would constitute an undecidable pair of formulas, rendering P incomplete. Hence, it is safe to replace the prerequisite of ω-consistency in sentence VI with the following requirement:

"¬Wid(χ) is unprovable in P $\cup$ χ."

In his last sentence, set in parentheses, Gödel points out an interesting fact. Consistency merely states that a formula φ can never be proved alongside its negation ¬φ, while correctness ensures that no substantively false formulas are provable. Thus, a formal system may claim that for some formula φ, both φ and ¬φ can be derived from the axioms, even though this is not the case. The system could still be consistent in the sense of Definition 1.5 but would obviously no longer be correct.

> Wir haben uns in dieser Arbeit im wesentlichen auf das System P beschränkt und die Anwendung auf andere Systeme nur angedeutet. In voller Allgemeinheit werden die Resultate in einer demnächst erscheinenden Fortsetzung ausgesprochen und bewiesen werden. In dieser Arbeit wird auch der nur skizzenhaft geführte Beweis von Satz XI ausführlich dargestellt werden.

> We have limited ourselves in this paper essentially to the system P and have only indicated the applications to other systems. The results will be expressed and proved in full generality in a sequel to appear shortly. Also in that paper, the proof of Theorem XI, which has only been sketched here, will be presented in detail.

The second part Gödel mentioned was never published. Most mathematicians found the proof sketch so convincing that hardly anyone doubted its correctness.

At this juncture, we have reached the end of Gödel's article. Indeed, the preceding pages demanded substantial mental effort, and understanding any aspect of the proof on the spot is quite challenging. However, Gödel has rewarded us generously: The two incompleteness theorems undoubtedly belong to the most exciting theorems ever proved in mathematics. They show us limits we cannot overcome, thus standing on par with Einstein's theory of relativity or Heisenberg's uncertainty principle in physics.

It is inherent in most mathematicians to strive for completeness, making the incompleteness theorems a persistent thorn in their side without the chance of removal. Nevertheless, we shouldn't perceive Gödel's theorems in a too negative light. Gödel undeniably demonstrated that mathematics couldn't do everything, which, in the grand scheme, may ultimately be a good thing.

(Eingelangt: 17. XI. 1930.)

———————

(Received: 17. XI. 1930.)

———————

7 Epilogue

In the 1930s, the reactions to Gödel's work varied greatly. It is reported that Hilbert initially reacted with anger [103]. However, he did not deny reality for long. In the years to come, he diligently worked out many proof steps that Gödel had merely sketched.

Zermelo could not come to terms with the incompleteness theorems and continued to reject their contents until his death. Several times he believed to have identified a flaw in Gödel's proof [108], but none of his objections stood the test of time.

For John von Neumann, logic would never be the same again. After delivering several lectures on the incompleteness theorems in the 1930s, he soon shifted his focus to a different field of research. He became involved in the construction of electronic calculating machines and made significant achievements in this area. In 1946, he introduced a fundamental concept for the internal workings of microcomputers, and the *von Neumann architecture* remains the basis for the construction of most modern computer systems to this day [75].

Bertrand Russell also largely withdrew from logic and shifted his attention towards socio-political and philosophical topics.

Gödel remained committed to mathematics and began working intensely on set theoretical problems. During that period, crucial questions revolved around the veracity of the continuum hypothesis (CH) and the axiom of choice (AC). As it gradually emerged that neither CH nor AC could be proved or disproved within Zermelo-Fraenkel set theory, Gödel became utterly absorbed. In the ensuing years, he made groundbreaking discoveries in this field, advancing our comprehension of sets and classes in unprecedented ways [33, 34]. Alas, we can no longer open this fascinating chapter here, so this is where our journey comes to an end.

Bibliography

[1] Ackermann, W.: Zur Axiomatik der Mengenlehre. In: *Mathematische Annalen* 131 (1956), August, Nr. 4, S. 336–345

[2] Banach, S.; Tarski, A.: Sur la décomposition des ensembles de points en parties respectivement congruentes. In: *Fundamenta Mathematicae* 6 (1924), S. 244–277

[3] Bedürftig, T.; Murawski, R.: *Philosophie der Mathematik*. Berlin: Walter de Gruyter Verlag, 2010

[4] Bernays, P.: *Axiomatische Untersuchung des Aussagen-Kalküls der „Principia Mathematica "*, Universität Göttingen, Habilitation, 1918

[5] Bernays, P.: Axiomatische Untersuchung des Aussagen-Kalküls der „Principia Mathematica ". In: *Mathematische Zeitschrift* 25 (1926), S. 305–320

[6] Cantor, G.: Über die Ausdehnung eines Satzes aus der Theorie der trigonometrischen Reihen. In: *Mathematische Annalen* 5 (1872), S. 123–132

[7] Cantor, G.: Über eine elementare Frage der Mannigfaltigkeitslehre. In: *Jahresbericht der deutschen Mathematiker-Vereinigung. Erster Band (1890–91)* 1 (1892), Nr. 4, S. 75–78

[8] Carnap, R.: Die logizistische Grundlegung der Mathematik. In: Carnap, R. (Hrsg.); Reichenbach, H. (Hrsg.): *Bericht über die 2. Tagung für Erkenntnislehre der exakten Wissenschaften Königsberg 1930*. Leipzig: Felix Meiner Verlag, 1931, S. 91–105

[9] Carnap, R.; Reichenbach, H.: *Bericht über die 2. Tagung für Erkenntnislehre der exakten Wissenschaften Königsberg 1930*. Leipzig: Felix Meiner Verlag, 1931

[10] Coffa, J. A.: *The Semantic Tradition from Kant to Carnap: To the Vienna Station*. Cambridge: Cambridge University Press, 1993

[11] Cohen, P.: The Independence of the Continuum Hypothesis. In: *Proceedings of the National Academy of Sciences of the United States of America* Bd. 50. Washington, DC: National Academy of Sciences, 1963, S. 1143–1148

[12] Coutura, L. (Hrsg.): *Opuscules et fragments inédits de Leibniz*. Hildesheim: Georg Olms Verlag, 1966

[13] Davis, M.: *The Undecidable*. Mineola, NY: Dover Publications, 1965

[14] Dawson, J. W.: *Kurt Gödel. Leben und Werk*. Berlin, Heidelberg, New York: Springer-Verlag, 1999

[15] Dedekind, R.: *Was sind und was sollen die Zahlen?* Braunschweig, 1918

[16] Dedekind, R.: What are numbers and what should they be? (1995)

[17] Ebbinghaus, H.-D.; Peckhaus, V.: *Ernst Zermelo: An Approach to His Life and Work*. Berlin, Heidelberg, New York: Springer-Verlag, 2007

[18] Einstein, A.: Zur Elektrodynamik bewegter Körper. In: *Annalen der Physik* 17 (1905), S. 891–921

[19] Euclid; Heath, T.L. (Hrsg.): *The thirteen books of Euclid's elements (3 vols)*. 2. Dover Publications, 1956 (Dover Books on Mathematics)

[20] Fraenkel, A.: Zu den Grundlagen der Cantor-Zermeloschen Mengenlehre. In: *Mathematische Annalen* 86 (1922), S. 230–237

[21] Frege, G.: *Begriffsschrift. Eine der arithmetischen nachgebildeten Formelsprache*. Ditzingen: Verlag Louis Nebert, 1879

[22] Frege, G.: *Die Grundlagen der Arithmetik – Eine logisch mathematische Untersuchung über den Begiff der Zahl*. Breslau: Wilhelm Koebner Verlag, 1884

[23] Frege, G.: *Grundgesetze der Arithmetik, Begriffsschriftlich abgeleitet*. Bd. 1. Jena: Verlag Hermann Pohle, 1903

[24] Frege, G.: *Grundgesetze der Arithmetik, Begriffsschriftlich abgeleitet*. Bd. 2. Jena: Verlag Hermann Pohle, 1903

[25] Frege, G.; Austin, J. L. (Hrsg.): *The Foundations of Arithmetic: A Logico-Mathematical Enquiry Into the Concept of Number*. Evanston, Ill.: Northwestern University Press, 1953

[26] Frege, G.: *Grundgesetze der Arithmetik*. Bd. 1. Hildesheim: Verlag Olms, 1962

[27] Frege, G.: *Grundgesetze der Arithmetik*. Bd. 2. Hildesheim: Verlag Olms, 1962

[28] Frege, G.: Begriffsschrift, a formula language, modeled upon that of arithmetic, for pure thought. In: Heijenoort, J. van (Hrsg.): *From Frege to Gödel : A Source Book in Mathematical Logic, 1879-1931 (Source Books in the History of the Sciences)*. Cambridge, MA: Harvard University Press, 2002, S. 1–82

[29] Frege, G.; P. A. Evert, C. W. M. Rossberg R. M. Rossberg (Hrsg.): *Basic Laws of Arithmetic. Volumes I and II*. Oxford: Oxford University Press, 2013

[30] Gabriel, G.; Kambartel, F.; Thiel, C.: *Philosophische Bibliothek*. Bd. 321: *Gottlob Freges Briefwechsel mit D. Hilbert, E. Husserl, B. Russell sowie ausgewählte Einzelbriefe Freges*. Hamburg: Felix Meiner Verlag, 1980

[31] Gödel, K.: Einige metamathematische Resultate über Entscheidungsdefinitheit und Widerspruchsfreiheit. In: *Anzeiger der Akademie der Wissenschaften in Wien* 67 (1930), S. 214–215. – Nachgedruckt in [36]

[32] Gödel, K.: Die Vollständigkeit der Axiome des logischen Funktionenkalküls. In: *Monatshefte für Mathematik* 37 (1930), S. 349–360

[33] Gödel, K.: The Consistency of the Axiom of Choice and of the Generalized Continuum-Hypothesis. In: *Proceedings of the U.S. National Academy of Sciences* Bd. 24, 1938, S. 556–557

[34] Gödel, K.: What is Cantor's Continuum Problem? In: *American Mathematical Monthly* 54 (1947), S. 515–525

[35] Gödel, K.: On Formally Undecidable Propositions of Principia Mathematica and Related Systems I. In: Heijenoort, J. van (Hrsg.): *From Frege to Gödel : A Source Book in Mathematical Logic, 1879-1931 (Source Books in the History of the Sciences)*. Cambridge, MA: Harvard University Press, 1977, S. 596–616

Bibliography

[36] Gödel, K.: *Collected Works I. Publications 1929–1936.* New York: Oxford University Press, 1986

[37] Gödel, K.: *Collected Works V. Correspondence, H-Z.* New York: Oxford University Press, 2003

[38] Graßmann, H.: *Lehrbuch der Arithmetik für höhere Lehrveranstaltungen.* Berlin: Enslin Verlag, 1861

[39] Heisenberg, W.: Über den anschaulichen Inhalt der quantentheoretischen Kinematik und Mechanik. In: *Zeitschrift für Physik* 43 (1927), März, Nr. 3, S. 172–198

[40] Heuser, H.: *Lehrbuch der Analysis I.* Wiesbaden: Teubner Verlag, 2006

[41] Heyting, A.: Die intuitionistische Grundlegung der Mathematik. In: Carnap, R. (Hrsg.); Reichenbach, H. (Hrsg.): *Bericht über die 2. Tagung für Erkenntnislehre der exakten Wissenschaften Königsberg 1930.* Leipzig: Felix Meiner Verlag, 1931, S. 106–115

[42] Hilbert, D.: Axiomatisches Denken. In: *Mathematische Annalen* 78 (1918), S. 405–415

[43] Hilbert, D.: Die logischen Grundlagen der Mathematik. In: *Mathematische Annalen* 88 (1923), S. 151–165

[44] Hilbert, D.: Über das Unendliche. In: *Mathematische Annalen* 95 (1926), Nr. 1, S. 161–190

[45] Hilbert, D.: Die Grundlagen der Mathematik. In: *Abhandlungen aus dem mathematischen Seminar* VI (1928), S. 80

[46] *Kapitel* On the Infinite. In: Hilbert, D.: *Philosophy of Mathematics: Selected Readings.* Cambridge University Press, 1984

[47] Hilbert, D.: *Die Hilbert'schen Probleme.* Frankfurt: Verlag Harri Deutsch, 1998 (Ostwalds Klassiker)

[48] Hilbert, D.: The foundations of mathematics. In: Heijenoort, J. van (Hrsg.): *From Frege to Gödel : A Source Book in Mathematical Logic, 1879-1931 (Source Books in the History of the Sciences).* Cambridge, MA: Harvard University Press, 2002, S. 464–479

[49] Hilbert, D.: From Kant to Hilbert: A Source Book in the Foundations of Mathematics. II (2005), S. 1105–1114

[50] Hilbert, D.: From Kant to Hilbert: A Source Book in the Foundations of Mathematics. II (2005), S. 1096–1104

[51] Hilbert, D.: From Kant to Hilbert: A Source Book in the Foundations of Mathematics. II (2007)

[52] Hilbert, D.; Ackermann, W.: *Grundzüge der theoretischen Logik.* 1. Auflage. Berlin, Heidelberg: Springer-Verlag, 1928

[53] Hilbert, D.; Ackermann, W.: *Grundzüge der theoretischen Logik.* 2. Auflage. Berlin, Heidelberg: Springer-Verlag, 1938

[54] Hilbert, D.; Ackermann, W.: *Grundzüge der theoretischen Logik.* 4. Auflage. Berlin, Heidelberg: Springer-Verlag, 1958

[55] Hilbert, D.; Bernays, P.: *Die Grundlehren der mathematischen Wissenschaften in Einzeldarstellungen.* Bd. 40: *Grundlagen der Mathematik – Band I.* Berlin, Heidelberg: Springer-Verlag, 1934

[56] Hilbert, D.; Bernays, P.: *Die Grundlehren der mathematischen Wissenschaften in Einzeldarstellungen.* Bd. 50: *Grundlagen der Mathematik – Band II.* Berlin, Heidelberg: Springer-Verlag, 1939

[57] Hilbert, D.; Bernays, P.: *Die Grundlehren der mathematischen Wissenschaften in Einzeldarstellungen.* Bd. 40: *Grundlagen der Mathematik – Band I.* 2. Aufl. Berlin, Heidelberg: Springer-Verlag, 1968

[58] Hoffmann, D. W.: *Software-Qualität.* Berlin: Springer-Verlag, 2008

[59] Hoffmann, D. W.: *Grenzen der Mathematik. Eine Reise durch die Kerngebiete der mathematischen Logik.* Heidelberg: Spektrum Akademischer Verlag, 2011

[60] Hoffmann, M.: Axiomatisierung zwischen Platon und Aristoteles. In: *Zeitschrift für philosophische Forschung* 58 (2004), S. 224–245

[61] Hofstadter, D. R.: *Gödel, Escher, Bach: Ein endloses geflochtenes Band.* Stuttgart: Klett-Cotta, 2006

[62] Kamareddine, F. D.; Laan, T.; Nederpelt, R.: *Applied logic series.* Bd. 29: *A Modern Perspective on Type Theory: From its Origins until Today.* Berlin, Heidelberg, New York: Springer-Verlag, 2004

[63] Kelley, J. L.: *General Topology.* New York: Van Nostrand Reinhold, 1955

[64] Kmhkmh: *Portraitphoto von Paul Erdős.* http://creativecommons.org/licenses/by/3.0/. – Creative Commons License 3.0, Typ: Attribution-ShareAlike

[65] Leibniz, G. W.; Strack, C. (Übersetzer): *Leibniz sogenannte Monadologie und Principes de la nature et de la grâce fondés en raison.* Berlin: Walter de Gruyter Verlag, 1967

[66] Löb, M. H.: Solution of a problem of Leon Henkin. In: *Journal of Symbolic Logic* 20 (1955), S. 115–118

[67] Łukasiewicz, L.; Tarski, A.: Untersuchungen über den Aussagenkalkül. In: *Comptes Rendus des séances de la Société des Sciences et des Lettres de Varsovie* 23 (1930), S. 30–50

[68] Meltzer, B.: *On Formally Undecidable Propositions of Principia Mathematica and Related Systems.* Dover Publications, 1962. – Translation of the German original by Kurt Gödel

[69] *Kapitel* On Formally Undecidable Propositions of Principia Mathematica and Related Systems I. In: Mendelson, E.: *The Undecidable.* Mineola, NY: Dover Publications, 1965. – Translation of the German original by Kurt Gödel

[70] Mendelson, E.: *Introduction to Mathematical Logic.* 4th edition. Boca Raton, FL: Chapman & Hall, CRC Press, 1997

[71] Morse, A. P.: *A Theory of Sets.* New York: Academic Press, 1965

[72] Neumann, J. von: Die formalistische Grundlegung der Mathematik. In: Carnap, R. (Hrsg.); Reichenbach, H. (Hrsg.): *Bericht über die 2. Tagung für Erkenntnislehre der exakten Wissenschaften Königsberg 1930.* Leipzig: Felix Meiner Verlag, 1931, S. 116–121

[73] Neumann, J. von: *The Computer and the Brain.* New Haven: Yale University Press, 1958

[74] Neumann, J. von: The Formalist Foundations of Mathematics. In: Benacerraf, P. (Hrsg.); Putnam, H. (Hrsg.): *Philosophy of Mathematics.* Cambridge: Cambridge University Press, 1984, S. 61–65

Bibliography

[75] Neumann, J. von: First Draft of a Report on the EDVAC. In: *IEEE Annals of the History of Computing* 15 (1993), Nr. 4, S. 27–75

[76] Peano, G.: *Calcolo geometrico secondo l'Ausdehnungslehre di H. Grassmann.* Torino: Fratelli Bocca, 1888

[77] Peano, G.: *Arithmetices principia, nova methodo exposita.* Torino: Fratelli Bocca, 1889

[78] Peano, G.: Principii di logica matematica. In: *Rivista di matematica* (1891)

[79] Peano, G.: The principles of arithmetic, presented by a new method. In: Heijenoort, J. van (Hrsg.): *From Frege to Gödel : A Source Book in Mathematical Logic, 1879-1931 (Source Books in the History of the Sciences).* Cambridge, MA: Harvard University Press, 1977, S. 83–97

[80] Petzold, C.: *The Annotated Turing: A Guided Tour Through Alan Turing's Historic Paper on Computability and the Turing Machine.* New York: John Wiley and Sons, 2008

[81] Poincaré, H.: Les Mathématiques et la Logique. In: *Revue de Métaphysique et de Morale* 14 (1906), Nr. 3, S. 294–317

[82] Prince, H.: *The Annotated Gödel: A Reader's Guide to his Classic Paper on Logic and Incompleteness.* Homebred Press, 2022

[83] Ramsey, F. P.: The foundations of mathematics. In: *Proceedings of the London Mathematical Society* 25 (1925), S. 338–384

[84] Reichhalter, M.: *Cantor – Frege – Zermelo. Grundzüge der Entwicklung der Mengenlehre.* Saarbrücken: VDM Verlag Dr. Müller, 2010

[85] Richard, J.: Les principes des mathématiques et le problème des ensembles. In: *Revue générale des sciences pures et appliquées* 16 (1905), S. 541–543

[86] Richard, J.: Lettre Monsieur le Rédacteur de la Revue Générale des Sciences. In: *Acta Mathematica* 30 (1906), S. 295–296

[87] Richard, J.: The principles of mathematics and the problem of sets. In: Heijenoort, J. van (Hrsg.): *From Frege to Gödel : A Source Book in Mathematical Logic, 1879-1931 (Source Books in the History of the Sciences).* Cambridge, MA: Harvard University Press, 1977, S. 142–144

[88] Rose, N. J.: *Mathematical Maxims and Minims.* Raleigh, North Carolina: Rome Press Inc., 1988

[89] Rosser, J. B.: Extensions of Some Theorems of Gödel and Church. In: *Journal of Symbolic Logic* 1 (1936), S. 87–91

[90] Russell, B.: *The principles of mathematics.* Cambridge: Cambridge University Press, 1903

[91] Russell, B.: Mathematical Logic as Based on the Theory of Types. In: *American Journal of Mathematics* 30 (1908), July, Nr. 3, S. 222–262

[92] Russell, B.: *The principles of mathematics.* 2nd edition. Cambridge: Cambridge University Press, 1937

[93] Russell, B.: *The Autobiography of Bertrand Russell.* London: George Allen and Unwin Ltd., 1967

[94] Russell, B.: Letter from Russell to Frege. In: Heijenoort, J. van (Hrsg.): *From Frege to Gödel : A Source Book in Mathematical Logic, 1879-1931 (Source*

Books in the History of the Sciences). Cambridge, MA: Harvard University Press, 1977, S. 124–125

[95] Singh, S.: *Fermats letzter Satz*. München: Deutscher Taschenbuch Verlag, 2000

[96] *Kapitel* The incompleteness theorems. In: Smoryński, C.: *Handbook of Mathematical Logic*. Amsterdam: North-Holland Publishing, 1977

[97] Turing, A. M.: On computable numbers with an application to the Entscheidungsproblem. In: *Proceedings of the London Mathematical Society* 2 (1936), Juli – September, Nr. 42, S. 230–265

[98] Weber, H.: Leopold Kronecker. In: *Jahresbericht der Deutschen Mathematiker-Vereinigung* 2 (1893), S. 19

[99] Whitehead, A. N.; Russell, B.: *Principia Mathematica. Volume I*. London: Merchant Books, 1910

[100] Whitehead, A. N.; Russell, B.: *Principia Mathematica. Volume II*. 2nd edition. London: Merchant Books, 1927

[101] Wikipedia: *Gödel's incompleteness theorems*. https://en.wikipedia.org/wiki/Gödel%27s_incompleteness_theorems

[102] Wiles, A.: Modular Elliptic Curves and Fermat's last theorem. In: *Annals of Mathematics* 141 (1995), S. 443–551

[103] Yourgrau, P.: *Gödel, Einstein und die Folgen: Vermächtnis einer ungewöhnlichen Freundschaft*. München: C. H. Beck, 2005

[104] Zermelo, E.: Beweis, dass jede Menge wohlgeordnet werden kann. In: *Mathematische Annalen* 59 (1904), S. 514–516

[105] Zermelo, E.: Neuer Beweis für die Möglichkeit einer Wohlordnung. In: *Mathematische Annalen* 65 (1908), S. 107–128

[106] Zermelo, E.: Untersuchungen über die Grundlagen der Mengenlehre I. In: *Mathematische Annalen* 65 (1908), S. 261–281

[107] Zermelo, E.: über Grenzzahlen und Mengenbereiche. In: *Fundamenta Mathematicae* 16 (1930), S. 29–47

[108] Zermelo, E.: über Stufen der Quantifikation und die Logik des Unendlichen. In: *Jahresbericht der Deutschen Mathematiker-Vereinigung* 41 (1932), S. 85–88

[109] Zermelo, E.: Investigations in the foundations of set theory I. In: Heijenoort, J. van (Hrsg.): *From Frege to Gödel : A Source Book in Mathematical Logic, 1879-1931 (Source Books in the History of the Sciences)*. Cambridge, MA: Harvard University Press, 1977, S. 199–215

[110] Zermelo, E.: A New Proof of the Possibility of a Well-Ordering. In: Heijenoort, J. van (Hrsg.): *From Frege to Gödel : A Source Book in Mathematical Logic, 1879-1931 (Source Books in the History of the Sciences)*. Cambridge, MA: Harvard University Press, 1977, S. 183–198

[111] Zermelo, E.: Proof that every set can be well-ordered. In: Heijenoort, J. van (Hrsg.): *From Frege to Gödel : A Source Book in Mathematical Logic, 1879-1931 (Source Books in the History of the Sciences)*. Cambridge, MA: Harvard University Press, 1977, S. 139–141

[112] *Kapitel* Bericht an die Notgemeinschaft der Deutschen Wissenschaft über meine Forschungen betreffend die Grundlagen der Mathematik. In: Zermelo,

E.: *Ernst Zermelo. Gesammelte Werke.* Berlin, Heidelberg, New York: Springer-Verlag, 2010, S. 432–434

Image Credits

Page 25   John von Neumann
              `commons.wikimedia.org/wiki/File:JohnvonNeumann-LosAlamos.gif`
Page 3   Euklid von Alexandria
              `commons.wikimedia.org/wiki/File:Euklid.jpg`
Page 2   Elemente von Euklid (Fragment)
              `commons.wikimedia.org/wiki/File:P._Oxy._I_29.jpg`
Page 4   David Hilbert
              `commons.wikimedia.org/wiki/File:Hilbert.jpg`
Page 19   Gottfried Wilhelm Leibniz
              `commons.wikimedia.org/wiki/File:Gottfried_Wilhelm_Leibniz_c1700.jpg`
Page 21   Leopold Kronecker
              `commons.wikimedia.org/wiki/File:Leopold_Kronecker.jpg`
Page 22   Georg Cantor
              `commons.wikimedia.org/wiki/File:Georg_Cantor2.jpg`
Page 27   Kurt Gödel
              `commons.wikimedia.org/wiki/File:Kurt_gödel.jpg`
Page 34   Gottlob Frege
              `commons.wikimedia.org/wiki/File:Young_frege.jpg`
Page 36   Begriffsschrift (title page)
              `commons.wikimedia.org/wiki/File:Begriffsschrift_Titel.png`
Page 42   Giuseppe Peano
              `commons.wikimedia.org/wiki/File:Giuseppe_Peano.jpg`
Page 53   Richard Dedekind
              `commons.wikimedia.org/wiki/File:Dedekind.jpeg`
Page 57   Bertrand Russell
              `commons.wikimedia.org/wiki/File:Russell1907-2.jpg`
Page 80   Bertrand Russell
              `commons.wikimedia.org/wiki/File:FourAnalyticPhilosophers.JPG`
Page 86   Ernst Zermelo
              `commons.wikimedia.org/wiki/File:Ernst_Zermelo.jpeg`
Page 114   Goldbach's conjecture
              `commons.wikimedia.org/wiki/File:Goldbach-1000000.png`
Page 115   Pierre de Fermat
              `commons.wikimedia.org/wiki/File:Pierre_de_Fermat.jpg`
Page 211   Hermann Graßmann
              `commons.wikimedia.org/wiki/File:Hgrassmann.jpg`
Page 212   Rózsa Péter
              `commons.wikimedia.org/wiki/File:RozsaPeter.jpg`
Page 235   Joseph Louis François Bertrand
              `commons.wikimedia.org/wiki/File:Joseph_bertrand.jpg`
Page 236   Pafnuty Lvovich Chebyshev
              `commons.wikimedia.org/wiki/File:Chebyshev.jpg`
Page 236   Srinivasa Ramanujan
              `commons.wikimedia.org/wiki/File:Ramanujan.jpg`
Page 236   Paul Erdős
              `commons.wikimedia.org/wiki/File:Erdos_head_budapest_fall_1992.jpg`

All clipart images are from `www.openclipart.org`.

© The Editor(s) (if applicable) and The Author(s), under exclusive license to Springer-Verlag GmbH, DE, part of Springer Nature 2024
D. W. Hoffmann, *Gödel's Incompleteness Theorems*,
https://doi.org/10.1007/978-3-662-69550-0

Name Index

A

Abbe, Ernst, 33
Ackermann, Wilhelm, 106, 276
Aristotle, 2

B

Banach, Stefan, 101
Bernays, Paul, 29, 84, 106, 172, 212, 358
Bertrand, Joseph, 235
Boole, George, 35
Borel, Émile, 100
Brouwer, Luitzen, 22
Burali-Forti, Cesare, 64, 73

C

Cantor, Georg, 21, **22**, 60, 86
Carnap, Rudolf, 1, 26
Chebyshev, Pafnuty Lvovich, 236
Cohen, Paul Joseph, 101

D

De Morgan, Augustus, 35
Dedekind, Richard, 41, **53**
Dirichlet, Peter Gustav, 3
Dubislav, Walter, 26

E

Einstein, Albert, 361, I
Erdős, Paul, 236
Euclid of Alexandria, 2
Eudoxos of Knidos, 2
Euler, Leonhard, 113

F

Feigl, Herbert, 1
Fermat, Pierre de, 113, **115**, 351

Fraenkel, Abraham, 104
Frege, Gottlob, 21, 33, **34**

G

Gauss, Carl Friedrich, 3
Gödel, Kurt, 1, **27**, 106, 363
Goldbach, Christian, 113, 350
Graßmann, Hermann, 210, **211**
Grelling, Kurt, 1

H

Hahn, Hans, 1, 26
Heisenberg, Werner, 26, I
Heuser, Harro, 2
Heyting, Arend, 1, 26
Hilbert, David, 3, **4**, 29, 87, 276, 358, 363
Hofstadter, Douglas, I

K

Kelley, John, 106
Klein, Felix, 87
Kronecker, Leopold, 21, 41

L

Leibniz, Gottfried Wilhelm, **19**, 43, 154, 161, 185
Löb, Martin Hugo, 359
Łukasiewicz, Jan, 38

M

Morse, Anthony, 106

N

Neumann, John von, **25**, 106, 363

P

Peano, Giuseppe, **42**, 100
Planck, Max, 86
Plato, 2
Poincaré, Henri, 100, 101
Péter, Rózsa, 212, **212**

R

Ramanujan, Srinivasa, 236
Ramsey, Frank Plumpton, 76
Reichenbach, Hans, 26
Richard, Jules, 132
Riemann, Bernhard, 3
Rosser, J. Barkley, 137, 330
Russell, Bertrand, **56, 57**, 363

S

Schoenflies, Arthur Moritz, 87
Scholz, Arnold, 26

T

Tarski, Alfred, 101
Turing, Alan, 278

W

Waismann, Friedrich, 1
Whitehead, Alfred, 58
Wiles, Andrew, 113

Z

Zermelo, Ernst, **86**, 363

Name	Born	Died
Wilhelm Ackermann	1896	1962
Paul Bernays	1888	1977
George Boole	1815	1864
Luitzen Brouwer	1881	1966
Cesare Burali-Forti	1861	1931
Georg Cantor	1845	1918
Rudolf Carnap	1891	1970
Paul Cohen	1934	2007
Richard Dedekind	1831	1916
Abraham Fraenkel	1891	1965
Gottlob Frege	1848	1925
Kurt Gödel	1906	1978
David Hilbert	1862	1943
Leopold Kronecker	1823	1891
John von Neumann	1903	1957
Giuseppe Peano	1858	1932
Rosza Peter	1905	1977
Barkley Rosser	1907	1989
Bertrand Russell	1872	1970
Alan Turing	1912	1954
Alfred Whitehead	1861	1947
Ernst Zermelo	1871	1953

Index

A

Absolute consistency proof, 17
AC, see Axiom of choice
Actual infinite, 21
α function, 220
AND connective, see Conjunction
Antinomy
    Cantor's, 60, 64
    Richard's, 132
    Russell's, 65
        logical, 65
        set-theoretical, 68
    semantic, 134
Architecture
    von Neumann, 363
Arithmetic
    formula, 110, **318**
    interpretation, 318
    Peano, 111, 317
        semantics, 318
        syntax, 318
    representation, 318
    term, 318
Arithmetices Principia, 43
    axioms, 46
Arithmetization
    of syntax, 117, 200
Asymmetric relation, 91
Atomic formula
    of predicate logic, 331
Axiom, 1
    of choice, **99**, 106, 313
    of comprehension, 160, 271
    of definiteness, 103, I
    of empty set, 102
    of foundation, 106
    of limitation, 105
    of pairing, 102
    of power set, 102
    of reducibility, 76
    of separation, 103
    of union, 102
Axiomatic
    method, 1
    set theory, 86
Axiomatizability
    finite, 106
Axioms
    of *Arithmetices Principia*, 46
    of *Begriffsschrift*, 37
    categorical, 105, 157
    Peano, 46, 47
    of *Principia Mathematica*, 80
    of system B, 8
    of system E, 11
    of system P, 156
    Zermelo's, 102

B

B (formal system)
    axioms, 8
    formula, 7
    inference rules, 8
    proof, 8
    syntax, 7
Base type, 75
Basic law of course-of-values, 40
Begriffsschrift, 34
    axioms, 37
Bertrand's postulate, 236
β function, 220, 322
Bound variable, 147
Bounded renaming, 184
Burali-Forti paradox, 65

C

Calculus ratiocinator, 20
Cantor's
    antinomy, 60, 64
    theorem
        contemporary, 63, 87
        historic, 60, 62

Cardinal number, 65, 88, 132
Cardinality, 60, 88
Categorical
    axioms, 105, 157
CH, *see* Continuum hypothesis
Chain inference, 165
Characteristica universalis, 20, 43
Choice function, 95
Class, 106
    sign, 148, 279, 292, 300
Closed formula, 147, 332
Closure
    universal, 181
Collision, 84, 150
Collision-free substitution, 150
Completeness, 15, 18
    theorem, 26, 335
Completion
    of a formal system, 116
Comprehension
    axiom, 160, 271
    schema
        general, 103
Computable function, 213
Conceptual scope, 41
Conjecture
    Goldbach's, 113, 351
Conjunction, 145
Consequence
    logical, 153
Consistency, 15
    ω-, 136, 295
    proof
        absolute, 17
        relative, 16
Constant, 333
Continuum hypothesis, 87
    generalized, 88, 313
Correctness, 15
Course-of-values, 41
    basic law of, 40

D

Decision
    problem
        semantic, 276
        syntactic, 275
    procedure, 274
Dedekind's
    isomorphism theorem, **50**, 55, 157
    recursion theorem, 53
Deduction theorem, 180
Definiteness
    axiom, 103
Definition
    non-predicative, 101
Detachment
    law of, 10
    rule, 38
Diagonal element, 61, 125
Diagonalization, **61**, 129, 133
Disjunction, 145
Domain, 74, **332**
Dot notation
    Peano's, 45
DT, *see* Deduction theorem

E

E (formal system)
    axioms, 11
    formula, 10
    inference rules, 11
    proof, 10
    semantics, 12
    syntax, 10
    term, 10
Elementary formula, **144**, 248
Elements
    Euclid's, 2
Empty set
    axiom, 102
Equality
    in system P, 154
    theorems, 185
Equisatisfiability, 339
Equivalence
    operator, 145
    in system P, 153
Euclid's *Elements*, 2
Exchange rule, 173
Excluded middle
    law of, 81
Existential quantification, 145
Extensionality principle, 161

F

Fermat's Last Theorem, 113, 351

Index

Finite
    axiomatizability, 106
    means, 25
First incompleteness theorem, 26, 27, 136, **315**
First-order predicate, 74
    logic, 26, 154, 317, **330**
        semantics, 333
        syntax, 331
        with equality, 332
Formal
    proof, 9
    system, 6, 9
        complete, 15, 18
        consistent, 15
        correct, 15
        negation complete, 15
Formalism, 1
Formula, 331
    arithmetic, 110, **318**
    closed, 147, 332
    elementary, **144**, 248
    open, 147, 332
    of Peano arithmetic, 111
    of predicate logic, 332
    satisfiable, 334
    of system B, 7
    of system E, 10
    of system P, 144, 145, 248
    universally valid, 26, 334
    unsatisfiable, 334
Formulario Mathematica, 43
Formulario project, 43
Foundation
    axiom, 106
Free variable, 82, 147
Function
    α, 220
    β, 220, 322
    computable, 213
    of predicate logic, 331
    primitive-recursive, 209, 213
    projection, 214, 286
    recursive, 213
    representable
        semantically, 280
        syntacially, 282
    successor, 214, 286, 343
    zero, 213, 286
Function calculus
    restricted, 330
        in the extended sense, 339

G

γ-set, 96
GCH, *see* Generalized continuum hypothesis
General comprehension schema, 103
Generalization, 145
    rule, 83, 162
Generalized continuum hypothesis, 88, 313
Goldbach's Conjecture, 113, 351
Gödel number, 119, 200, 204
Gödel's
    completeness theorem, 26, 335
    incompleteness theorem
        first, 26, 27, 136, **315**
        second, 28, 137, **352**
Gödel-Rosser theorem, 330
Gödelization, 119

H

Higher-order predicate logic, 65, 154
Hilbert's program, **20**, 137, 359
Hypothesis, 178

I

Implication, 145
Incompleteness
    of arithmetic, 315
    of system P, 296
    theorem
        first, 26, 27, 136, **315**
        second, 28, 137, **352**
Individual
    object, 74
    predicate, 74
    set, 332
Inference rules
    of system B, 8
    of system E, 11
    of system P, 156
Infinity
    actual, 21
Initial segment
    of well-ordered sets, 96

Interpretation, 332
    arithmetic, 318
    of system P, 151
Intuitionism, 1, 305
Irreflexive relation, 91
Isomorphism, 55
    theorem
        Dedekind's, **50**, 55, 157

K

Klassenzeichen, 123

L

Law
    of detachment, 10
    of excluded middle, 81
    Leibniz', 155, 158, 185
Leibniz's law, 155, 158, 185
Liar's paradox, 130
Limitation
    axiom, 105
Linear order, 91
Logic
    first-order, 26, 154, 317, **330**
        semantics, 333
        syntax, 331
        with equality, 332
    higher-order, 65, 154
    second-order, 65
Logical
    calculus, *see* Formal system
    consequence, 153
Logicism, 1, **33**, 34

M

Main result (Theorem VI), 293
Manifold, 21
MB, *see* Modus barbara
Metamathematics, 13
Model, 16, 152
    relation, 12
        of P, 151
        of PA, 318
        of PL1, 333
Modus
    barbara, 165
    ponens, 10, 38, 162

Morse-Kelley set theory, 106
MP, *see* Modus ponens

N

Natural numbers
    in system P, 155
NBG set theory, 106, 314
Negation, 145
    completeness, 15
Non-predicative definition, 101
Notation
    Polish, 38
Number
    cardinal, 65, 88, 132
    ordinal, 21, 132
    sequence
        Zermelo's, 104
Numeric theorems, 190

O

ω-consistency, 136, 295
Open formula, 147, 332
OR connective, *see* Disjunction
Order
    linear, 91
    partial, 91
    total, 91
    well-, 92
Ordinal number, 21, 132

P

P (formal system)
    atomic formula, 144, 248
    axioms, 156
    equality, 154
    equivalence, 153
    formula, 145
    incompleteness, 296
    inference rules, 156
    interpretation, 151
    model relation, 151
    natural numbers, 155
    primitive sign, 141
    proof, 162
    semantics, 150
    substitution, 148
    syntax, 140

Index

term, 143
PA, *see* Peano arithmetic
Pairing
    axiom, 102
Paradox
    Burali-Forti, 65
    Liar's, 130
    semantic, 77
Partial order, 91
Peano
    arithmetic, 111, 317
        formula, 111
        semantics, 318
        syntax, 318
    axioms, 46, 47
PL, *see* Predicate logic
PL1, *see* First-order predicate logic
Platonism, 22
PM, *see* Principia Mathematica
Polish notation, 38
Postulate
    Bertrand's, 236
Power set, 62
    axiom, 102
Predicate
    first-order, 74
    satisfiable, 67
    second-order, 74
    unsatisfiable, 67
Predicate logic
    atomic formula, 331
    first-order, 26, 154, 317, **330**
        semantics, 333
        syntax, 331
        with equality, 332
    formula, 332
    function, 331
    higher-order, 65, 154
    predicate, 331
    second-order, 65
    semantics, 332
    syntax, 330
    theorem, 182
    variable, 331
Primitive
    propositions, 78
    recursion, 210
    sign
        of system P, 141
Primitive-recursive

function, 209, 213
relation, 217
set, 217
Principia Mathematica, 30, **56**, 77
    axioms, 80
Principle
    of contradiction, 19
    of excluded middle, 22, 23
    of extensionality, 161
    of sufficient reason, 19
Projection function, 214, 286
Proof
    consistency
        absolute, 17
        relative, 16
    formal, 9
    in system B, 8
    in system E, 10
    in system P, 162
    template, 164
    theory, 6
Propositional
    theorems, 166
    variable, 333
Provability relation, 12

Q

Quantor, 147
    scope, 147

R

Ramified type theory, 73, 76
Recursion
    primitive, 210
    theorem
        Dedekind's, 53
Recursive function, 213
Reducibility
    axiom, 76
Relation
    asymmetric, 91
    irreflexive, 91
    model, 12
    of PA, 318
    of PL1, 333
    primitive-recursive, 217
    provability, 12
    representable

semantically, 280
syntactically, 282
sign, **148**, 292
transitive, 91
Relative consistency proof, 16
Renaming
bounded, 184
Replacement schema, 104
Representable
function
semantically, 280
syntatically, 282
relation
semantically, 280
syntactically, 282
Representation
arithmetic, 318
Restricted function calculus, 330
in the extended sense, 339
Rewriting system, 8
Richard's antinomy, 132
Rossers trick, 330
Rule
of detachment, 38
generalization, 83, 162
of substitution, 38
Russell's antinomy, 65
logical, 65
set-theoretical, 68

S

Satisfiable
formula, 334
predicate, 67
Schema
of replacement, 104
Scope
of a quantor, 147
Second incompleteness theorem, 28, 137, **352**
Second-order
predicate, 74
logic, 65
Self reference, 73, 130
Semantic
antinomy, 134
decision problem, 276
paradox, 77
Semantically

representable
function, 280
relation, 280
Semantics
of predicate logic, 332
of system E, 12
of system P, 150
Sentential formula, 148
Separation
axiom, 103
Set
primitive-recursive, 217
Set theory
axiomatic, 86
Morse-Kelley, 106
NBG, 106, 314
von Neumann, 106
Simple type, 75
theory, 73
Soundness, *see Correctness*
Substitution
collision, 84
collosion-free, 150
rule, 38
in system P, 148
Successor function, 214, 286, 343
Syntactic decision problem, 275
Syntactically
representable
function, 282
relation, 282
Syntax
arithmetization of, 117, 200
of predicate logic, 330
of system B, 7
of system E, 10
of system P, 140
System
B
axioms, 8
formula, 7
inference rules, 8
proof, 8
syntax, 7
E
axioms, 11
formula, 10
inference rules, 11
proof, 10
semantics, 12

Index

syntax, 10
term, 10

P
atomic formula, 144, 248
axioms, 156
equality, 154
equivalence, 153
formula, 145
incompleteness, 296
inference rules, 156
interpretation, 151
model relation, 151
natural numbers, 155
primitive sign, 141
proof, 162
semantics, 150
substitution, 148
syntax, 140
term, 143

T

Term, 247, 331
   arithmetic, 318
   of system E, 10
   of system P, 143
Tertium non datur, 22, 81
Theorem, 9
   Cantor's
      contemporary, 63, 87
      historic, 60, 62
   equality, 185
   Gödel-Rosser, 330
   numeric, 190
   of predicate logic, 182
   propositional, 166
   V, 279, 284
      semantic variant, 286
   VI, 293
   well-ordering, 90, 94
Theory axioms, 2
Total order, 91
Transfinite, 103
Transitive
   relation, 91
Type
   base, 75
   elevation, 150
   simple, 75
   theory
      ramified, 73, 76
      simple, 73

U

Undecidability, 110
Union
   axiom, 102
Universal
   closure, 181
   quantification, 145
Universally valid formula, 26, 334
Universe, 74, **332**
Unsatisfiable
   formula, 334
   predicate, 67

V

Variable
   bound, 147
   free, 82, 147
   of predicate logic, 331
Vicious circle principle, 74, 76, 102
Von Neumann
   architecture, 363
   set theory, 106

W

Well-ordering, 92
   theorem, 90, 94

Z

Zermelo's
   axiom system, 102
   number sequence, 104
Zermelo-Fraenkel
   set theory, 30, 106
      with axiom of choice, 106
Zero function, 213, 286